计算机

科学与技术丛书

Dart语言实战

基于Angular框架的Web开发

刘仕文 ◎ 编著
Liu Shiwen

DART LANGUAGE IN ACTION
WEB DEVELOPMENT BASED ON ANGULAR FRAMEWORK

清華大學出版社
北京

内 容 简 介

本书系统且详尽地阐述编程语言 Dart 的基础知识，以及用于 Web 开发的 Dart 版 Angular 框架。

全书共 17 章，分 4 部分。第一部分（第 1～12 章）介绍开发环境的搭建、变量、内置类型、函数、运算符、流程控制语句、类、异常、泛型、库、异步、Isolate、扩展方法等基础知识；第二部分（第 13 章）主要介绍服务端的开发；第三部分（第 14～16 章）详细介绍 Angular 框架的使用和材质化组件库；第四部分（第 17 章）介绍数据库的配置与连接，并将前三部分的知识应用于项目实战。

本书适合 Dart 从入门到精通阶段的读者参考学习，所有 Dart 初学者、Angular 编程爱好者、Flutter 开发者等均可选择本书作为软件开发的实战指南或参考工具书。应用型高校计算机相关专业、培训机构也可选择本书作为 Dart 编程语言的教材或参考书。

图书在版编目（CIP）数据

Dart 语言实战：基于 Angular 框架的 Web 开发/刘仕文编著. —北京：清华大学出版社，2021.6
（计算机科学与技术丛书）
ISBN 978-7-302-57280-0

Ⅰ. ①D… Ⅱ. ①刘… Ⅲ. ①程序语言—程序设计 Ⅳ. ①TP312

中国版本图书馆 CIP 数据核字（2021）第 005028 号

责任编辑：赵佳霓
封面设计：吴　刚
责任校对：时翠兰
责任印制：沈　露

出版发行：清华大学出版社
网　　址：http://www.tup.com.cn，http://www.wqbook.com
地　　址：北京清华大学学研大厦 A 座　　**邮　　编**：100084
社 总 机：010-62770175　　**邮　　购**：010-83470235
投稿与读者服务：010-62776969，c-service@tup.tsinghua.edu.cn
质量反馈：010-62772015，zhiliang@tup.tsinghua.edu.cn
课件下载：http://www.tup.com.cn，010-83470236
印 装 者：大厂回族自治县彩虹印刷有限公司
经　　销：全国新华书店
开　　本：186mm×240mm　　**印　　张**：28　　**字　　数**：621 千字
版　　次：2021 年 7 月第 1 版　　**印　　次**：2021 年 7 月第1 次印刷
印　　数：1～1500
定　　价：109.00 元

产品编号：088043-01

前言
FOREWORD

Dart 是由谷歌公司推出的现代化编程语言，最初知道它是因为 Angular 框架推出了 Dart 版本。抱着好奇访问了 Dart 的官网，它以简单、高效、可扩展为开发目标，将强大的新语言特性与熟悉的语言构造组合成清晰、可读的语法，并提供很多语法糖来保证以更少的代码量完成指定功能。

Dart 不仅仅是一门语言，Dart 的各种开源项目和 Pub 包管理工具帮助开发人员变得更有生产力。例如，开发者可以使用 Pub 获得与 JavaScript 互操作的能力、Web UI 框架、单元测试库、用于游戏开发的库及使用 Dart 语言开发的 Flutter 跨平台移动 UI 框架等。

本书主要内容：

第 1 章搭建开发环境，安装与配置编辑器。

第 2 章主要介绍内置类型，包括数字、字符串、布尔、List 集合、Set 集合、Map 集合及符文类型的定义及使用。

第 3 章讲解函数的定义，主要包含可选参数、匿名函数、回调函数及将函数作为对象传递。

第 4 章介绍运算符，包含算术运算符、关系运算符、赋值运算符、逻辑运算符、位运算符、条件运算符及类型测试运算符。

第 5 章介绍流程控制语句，包含分支语句 if、switch；循环语句 for、while、do-while；跳转语句 break、continue。

第 6 章讲解类，包含类的定义、属性、构造函数、方法、接口、Mixin 及枚举类。

第 7 章讲解异常，包含异常的抛出、异常的捕获、自定义异常。

第 8 章讲解泛型，包含使用集合 List、Set、Map 提供的泛型接口，以及自定义泛型类和方法。

第 9 章介绍库，包含库的声明、导入、核心库、数学库、数据转换库、输入输出库。

第 10 章介绍异步，包含 Future、Stream 及生成器函数。

第 11 章介绍多线程实践途径 Isolate，包含 Isolate 的含义、事件循环、消息传递及不同 Isolate 间相互通信。

第 12 章是扩展阅读，包含可调用类的声明、扩展方法、类型定义、元数据及注释。

第 13 章介绍服务端开发，包含基础的 HTTP 请求与响应、shelf 框架的使用及使用路由包定义服务的 API。

第 14 章介绍 Angular 框架的基础知识，包含项目结构、数据绑定、内置指令、模板引用变量、服务、子组件及表单。

第 15 章介绍 Angular 框架的高级知识，包含属性指令、组件样式、依赖注入、生命周期挂钩、管道、路由、结构指令、HTTP 连接及项目部署。

第 16 章介绍材质化组件库 angular_components，该库包含表单、业务流及布局中常用的组件。

第 17 章是项目实战，介绍数据库的安装与连接，以及通过用于时间规划的项目 Deadline 来温习本书所学的知识点。

扫描下方二维码可下载本书源代码。读者可直接扫描书中二维码观看本书配套视频教程。

本书源代码

在学习本书之前，读者应当具备使用 C 或 Java 等语言的编程经验。本书所涉及的示例代码均可在 Dart SDK 2.7 或更高版本中运行。

刘仕文

2021 年 5 月

目录

CONTENTS

第一部分

第三部分

第四部分

第 一 部 分

第 1 章

简　　介

1.1　概述

Dart 是由 Google 开发的计算机编程语言，它的语法类似 C 语言。可以使用 Dart 编写命令行脚本、服务器端应用、Web 应用和移动端应用。

Web 应用：常采用 AngularDart 框架开发 Web 应用程序，它是 Angular 框架的 Dart 版本。Dart Web 采用编译器将 Dart 代码编译成 JavaScript，使得 Dart Web 应用程序可在浏览器中运行。在开发应用时采用 dartdevc 编译器，部署应用时采用 dart2js 编译器。

移动端应用：采用 Flutter 框架可以编写 iOS 和 Android 应用程序。

1.2　环境安装与配置

10min

Dart SDK 支持 Windows、Mac OS、Linux 3 种操作系统，这里提供两个资源站供 SDK 下载：

官网：dart.dev/get-dart

国内：dartlang.tech

安装包分为 3 种类型：稳定版、测试版和开发版，推荐使用稳定版。

首先请根据相应平台下载文件包，默认为 zip 压缩格式文件。解压 zip 文件，解压后的目录里包含所需的 dart-sdk 目录。

1.2.1　Windows 用户

第 1 步：复制 dart-sdk 目录到 C:\Program Files 目录下。

第 2 步：添加 dart-sdk\bin 路径到 Path 环境变量，这样在任何目录下都可以执行 Dart SDK 提供的工具命令。

以 Windows 10 为例，右击屏幕左下角的“开始”按钮，在弹出的菜单中选择“系统”，弹出如图 1-1 所示的设置对话框，单击右侧的系统信息即可弹出如图 1-2 所示的 Windows 系统对话框。

图 1-1 设置对话框

图 1-2 Windows 系统对话框

选择左边的"高级系统设置",打开如图 1-3 所示的系统属性对话框。

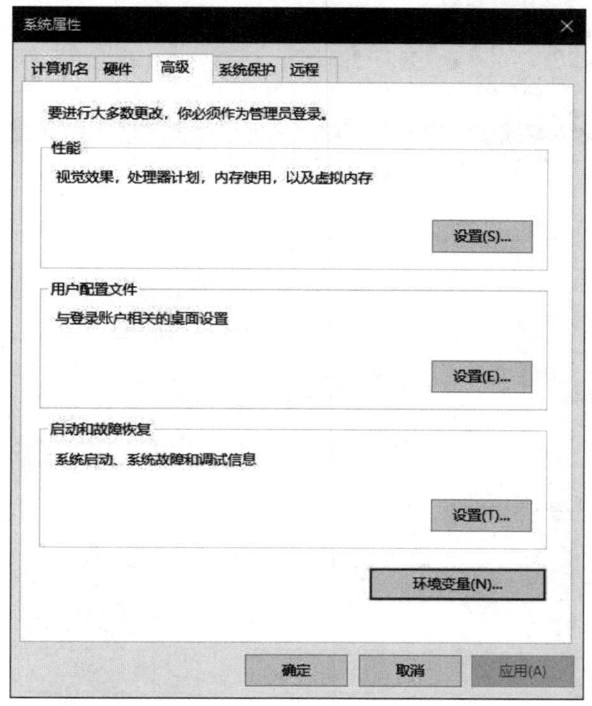

图 1-3　系统属性对话框

在如图 1-3 所示的系统属性对话框中,首先选择"高级",然后单击"环境变量"按钮打开环境变量对话框,如图 1-4 所示。可以在用户变量或系统变量中添加环境变量,一般情况下在用户变量中添加,双击 Path 变量条目进入如图 1-5 所示的编辑环境变量对话框。

在图 1-5 所示的编辑环境变量对话框中单击右侧的"新建"按钮,输入值 C:\Program Files\dart-sdk\bin,最后单击"确定"按钮即可完成编辑。

第 3 步:单击屏幕左下角"开始"按钮,在弹出菜单中找到"Windows 系统"文件夹,单击该文件夹下的"命令提示符"菜单。在命令提示符中输入 dart 命令。出现如图 1-6 所示信息表示环境配置成功。

1.2.2　Mac 用户

下载 Mac OS 版本的安装包到计算机,这里默认放在下载目录里,如图 1-7 所示。

双击文件名即可解压,得到 dart-sdk 文件夹。通常会将安装目录放在/usr/local 目录下,因此首先打开终端输入命令 sudo mv,再将 dart-sdk 目录拖到终端,这里 mv 和路径间有空格,如图 1-8 所示。

图 1-4　环境变量对话框

图 1-5　编辑环境变量对话框

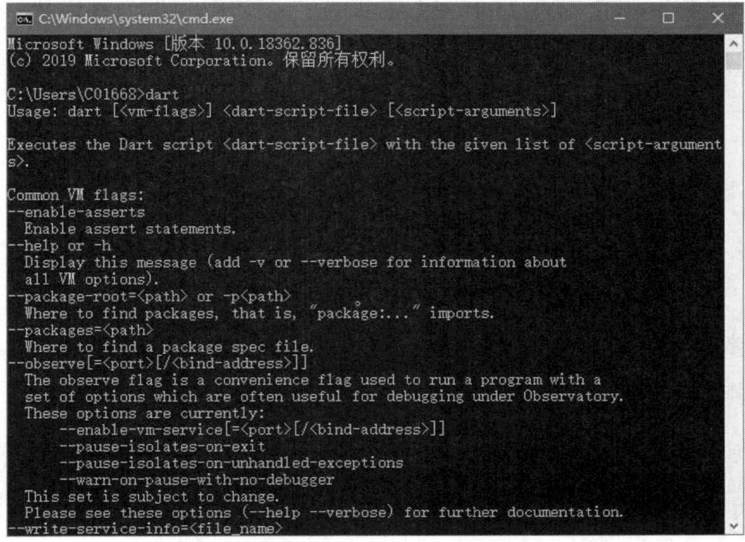

图 1-6　命令提示符对话框

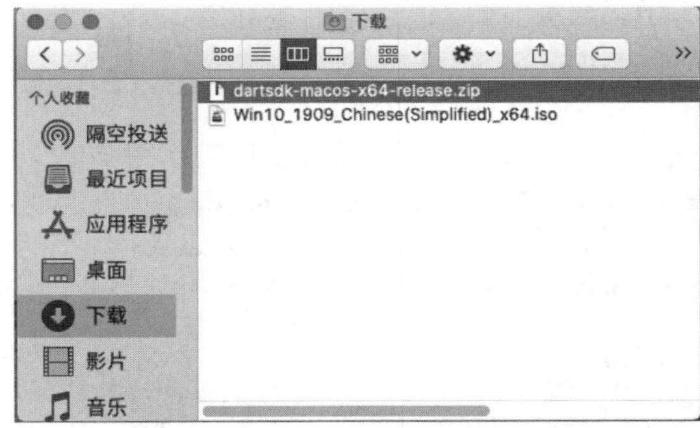

图 1-7　安装包存放目录

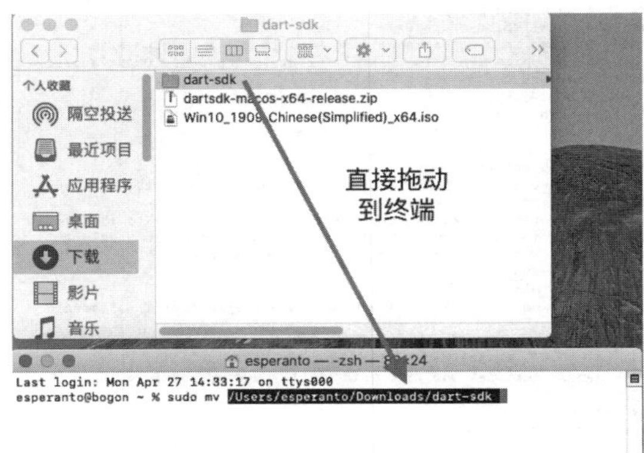

图 1-8　移动文件夹

在终端中继续输入/usr/local，这里/usr 前面需要空格，然后回车即可执行命令，如图 1-9 所示。

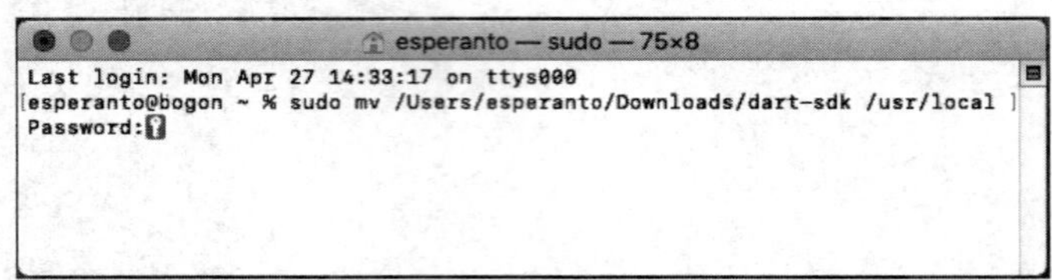

图 1-9　执行移动目录命令

接下来配置环境变量。

bash 终端配置代码如下：

```
echo 'export PATH = /usr/local/dart - sdk/bin: $ {PATH}' >> ~/.bash_profile
source ~/.bash_profile
```

zsh 终端配置代码如下：

```
echo 'export PATH = /usr/local/dart - sdk/bin: $ {PATH}' >> ~/.zshrc
source ~/.zshrc
```

在终端中输入 dart 命令，出现如图 1-10 所示信息表示配置成功。

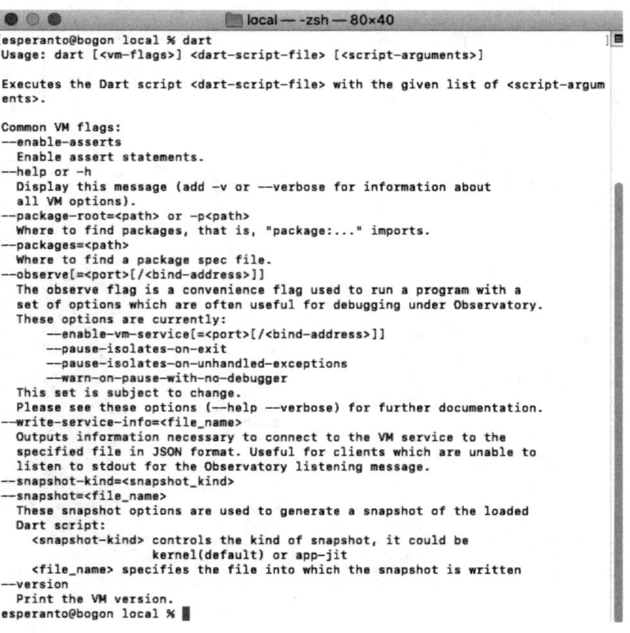

图 1-10　执行 dart 命令

1.3 IntelliJ IDEA 的安装与配置

12min

在官网 https://www.jetbrains.com/idea/download 页面有所需的下载链接，如图 1-11 所示。IntelliJ IDEA 支持 Windows、Mac、Linux 3 种操作系统，并且提供 Ultimate 和 Community 两种版本，选择 Community 版就可以满足需求。

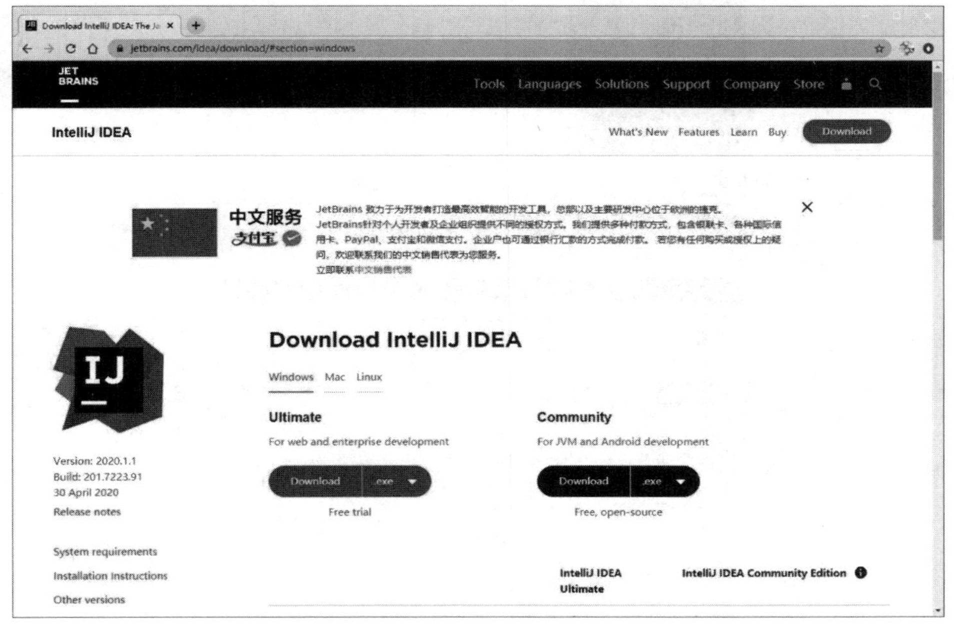

图 1-11 编辑器 IDEA 下载页面

1.3.1 Windows 用户

双击安装包，进入安装界面，如图 1-12 所示。

单击 Next 按钮，进入如图 1-13 所示安装目录对话框，通常采用默认安装路径。

单击 Next 按钮，进入如图 1-14 所示安装选项对话框。在 Create Desktop Shortcut 下勾选启动器，这里选择的是 64-bit launcher。在 Update PATH variable 下勾选 Add launchers dir to the PATH。

单击 Next 按钮，进入如图 1-15 所示的选择启动菜单目录对话框，这里采用默认菜单目录即可，单击 Install 按钮，程序将自动安装。

安装完成后如图 1-16 所示，选择 Reboot now 选项后单击 Finish 按钮，系统将重启。

系统重启后，单击桌面上的 IntelliJ IDEA 编辑器启动图标，弹出如图 1-17 所示的编辑器主题选择对话框，根据习惯选择相应主题，然后单击对话框左下角的 Skip Remaining and

Set Defaults 按钮。

图 1-12　编辑器安装界面

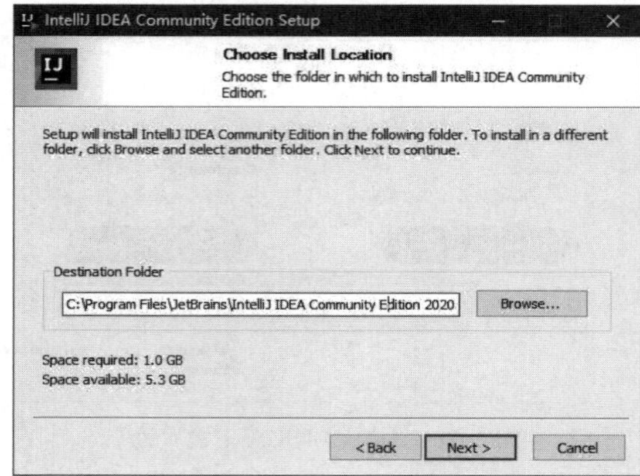

图 1-13　安装目录对话框

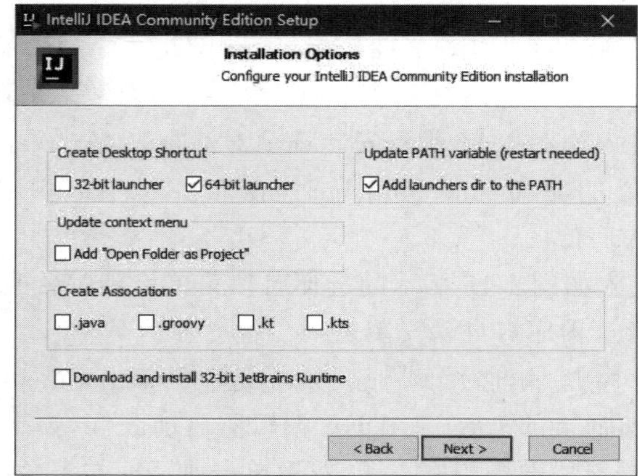

图 1-14　安装选项对话框

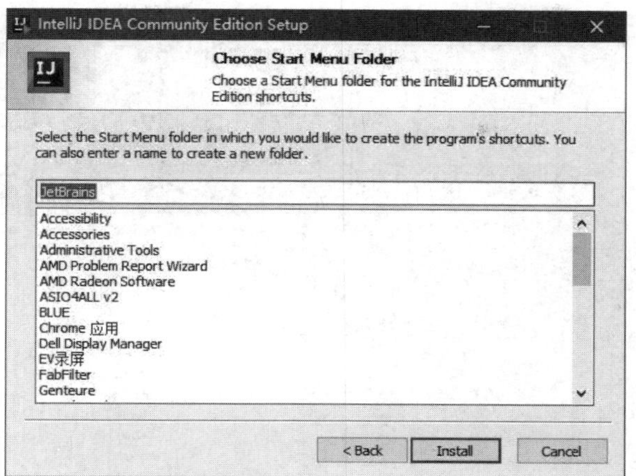

图 1-15 选择启动菜单目录对话框

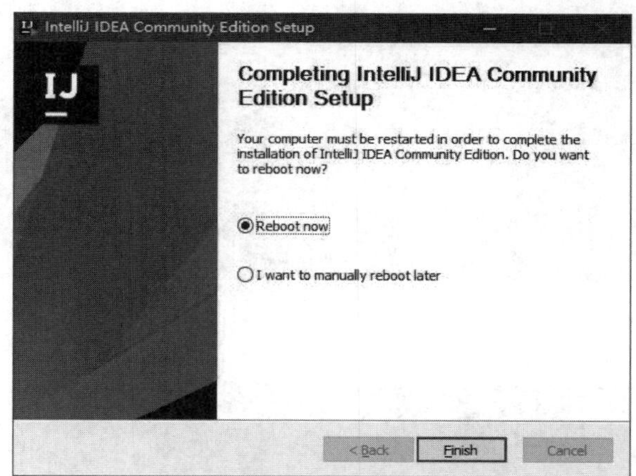

图 1-16 安装完成对话框

进入如图 1-18 所示的编辑器欢迎界面对话框，选择对话框右下角的 Configure 下拉菜单，在菜单中选择 Plugins 选项，弹出插件安装对话框。

在插件对话框中选择 Marketplace 选项卡，在搜索框中输入 Dart，单击 Dart 插件右边的 Install 按钮安装该插件，如图 1-19 所示。

安装完成后，重启编辑器，在欢迎界面选择 Create New Project 选项来创建新项目。弹出如图 1-20 所示的项目创建对话框，单击左侧列表中的 Dart 选项，在右侧标题为 Dart SDK path 的路径选择框中选择 Dart SDK 的安装路径，即 C:\Program Files\dart-sdk。在 Generate sample content 下会罗列一些 Dart 项目创建模板。

实际上此时的模板并不够用，因此还需要安装第三方软件包 stagehand。在 Dart SDK 中包含 pub 工具，该工具可用于获取 Dart 项目依赖的第三方软件包和工具。

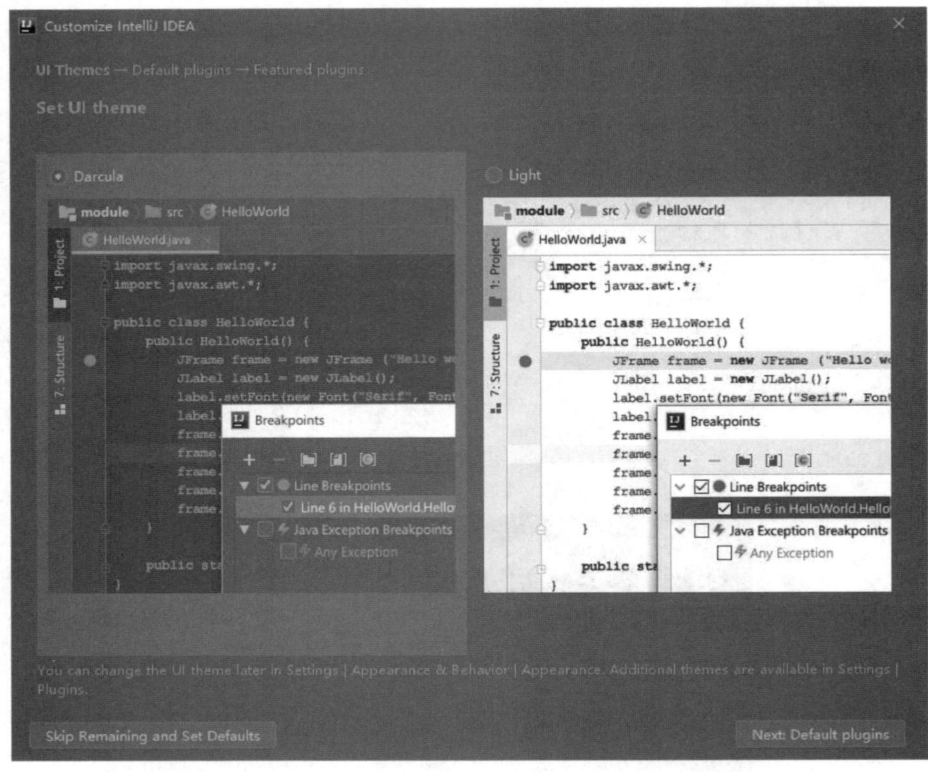

图 1-17　编辑器主题选择对话框

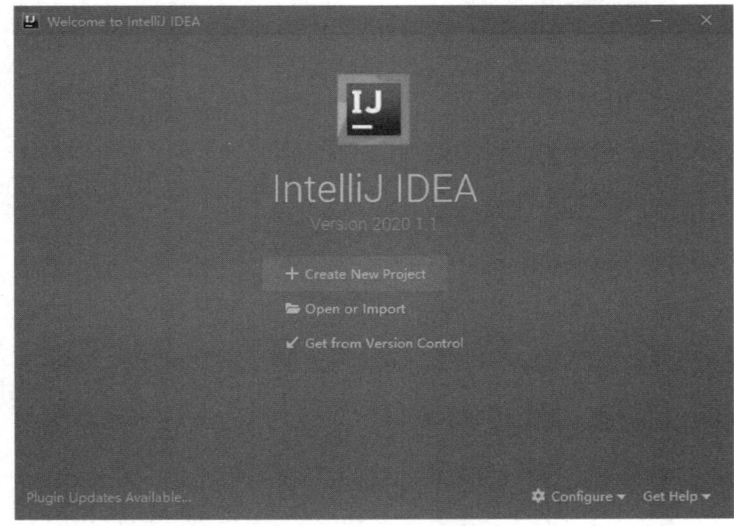

图 1-18　编辑器欢迎界面对话框

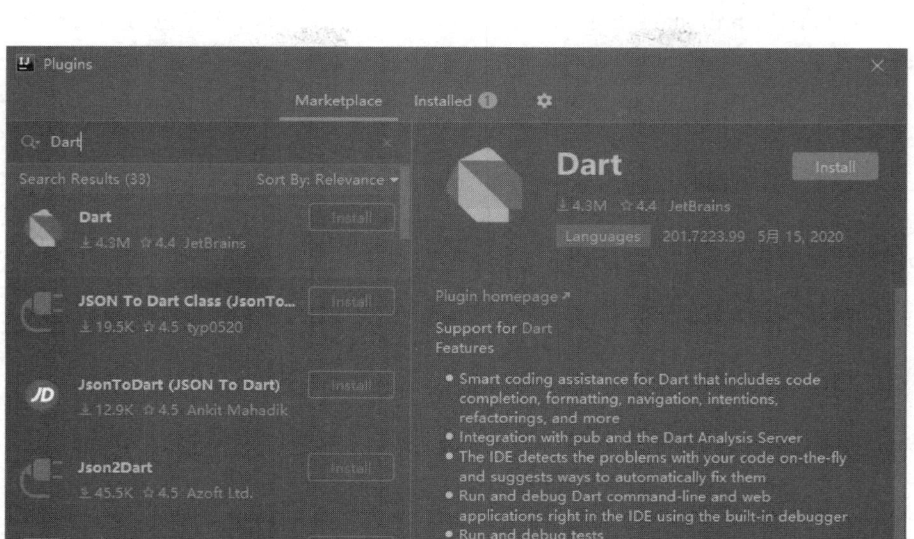

图 1-19　插件安装对话框

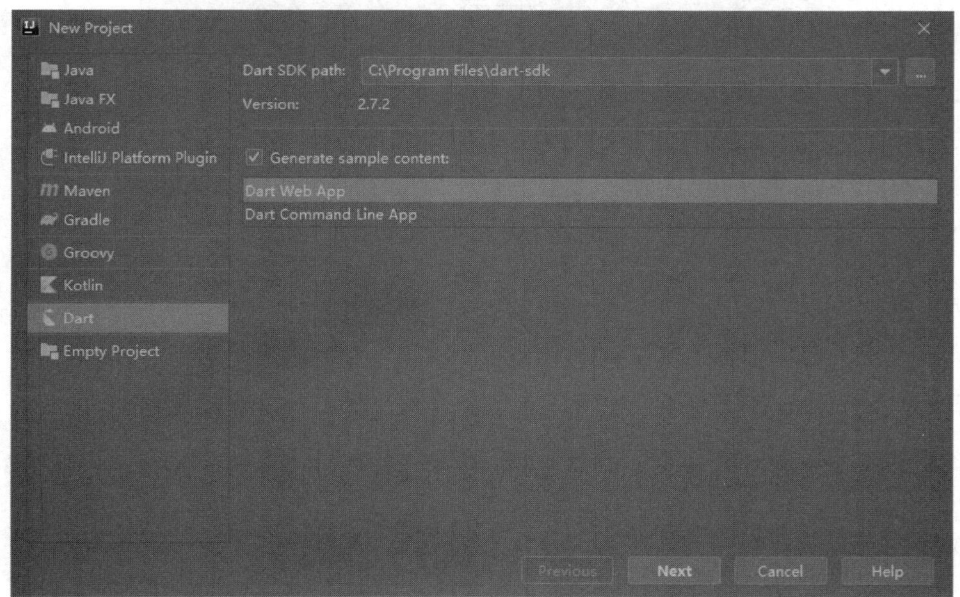

图 1-20　项目创建对话框

pub 工具默认从 pub. dartlang. org 上获取软件包，通常获取软件包的速度较慢，因此需要为 pub 工具更换镜像源。

上海交通大学镜像地址：https://dart-pub. mirrors. sjtug. sjtu. edu. cn

清华大学镜像地址：https://mirrors. tuna. tsinghua. edu. cn/dart-pub

打开环境变量对话框，打开步骤参见环境安装与配置。在该对话框中单击用户变量下的"新建"按钮，弹出"新建用户变量"对话框。变量名填入值 PUB_HOSTED_URL，这里选择清华大学的镜像源，因此变量值填入 https://mirrors. tuna. tsinghua. edu. cn/dart-pub。

打开命令提示符，输入如下命令：

```
pub global activate stagehand
```

回车执行该命令，如果结果如图 1-21 所示即表示 stagehand 软件包安装成功。

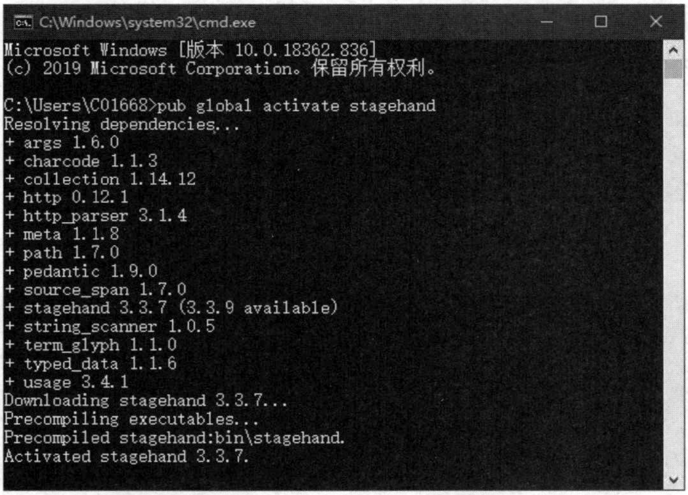

```
C:\Windows\system32\cmd.exe
Microsoft Windows [版本 10.0.18362.836]
(c) 2019 Microsoft Corporation。保留所有权利。

C:\Users\C01668>pub global activate stagehand
Resolving dependencies...
+ args 1.6.0
+ charcode 1.1.3
+ collection 1.14.12
+ http 0.12.1
+ http_parser 3.1.4
+ meta 1.1.8
+ path 1.7.0
+ pedantic 1.9.0
+ source_span 1.7.0
+ stagehand 3.3.7 (3.3.9 available)
+ string_scanner 1.0.5
+ term_glyph 1.1.0
+ typed_data 1.1.6
+ usage 3.4.1
Downloading stagehand 3.3.7...
Precompiling executables...
Precompiled stagehand:bin\stagehand.
Activated stagehand 3.3.7.
```

图 1-21　软件包安装对话框

完成软件包安装后，重启编辑器，继续创建项目，此时在项目生成模板列表下就会出现足够的项目模板，如图 1-22 所示。到此就完成了编辑器的安装与配置。

1.3.2　Mac 用户

编辑器安装与 Windows 用户类似，这里不再赘述。需要特别说明的是 pub 工具镜像源的更换，以清华大学镜像源为例。

bash 终端配置代码如下：

```
echo 'export PUB_HOSTED_URL = https://mirrors.tuna.tsinghua.edu.cn/dart - pub' >> ~/.bash
_profile
source ~/.bash_profile
```

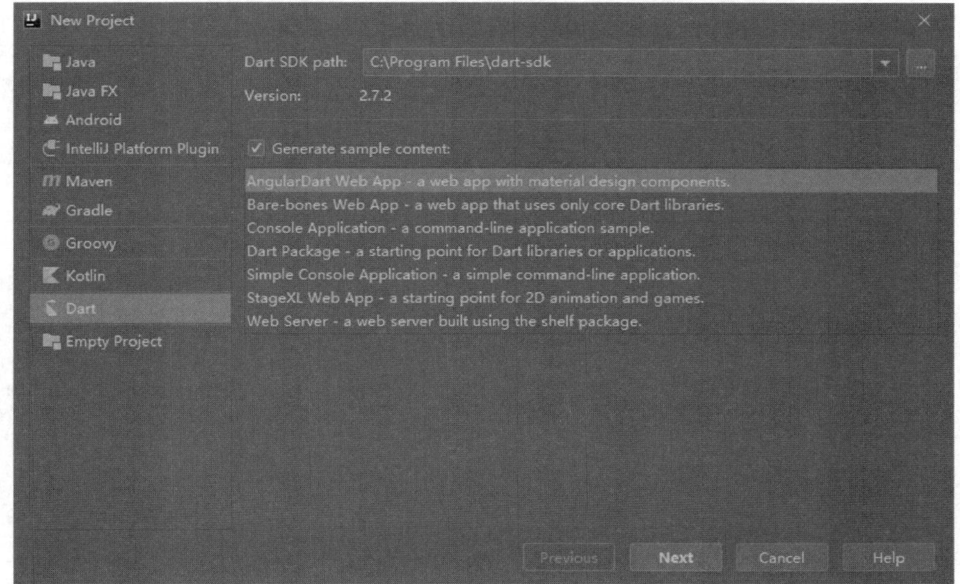

图 1-22　使用项目模板

zsh 终端配置代码如下：

```
echo 'export PUB_HOSTED_URL=https://mirrors.tuna.tsinghua.edu.cn/dart-pub' >> ~/.zshrc
source ~/.zshrc
```

第 2 章 变量和内置类型

创建命令行应用程序，将项目命名为 chapter2。本章所有文件都在该项目的 bin 文件夹中创建与运行。

11min

2.1 标识符

标识符就是为变量、方法、枚举、类、接口等指定的名字。在 Dart 语言中标识符的命名风格有 3 种：小驼峰命名法、大驼峰命名法、下画线命名法。

2.1.1 小驼峰命名法

小驼峰命名法的单词首字母大写，其余字母小写，但第一个单词除外。如：lowerCamelCase，适用于变量、参数、常量、枚举值及类成员等。

2.1.2 大驼峰命名法

大驼峰命名法又称帕斯卡命名法，单词首字母大写，其余字母小写。如：UpperCamelCase，适用于类名、枚举类型、类型定义、类型参数及扩展名等。

2.1.3 下画线命名法

下画线命名法的单词全部小写，单词间使用下画线连接。如：lowercase_with_underscores，适用于库名、包名、文件夹及源文件等。

如果对本节中出现的关键词不熟悉没有关系，这些关键词将在后续一一呈现且遵循上述规则。

2min

2.2 关键字

关键字是 Dart 中具有特殊用途的一系列词，应当避免将它们用作标识符。在这里仅需了解，我们会在后面的章节中陆续使用这些关键字。

表 2-1 所示的是保留字，它们不能用作标识符。

表 2-1 保留字

assert	break	case	catch	class	const	continue
default	do	else	enum	extends	false	final
finally	for	if	in	is	new	null
rethrow	return	super	switch	this	throw	true
try	var	void	while	with		

表 2-2 所示的是上下文关键字，它们仅在特定的位置具有含义，在所有地方都是有效标识符。

表 2-2 上下文关键字

async	hide	on	show	sync		

表 2-3 所示的是内置标识符，在大部分地方是有效标识符，不能用作类或类型名，也不能用作导入前缀。

表 2-3 内置标识符

abstract	as	covariant	deferred	dynamic	export	external
factory	function	get	implements	import	interface	library
mixin	operator	part	set	static	typedef	

表 2-4 所示的是受限制的保留字，与异步相关，在带有 async、async * 或 sync * 标识的函数体中不能使用 await 和 yield 作为标识符。

表 2-4 受限保留字

await	yield					

2.3 变量

12min

变量包含变量名和变量值，变量名属于标识符，故需遵守标识符的命名规范。赋给变量的一切值都是对象，每个对象都是一个类的实例。数字、函数、字符串及 null 等都是对象，所有对象都继承自 Object 类。变量存储的是对象的引用，当把对象赋给变量时，实际上是将对象的引用赋给变量。

变量的声明格式如下，中括号中的内容代表可选。

```
数据类型  变量名 [ = 初始值];
```

定义一个类型为 String 的变量 catName，且为其赋值。示例代码如下：

```
//chapter2/bin/example_01.dart
void main(){
  String catName = '千岁岁';
  print('变量 catName 运行时类型为 ${catName.runtimeType}');
}
```

名为 catName 的变量存储着 String 类型的对象的引用，该对象的值为千岁岁。

在 Dart 中可以使用 dynamic 或 var 声明变量的类型，它们都会根据赋值类型推断该变量的类型。其中 var 常用于声明局部变量，而不使用确定的类型，如上例中使用了 String 这一确定类型。示例代码如下：

```
//chapter2/bin/example_02.dart
void main(){
  dynamic dogName = '啵啵乐';
  print('变量 dogName 运行时类型为 ${dogName.runtimeType}');
}
```

如上例，名为 dogName 的变量将被推断为 String 类型。

2.3.1 默认值

在 Dart 中无论声明类型是什么，未初始化的变量都拥有一个默认值 null，即使其类型为数字。示例代码如下：

```
//chapter2/bin/example_03.dart
void main(){
  int sum;
  print(sum == null);
}
```

上例中运算符"＝＝"用于判断两个对象的值是否相等，它的返回值是布尔类型，即 true 或 false。

2.3.2 const 和 final

如果在赋值后不需要改变变量的值，那么可以使用 const 或 final 来修饰变量，const 和 final 都可以代替 var，被修饰的变量的类型由赋值对象的类型决定，也可以放在确定的类型前共同修饰变量。

final 修饰变量的格式如下：

```
final [数据类型] 变量名 = 初始值;
```

const 修饰变量的格式如下：

```
const [数据类型] 变量名 = 初始值;
```

final 修饰的变量只能设置一次值，除了在声明处赋值，还可以通过类的构造函数为 final 修饰的变量赋值，这一点在后面会介绍。示例代码如下：

```
//chapter2/bin/example_04.dart
void main(){
//不指定类型,赋值的类型作为推断类型的依据,这里为 String 类型
  final name = '猫猫村长';
//可放在确定的类型前,共同修饰变量
  final String nickName = '猫猫';
}
```

不能修改由 final 修饰的变量的值。

```
//错误,final 修饰的变量不能被修改
name = '猫猫';
```

const 修饰的变量应在声明处赋值，且值必须为编译时常量。用 const 修饰变量时可以直接使用常量为其赋值，也可以使用由 const 修饰的其他变量组成的表达式来赋值。示例代码如下：

```
//chapter2/bin/example_05.dart
void main(){
  //创建常量
  const pi = 3.14;
  const r = 6;
  //常量的表达式可由其他常量组成
  const l = 2 * pi * r;
}
```

const 不仅用于定义常量，也可以用来创建常量值，示例代码如下：

```
var  ls = const [];
```

没有使用 final 或 const 修饰的变量，其值可以被修改，即使变量引用过 const 修饰的对象，示例代码如下：

```
ls = [1,2,3];
```

2.4 数字

3min

数字在 Dart 中对应的类为 num，num 类中定义了+、-、*、/等基本运算符，num 类有两个子类：

(1) int：常被称为整型，整型值长度不超过 64 位，这依赖于平台，在 Dart VM 中，其取值可从 -2^{63} 到 $2^{63}-1$。当将 Dart 程序编译成 JavaScript 时将使用 JavaScript 中的数值范围，其取值可从 -2^{53} 到 $2^{53}-1$。

(2) double：64 位双精度浮点数，符合 IEEE 754 标准。

整型是不带小数点的数字，定义时可以使用 int 或 num 修饰变量。示例代码如下：

```
//chapter2/bin/bin/example_06.dart
void main(){
  //定义 int 型数字
  int i = 1;
  var j = 332131;
  num k = 1;
  print('k 的运行时类型：${k.runtimeType}');
}
```

如果数字中包含小数点，那么它就是浮点型。定义时可以使用 double 或 num 修饰变量。示例代码如下：

```
//定义 double 型数字
double pi = 3.14;
var r = 6.7;
num w = 1.2;
print('w 的运行时类型：${w.runtimeType}');
```

必要时，整型可以自动转为浮点型。示例代码如下：

```
//等同于 double l = 3.0;
double l = 3;
print('l:$l');
```

2.5 字符串

6min

Dart 字符串是 UTF-16 编码的字符序列，对应的类型为 String，可以使用单引号或双引号来创建字符串。示例代码如下：

```
//使用单引号创建字符串
var s1 = 'Today is sunny';
//使用双引号创建的字符串
var s2 = "Today is sunny";
```

也可以使用三重单引号或三重双引号来创建多行字符串。示例代码如下：

```
//使用三重单引号创建多行字符串
var s3 = '''Today
        is
        sunny''';
//使用三重双引号创建多行字符串
var s4 = """Today
        is
        sunny""";
```

插值操作可以将表达式的值放入字符串中,使用格式：${expr},expr 代表的是表达式。如果表达式是标识符,则可以省略{}。使用格式：$Identifier,Identifier 代表的是标识符。插值实际上是通过调用对象的 toString()方法获取该对象的字符串形式。示例代码如下：

```
//chapter2/bin/example_07.dart
void main(){
  var pi = 3.14;
  var r = 6;
  var s = 'QianSuiSui is a cat';
//插值表达式
  print('字符串 s: $ s');
  print('圆的周长 L: $ {2 * pi * r}');
}
```

可以通过多个相邻字符串组成一个新的字符串,也可以使用"+"运算符。示例代码如下：

```
//将相邻字符串拼接在一起
  var s5 = 'Today'
           'is'
           'sunny';
//通过使用"+"运算符将字符串拼接在一起
  var s6 = 'Today' + 'is' + 'sunny';
```

2.6 布尔

为了表示布尔值,Dart 中有一个 bool 类型,只有两个对象是 bool 类型：true 和 false。它们都是编译时常量。示例代码如下：

```
bool isEmpty = true;
bool isNull = false;
```

不能为 bool 类型的变量赋值 true 和 false 之外的常量，也不能用运算结果不是 true 或 false 的表达式为其赋值，否则会发生编译错误。示例代码如下：

```
//下列代码将发生编译错误
//使用字符串为 bool 变量赋值
bool isEmpty = 'AA';
//使用数字为 bool 变量赋值
bool isNull = 1;
```

当 Dart 程序需要一个布尔值时，只能向其提供 true 或 false。不能使用 if(非布尔值)或者 assert(非布尔值)这样的代码检查布尔值。应该显式地检查布尔值，示例代码如下：

```
//chapter2/bin/example_08.dart
void main(){
//检查是否为空字符串
  var name = '';
  print(name.isEmpty);

//检查是否小于或等于 0
  var topPoint = 0;
  print(topPoint <= 0);

//检查是否为 null
  var label;
  print(label == null);

//检查是否为 NaN
  var y = 0/0;
  print(y.isNaN);
}
```

13min

2.7 List 集合

集合是编程中常用的数据类型，它们相比普通类型更加丰富、更具扩展性。

List 集合中的元素是有序的，并且可以重复出现相同元素。在其他语言中会将数组(Array)单独拿出来作为一种类型，但在 Dart 中数组(Array)由 List 集合来表示。

List 集合中可以存放任意类型的对象，也可以通过“<>”符号指定泛型，确保存入该集合的对象都是泛型所指定的单一类型。

Dart 中 List 字面量看起来与 JavaScript 中的数组字面量一样。示例代码如下：

```
var list = [1,2,3];
```

在上述代码中 Dart 推断对象 list 的类型为 List<int>，因此只可以向该集合中添加 int 类型的对象，否则会报错。

上述代码也可以使用显式类型声明。示例代码如下：

```
List  list = [1,2,3];
```

或者显式指定泛型为 int：

```
List<int> list = [1,2,3];
```

当使用字面量初始化 List 集合时，如果初始化值不是同一类型，则 List 集合中的元素为动态类型，即任意类型都可以存放进该集合。示例代码如下：

```
var list = [1,2,'abc'];
    //添加 int 类型的数据
    list.add(33);
    //添加字符串类型的数据
    list.add('value');
    //添加布尔类型的数据
    list.add(true);
```

此时对象 list 的推动类型为 List<Object>，这样使用起来确实很方便。但是在处理集合中的对象时会增加难度，因此推荐使用泛型指定 List 集合中可存放的数据类型。

2.7.1 常用属性

(1) first：返回 List 集合中的第一个元素。

(2) last：返回 List 集合中的最后一个元素。

(3) length：返回 List 集合中元素的数量，返回值为 int 型。

(4) isEmpty：判断 List 集合中是否有元素，如果没有元素则返回 true，否则返回 false。

(5) iterator：返回迭代器(Iterator)对象，迭代器对象用于遍历集合，该属性是从 Iterable 类继承而来。

使用示例代码如下：

```
//chapter2/example_09.dart
void main(){
  var list = [63, 70.9, 'abc', true];
//打印第一个元素
  print('第一个元素：${list.first}');
```

```
  //打印最后一个元素
    print('最后一个元素: ${list.last}');
  //打印集合的长度
    print('List 集合的长度: ${list.length}');
  //打印运行时类型
    print('运行时类型: ${list.runtimeType}');
  //打印迭代器的类型
    print('迭代器的类型: ${list.iterator.runtimeType}');
  }
```

运行结果如下：

```
第一个元素:63
最后一个元素:true
List 集合的长度:4
运行时类型:List<Object>
迭代器的类型:ListIterator<Object>
```

2.7.2 常用方法

(1) 操作符[](int index)：返回 List 集合中给定索引处的对象。

(2) 操作符[]=(int index,E value)：将 List 集合中给定索引处 index 的值替换为 value。

(3) add(E value)：将 value 添加到 List 集合的末尾，并使 List 集合长度增加 1。

(4) insert(int index,E element)：在 List 集合指定位置 index 插入元素 element。

(5) remove(Object value)：从 List 集合中删除第一次出现 value 值的元素。

(6) removeAt(int index)：移除 List 集合指定位置的元素，并返回该元素。

(7) clear()：移除 List 集合中所有的元素，并将 List 集合的长度置为 0。

(8) sublist(int start,[int end])：返回 List 集合中位置 start 和 end 之间的元素的集合，包括 start 处的元素，但不包括 end 处的元素。

(9) indexOf(E element,[int start = 0])：按顺序查找 List 集合中的元素，返回第一次出现指定元素 element 的索引。如果集合中没有该元素，则返回 −1。可选参数 start 表示从 List 集合中指定索引处开始查找，一直查找到集合的最后一个元素，其默认值为 0。

List 使用基于 0 的索引，其中 0 是第一个元素的索引，list.length−1 是最后一个元素的索引。

使用示例代码如下：

```
//chapter2/bin/example_10.dart
void main() {
```

```
  var list = [63, 70.9, 'abc', true, 'abc'];
//打印下标为 2 的元素,即集合中的第 3 个元素
  print('元素 list[2]: ${list[2]}');
//修改 list[1]中的元素
  list[1] = false;
//打印修改后 list[1]中的元素
  print('元素 list[1]: ${list[1]}');
//向 list 添加元素
  list.add('new element');
  print('修改后的 list: $list');
//向 list 下标 1 处插入元素
  list.insert(1, 'insert element');
  print('插入新元素后的 list: $list');
//在 list 中查找元素,返回其第一次出现的下标
  print('abc 在 list 中第一次出现的下标: ${list.indexOf('abc')}');
//返回 list 下标从 2 到 5 的元素的集合
  print('list 中下标 2 到 5 的元素: ${list.sublist(2, 5)}');
//在 list 中查找元素,删除第一次出现 abc 的元素
  print('list 中移除 abc 元素: ${list.remove('abc')}');
//移除元素后的 list
  print('移除元素后的 list: $list');
//清除 list 中的元素
  list.clear();
}
```

运行结果如下：

```
初始 list 中的元素:[63, 70.9, abc, true, abc]
元素 list[2]:abc
元素 list[1]:false
修改后的 list:[63, false, abc, true, abc, new element]
插入新元素后的 list:[63, insert element, false, abc, true, abc, new element]
abc 在 list 中第一次出现的下标:3
list 中下标 2 到 5 的元素:[false, abc, true]
list 中移除 abc 元素:true
移除元素后的 list:[63, insert element, false, true, abc, new element]
```

若需要创建一个包含编译时常量的 List,可以在 List 字面量前添加 const：

```
var  constantList = const [1,2,3];
```

需要创建固定长度的 List 集合,需使用构造函数指定：

```
var  list = List(6);
```

8min

2.8 Set 集合

Set 集合中的元素无序且唯一，即不可重复。Set 的默认实现是 LinkedHashSet 类，该子类依据元素插入的顺序进行迭代。Dart 中提供了 Set 字面量和 Set 类型两种方式声明 Set 集合。

采用 Set 字面量来创建 Set 集合的代码如下：

```
var halogens = {'fluorine','chlorine','bromine','iodine','astatine'};
```

Dart 推断 halogens 变量是一个 Set < String >类型的集合，只可以向该集合添加 String 类型的对象。

可以在{}前面加上泛型参数来创建一个空的 Set 集合，也可以将{}赋值给声明类型为 Set 的变量。示例代码如下：

```
var cats = <String>{};
Set<String> dogs = {};
```

可以在 Set 字面量前添加 const 关键字创建一个 Set 类型的编译时常量。示例代码如下：

```
final constantSet = const {
  'fluorine',
  'chlorine',
  'bromine',
  'iodine',
  'astatine',
};
```

2.8.1 常用属性

(1) length：返回 Set 集合中元素的数量，返回值为 int 类型。

(2) iterator：返回迭代器(Iterator)对象，迭代器对象用于遍历集合。

(3) isEmpty：判断 Set 集合中是否有元素，如果有则返回 false，如果没有则返回 true。

使用示例代码如下：

```
//chapter2/bin/example_11.dart
void main(){
  var set = {'abc',1,true};
  //打印 Set 集合的长度
  print('Set 集合的长度：${set.length}');
  //打印迭代器的类型
  print(set.iterator.runtimeType);
```

```
  //判断 Set 集合是否为空
  print('Set 集合是否为空：${set.isEmpty}');
}
```

运行结果如下：

```
Set 集合的长度:3
_CompactIterator<Object>
Set 集合是否为空:false
```

2.8.2 常用方法

(1) add(E value)：将元素添加到 Set 集合。

(2) clear()：清除 Set 集合中的所有元素。

(3) remove(Object value)：移除 Set 集合中的指定元素。

使用示例代码如下：

```
//chapter2/bin/example_12.dart
void main(){
  var set = {'abc',1,true};
  //打印原 Set 集合
  print('原 Set 集合：$set');
  //添加元素到 Set 集合
  set.add(99);
  print('添加元素后的 Set 集合：$set');
  //移除 Set 集合中的元素
  set.remove(true);
  print('移除元素后的 Set 集合：$set');
  //清除 Set 集合中的所有元素
  set.clear();
}
```

运行结果如下：

```
原 Set 集合：{abc, 1, true}
添加元素后的 Set 集合：{abc, 1, true, 99}
移除元素后的 Set 集合：{abc, 1, 99}
```

2.9 Map 集合

8min

Map 集合是一种将键(key)和值(value)相关联的对象，key 和 value 都可以是任何对象。key 不可重复，但 value 可重复。Dart 中支持使用 Map 字面量和 Map 类型来构建 Map 对象。

采用 Map 字面量来创建 Map 集合的示例代码如下：

```
//chapter2/bin/example_13.dart
void main(){
//键和值都是 String 类型
  var gifts = {
    //Key:Value
    'first':'partridge',
    'second':'turtledoves',
    'fifth':'golden rings'
  };
//键是 int 型,值是字符串型
  var nobleGases = {
    2:'helium',
    10:'neon',
    10:'argon',
  };
}
```

Dart 将名为 gifts 的变量的类型推断为 Map < String,String >,将 nobleGases 推断为 Map < int,String >。

也可以使用 Map 构造函数来创建 Map。示例代码如下：

```
//chapter2/bin/example_14.dart
void main(){
  //用构造函数 Map()构建 Map 对象
  var gifts1 = Map();
  gifts['first'] = 'partridge';
  gifts['second'] = 'turtledoves';
  gifts['fifth'] = 'golden rings';

  var nobleGases1 = Map();
  nobleGases[2] = 'helium';
  nobleGases[10] = 'neon';
  nobleGases[18] = 'argon';
}
```

2.9.1 常用属性

(1) length：返回 Map 集合中的键值对数量。

(2) keys：返回 Map 集合中的所有键(key),返回值是可迭代对象。

(3) values：返回 Map 集合中的所有值(value),返回值是可迭代对象。

使用示例代码如下：

```
//chapter2/bin/example_15.dart
void main(){
  var map = {
    1: 'partridge',
    'second': 'turtledoves',
    'age': 32
  };
  //打印 Map 集合的键值对数量
  print('Map 集合的键值对数量: ${map.length}');
  //返回 Map 集合中的所有键
  print('Map 集合中的所有键: ${map.keys}');
  //返回 Map 集合中的所有值
  print('Map 集合中的所有值: ${map.values}');
}
```

运行结果如下：

```
Map 集合的键值对数量:3
Map 集合中的所有键:(1, second, age)
Map 集合中的所有值:(partridge, turtledoves, 32)
```

2.9.2 常用方法

(1) 操作符[](Object key)：获取给定 key 的值，如果该 key 不存在，则返回 null。

(2) 操作符[]=(K key, V value)：将值(value)与给定键(key)相关联，如果该键存在于 Map 集合中，则将对应值修改为 value。如果对应键不存在，则将键值对添加到 Map 集合。

(3) containsKey(Object key)：判断 Map 集合中是否包含指定键，如果包含则返回 true，否则返回 false。

(4) containsValue(Object value)：判断 Map 集合是否包含指定值，如果包含则返回 true，否则返回 false。

使用示例代码如下：

```
//chapter2/bin/example_16.dart
void main(){
  var map = {
    1: 'partridge',
    'second': 'turtledoves',
    'age': 32
  };
  //打印原 Map
  print('原 Map: $map');
```

```
  //获取 Map 集合中键为 age 的值
  print('Map 的键值对数量: ${map['age']}');
  //向 Map 集合中添加键值对
  map['name'] = 'Bob';
  print('添加键值对后的 Map: $map');
  //修改 Map 集合中键为 name 的值
  map['name'] = 'Jobs';
  print('修改键的值后的 Map: $map');
  //判断 Map 集合中是否包含键 age
  print('Map 集合中是否包含键 age: ${map.containsKey('age')}');
  //判断 Map 集合中是否包含值 32
  print('Map 集合中是否包含值 32: ${map.containsValue(32)}');
}
```

运行结果如下：

```
原 Map:{1: partridge, second: turtledoves, age: 32}
Map 的键值对数量:32
添加键值对后的 Map:{1: partridge, second: turtledoves, age: 32, name: Bob}
修改键的值后的 Map:{1: partridge, second: turtledoves, age: 32, name: Jobs}
Map 集合中是否包含键 age:true
Map 集合中是否包含值 32:true
```

在 Map 字面量前添加 const 关键字可以创建 Map 类型的编译时常量。示例代码如下：

```
final constantMap = const {
  2:'helium',
  10:'neon',
  18:'argon',
};
```

3min

2.10 符文

在 Dart 中符文对应的类型为 Runes。Unicode 为世界上所有书写系统中使用的每个字母、数字和符号定义了唯一的数值。由于 Dart 字符串是 UTF-16 代码单元的序列，因此在字符串中表示 Unicode 代码点需要特殊的语法。

表示 Unicode 代码点的常用形式是：\uXXXX，其中 XXXX 是 4 位十六进制数。例如：心脏字符是\u2665。要指定多于或少于 4 个十六进制数字，需将值放在大括号中。例如：笑的表情符号是\u{1f600}。

代码点(code point)：代码点是指编码字符集中，字符所对应的数字。有效范围从\u0000到\u10FFFF，其中\u0000 到\uFFFF 为基本字符，\u10000 到\u10FFFF 为增补字符。

代码单元(code unit)：代码单元对代码点进行编码得到的1或2个16位序列。其中基本字符的代码点直接用一个相同值的代码单元表示，增补字符的代码点用两个代码单元进行编码。

String类有几个属性可用于提取符文信息，使用codeUnits属性返回16位代码单元，使用runes属性获取字符串的符文。

为保证一致性，请在在线环境DartPad(https://dartpad.dev)中运行。示例代码如下：

```
//chapter2/bin/example_17.dart
void main() {
  //创建符文字符串
  var clapping = '\u{1f44f}';
  //打印符文
  print(clapping);
  //返回此字符串的 UTF-16 代码单元
  print(clapping.codeUnits);
  //返回此字符串的 Unicode 代码点
  print(clapping.runes.toList());
  //构建 Runes 对象
  Runes input = Runes(
      '\u2665 \u{1f605} \u{1f60e} \u{1f47b} \u{1f596} \u{1f44d}');
  //将 Runes 对象转换为 String 并打印
  print(String.fromCharCodes(input));
}
```

第 3 章 函　　数

Dart 是面向对象的语言，即使函数也是对象，对应的类为 Function。函数也是对象意味着可以将函数赋值给变量，或者将函数作为参数传递给另一个函数。函数有时又被称为方法，它们是等同的，只是叫法上有差异。

声明函数的格式如下：

```
[返回值类型] 函数名([参数列表]){
  //函数体
  return expr;
}
```

每个函数都必须有返回值，如果不提供返回值类型，则默认为 void。函数名是标识符，应当遵循标识符命名规范。参数列表中可能包含零个到多个参数，这取决于实际情况。函数体由变量声明、语句组成。当函数存在返回值时，函数体应当提供 return 语句，例如：return expr;。如果函数返回值的类型为 void，则无须提供 return 语句，等价于 return null;。

定义函数的示例代码如下：

```
bool isEven(int x){
  return x % 2  ==  0 ? true : false;
}
```

该函数返回值的类型为 bool，接收一个 int 类型的参数，函数体由一个 return 语句组成，返回语句的表达式是一个条件表达式，其返回值是布尔值。对于只有一个表达式的函数可以采用简写形式，示例代码如下：

```
bool isEven(int x)  => x % 2  ==  0 ? true : false;
```

=> expr;是{return expr;}的简写形式，常称为胖箭头语法。胖箭头(=>)与分号(;)之间只能是表达式而不能是语句，例如不能是 if 语句，但可以是条件表达式(expr1 ? expr2 : expr3)。

创建命令行应用程序，将项目命名为 chapter3。本章所有文件都在该项目的 bin 文件夹下创建与运行。

函数包含两种参数形式：必选参数和可选参数。参数列表先列出必选参数，再列出可选参数。

3.1 可选参数

可选参数分为命名参数和位置参数。在参数列表中只能选择其中一种作为可选参数，不可同时出现。

3.1.1 命名参数

定义函数时使用花括号({})包裹可选参数列表，使用{arg1,arg2,…}的形式来定义命名参数，示例代码如下：

```
void findAll({int currentPage, int pageSize}){ … }
```

调用函数时，采用 arg1:value1,arg2:value2 的形式来传递参数。即先提供参数名，接着是冒号(:)，冒号后边接着参数值：

```
findAll(currentPage: 1, pageSize: 10);
```

示例代码如下：

```
//chapter3/bin/example_01.dart
main(){
  //以下定义了一个包含必选参数和可选命名参数的函数
  void message(String from, String content, {DateTime time, String device})
{
    //调用该函数必须提供参数 from 和 content
    //因此无须做任何判断即可直接打印这两个参数
    print('来自: $ from,正文: $ content');
    //如果参数 time 不为空,则打印 time
    if (time != null) {
      print('时间: $ time');
    }
    //如果参数 device 不为空,则打印 device
    if (device != null) {
      print('发送设备: $ device');
    }
  }
  //调用 message 函数且提供两个必选参数
  message('jobs', 'hello');
  //调用 message 函数且提供可选参数 time
```

```
  message('jobs', 'hello', time: DateTime.now());
  //调用 message 函数且提供可选参数 time 和 device
  message('jobs', 'hello', time: DateTime.now(), device: 'phone');
}
```

3.1.2 位置参数

使用方括号([])包裹可选参数列表来定义位置参数：

```
void  message(String from,String content,[DateTime time,String device]){...}
```

在调用包含可选位置参数的函数时应当按照位置参数列表的顺序依次为参数赋值。示例代码如下：

```
//chapter3/bin/example_02.dart
void main(){
  //定义带可选位置参数 time 和 device 的函数
  void message(String from,String content,[DateTime time,String device]){
    print('来自：$from,正文：$content');
    if(time != null){
      print('时间：$time');
    }
    if(device != null){
      print('发送设备：$device ');
    }
  }
  //调用函数时不指定可选参数
  message('jobs','hello');
  //调用函数时指定可选位置参数 time
  message('jobs','hello',DateTime.now());
  //调用函数时指定可选位置参数 time 和 device
  message('jobs','hello',DateTime.now(),'phone');
}
```

3.1.3 默认参数值

在声明可选参数时可以使用等号运算符为可选参数指定默认值，默认值必须是编译时常量，没有指定默认值的参数将被赋值为 null。

为命名参数提供默认值的示例代码如下：

```
//chapter3/bin/example_03.dart
main(){
  //为可选命名参数 device 提供默认值
```

```
  void message(String from,String content,{DateTime time,String device = 'phone'}){
    print('来自:$from,正文:$content');
    if(time != null){
      print('时间:$time');
    }
    //参数 device 始终会被打印
    if(device != null){
      print('发送设备:$device');
    }
  }
  //调用函数时不指定可选参数
  message('jobs','hello');
  //调用函数且指定可选命名参数 time
  message('jobs','hello',time: DateTime.now());
  //调用函数且指定可选命名参数 time 和 device
  message('jobs','hello',time: DateTime.now(),device: 'pc');
}
```

为位置参数提供默认值的示例代码如下：

```
//chapter3/bin/example_04.dart
main(){
  //为可选位置参数 device 提供默认值
  void message(String from,String content,[DateTime time,String device = 'phone']){
    print('来自:$from,正文:$content');
    if(time != null){
      print('时间:$time');
    }
    //参数 device 始终会被打印
    if(device != null){
      print('发送设备:$device');
    }
  }
  //调用函数且不指定可选参数
  message('jobs','hello');
  //调用函数且指定可选命名参数 time
  message('jobs','hello',DateTime.now());
  //调用函数且指定可选命名参数 time 和 device
  message('jobs','hello',DateTime.now(),'pc');
}
```

3.2 main 函数

每个 Dart 程序都必须有一个 main 函数作为入口，main 函数的返回值为 void，它有一个 List < String >类型的可选参数。

使用命令行访问带参数的 main 函数的示例代码如下：

```
//chapter3/bin/example_05.dart
//带有参数列表的 main 函数
void main(List<String> args){
  //打印所有参数值
  print(args);
  var len = args.length;
  //打印参数数量
  print('参数数量:$len');
}
```

保存上述代码，文件名为 example_05.dart。打开编辑器的命令行工具 Terminal，运行命令并携带参数：

```
dart bin/bin/example_05.dart first 2 dog
```

运行结果如下：

```
[first, 2, dog]
参数数量:3
```

也可以省略参数：

```
void main(){
//函数体
}
```

4min

3.3 函数对象

可以将函数作为参数传递给另一个函数，下例中将函数 printElement 传递给 List 类型的对象 list 的 forEach 函数。示例代码如下：

```
//chapter3/bin/example_06.dart
void main(){
  //定义一个带有 int 类型参数的函数
  void printElement(int element) {
    //打印传入的参数值
    print(element);
  }
  //定义一个 List 类型的对象并赋初值
  var list = [1, 2, 3];
```

```
  //将 printElement 函数作为参数传递给 List 对象的 forEach 函数
  //forEach 函数会按迭代顺序将函数 printElement 应用于 List 集合的每个元素
  list.forEach(printElement);
}
```

也可以将函数赋值给一个变量。示例代码如下：

```
//chapter3/bin/example_07.dart
void main() {
  void printElement(int element) {
    print(element);
  }
  //将函数 printElement 赋值给变量 show
  var show = printElement;
  //像执行函数一样使用变量 show
  show(1);
}
```

3.4 匿名函数

5min

大多数方法都是有名字的，例如 main 或 printElement。也可以创建一个没有名字的函数，称为匿名函数。匿名函数常以回调函数或另一个函数的参数的形式出现。

匿名函数看起来与命名函数类似，在括号之间可以定义参数，参数之间用逗号分隔，后面大括号中的内容则为函数体，也可以有返回值。

```
([类型] [参数, … ]){
//函数体;
}
```

匿名函数常用于集合迭代中，下面的代码定义了只有一个参数 item 且没有参数类型的匿名函数。List 中的每个元素都会调用这个函数，函数体打印元素在集合中的下标和值。示例代码如下：

```
//chapter3/bin/example_08.dart
void main(){
  var list = ['apples', 'bananas', 'oranges'];
  //向 forEach 函数提供匿名函数
  list.forEach((item){
    print('${list.indexOf(item)}: $item');
  });
}
```

如果函数体内只有一行语句,则可以使用胖箭头语法:

```
list.forEach((item) => print('${list.indexOf(item)}: $item'));
```

匿名函数也可以作为另一个函数的返回对象,返回语句 return (num r) => 2 * pi * r; 中的返回值就是匿名函数。示例代码如下:

```
//chapter3/bin/example_09.dart
void main(){
  //定义一个返回值类型为 Function 的函数
  Function perimeter(){
    var pi = 3.14;
    //return 关键字后面跟着的是一个匿名函数
    return (num r) => 2 * pi * r;
  }
  //将函数赋值给变量
  var per = perimeter();
  var r = 9;
  var l = per(r);
  print('圆的周长: $l');
}
```

7min

3.5 语法作用域

Dart 是有词法作用域的语言,变量的作用域在写代码的时候就确定了。大括号内定义的变量只能在大括号内被访问,在大括号内可以访问大括号外的变量,与 Java 类似。嵌套函数中变量在多个作用域中使用的示例代码如下:

```
//chapter3/bin/example_10.dart
//定义一个顶层变量
String topLevel = 'top variable';
void main(){
  //在 main 函数中定义变量
  var insideMain = 'insideMain variable';
  void myFunction(){
    //在 myFunction 函数中定义变量
    var insideFunction = 'insideFunction variable';
    void nestedFunction(){
      //在嵌套函数中定义变量
      var insideNestedFunction = 'insideNestedFunction variable';
      print('$topLevel');
      print('$insideMain');
```

```
        print('$insideFunction');
        print('$insideNestedFunction');
      }
    }
  }
```

注意：nestedFunction()函数可以访问包括顶层变量在内的所有变量。

3.6　语法闭包

4min

语法闭包即一个函数对象，即使函数对象的调用在它原始作用域之外，依然能够访问在它词法作用域内的变量。

函数可以封闭定义到它作用域内的变量。下面的示例中，函数 makeAdder()捕获了变量 addBy。无论函数在什么时候返回，它都可以使用捕获的 addBy 变量。示例代码如下：

```
//chapter3/bin/example_11.dart
void main(){
  //返回一个函数，该函数的参数将与 addBy 相加
  Function makeAdder(num addBy){
    return (num i) => addBy + i;
  }
  //生成加 2 的函数
  var add2 = makeAdder(2);
  //生成加 4 的函数
  var add4 = makeAdder(4);
  print(add2(3));
  print(add4(3));
}
```

3.7　函数相等性测试

7min

函数相等性测试用来判断两个函数是否是同一个对象。

顶层函数、静态方法和实例方法相等性的测试示例代码如下：

```
//chapter3/bin/example_12.dart
//定义顶层函数
void foo(){}

//定义一个类
class A {
```

```
  //定义静态方法
  static void bar(){}
  //定义实例方法
  void baz(){}
}

void main() {
  var x;
  x = foo;
  //比较顶层函数是否相等
  print(foo == x);

  //比较静态方法是否相等
  x = A.bar;
  print(A.bar == x);

  //A 的实例#1
  var v = A();
  //A 的实例#2
  var w = A();
  var y = w;
  x = w.baz;
  //这两个闭包引用了相同的实例对象,因此它们相等
  print(y.baz == x);
  //这两个闭包引用了不同的实例对象,因此它们不相等
  print(v.baz != w.baz);
}
```

3min

3.8 返回值

所有的函数都有返回值,返回值本质上是对象。返回值类型可以是内置类型也可以是自定义类型,返回值类型放在函数名前面。示例代码如下:

```
//chapter3/bin/example_13.dart
void main(){
  //返回值类型为 num
  num area(num pi,num r) {
    //返回语句
    return pi * r * r;
  }
  print('圆面积为: ${area(3.14,6)}');
}
```

当函数名前有类型修饰时，函数体最后必须提供 return 语句。当函数名前没有类型修饰且未提供 return 语句时，最后一行默认为执行 return null;。示例代码如下：

```
//chapter3/bin/example_14.dart
void main() {
  //定义没有提供返回值类型的函数
  foo(){
    //函数体未提供 return 语句
    var a;
  }
  //判断函数 foo 的返回值是否为 null
  print(foo() == null);
}
```

3.9 回调函数

4min

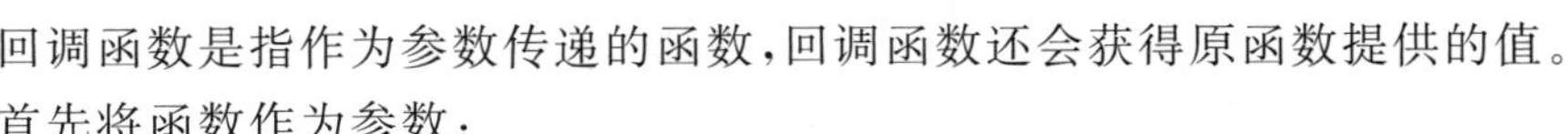

回调函数是指作为参数传递的函数，回调函数还会获得原函数提供的值。

首先将函数作为参数：

```
void printProgress({Function(int) callback}){...}
```

有条件地在函数内执行回调函数：

```
if(callback! = null) callback(progress);
```

接收原函数提供的参数：

```
printProgress(callback: (int progress){
  print('打印进度 : $progress');
});
```

完整的示例代码如下：

```
//chapter3/bin/example_15.dart
void main(){
  //定义 printProgress 函数，将回调函数 callback 作为可选参数
  void printProgress({Function(int) callback}){
    for (int progress = 0; progress <= 10; progress++){
      //判断回调函数是否为空
      if(callback!= null){
        //若提供则调用回调函数并传递循环变量 progress
        callback(progress);
```

```
            }
        }
    }
    //调用 printProgress 函数并提供匿名回调函数
    printProgress(callback: (int progress){
      print('打印进度 : $progress%');
    });
}
```

第 4 章 运 算 符

表 4-1 所示的是 Dart 语言定义的运算符，可以覆写其中一部分运算符。

表 4-1 运算符

运算符名称	操 作 符
一元后缀运算符	expr++ expr-- () [] . ?..
一元前缀运算符	-expr !expr ~expr ++expr await expr
乘除	* / % ~/
加减	+ -
位运算符	<< >> >>>
二进制与	&
二进制异或	^
二进制或	|
关系和类型测试	>= > <= < as is is!
相等判断	== !=
逻辑与	&&
逻辑或	||
空判断	??
条件表达式	expr1 ? expr2 : expr3
级联	..
赋值	= *= /= += -= &= ^=

运算符存在优先级，就像在数学中常说的先乘除后加减一样。在运算符表中，在同一列中优先级依次降低，即第一行优先级最高，最后一行优先级最低。在同一行中从左到右优先级依次降低，即最左边的优先级最高，最右端的优先级最低。

创建命令行应用程序，将项目命名为 chapter4。本章所有文件都在该项目的 bin 文件夹下创建与运行。

4.1 算术运算符

表 4-2 所示的是 Dart 语言中的算术运算符，表中的 expr 代表的是表达式，var 代表的是变量。

表 4-2 算术运算符

运 算 符	说 明
+	加
-	减
-expr	负号,也可以对表达式符号取反
*	乘
/	除
~/	除,返回正整数部分
%	求模
++var	先自加再进行运算
var++	先运算再进行自加
--var	先自减再进行运算
var--	先运算再进行自减

常见算术运算符使用示例代码如下:

```
//chapter4/bin/example_01.dart
void main(){
  //加号运算
  print('1 + 3 = ${1 + 3}');
  //减号运算
  print('8 - 6 = ${8 - 6}');
  var a = 6;
  //负号运算
  print('-a= ${-a}');
  //对表达式符号取反
  print('-(-a) = ${-(-a)}');
  //乘号运算
  print('6 * 7 = ${6 * 7}');
  //除号运算
  print('8/2 = ${8/2}');
  //除并取整运算
  print('9~/2 = ${9~/2}');
  //除并取余运算
  print('9 % 2 = ${9 % 2}');
}
```

自增和自减需要特别注意。示例代码如下:

```
//chapter4/bin/example_02.dart
void main(){
```

```
    var a = 10;
    var b = 20;
    var c = 30;
    var d = 40;
    var e,f,g,h;
    //a 先自加再赋值给 e
    e = ++a;
    print('e: $ e a: $ a');
    //b 先赋值给 f 再自加
    f = b++;
    print('f: $ f b: $ b');
    //c 先自减再赋值给 g
    g = --c;
    print('g: $ g c: $ c');
    //d 先赋值给 h 再自减
    h = d--;
    print('h: $ h d: $ d');
}
```

4.2 关系运算符

11min

表 4-3 中的关系运算符常用来比较两个对象是否相等或大小关系。关系表达式属于布尔表达式，所以它们的返回值均为布尔值，即 true 或 false。

表 4-3 关系运算符

运算符	说明
==	相等
!=	不等
>	大于
<	小于
>=	大于或等于
<=	小于或等于

要测试两个对象是否表示同样的值，例如数值、布尔值或字符串值等，需使用==运算符。有时需要知道两个对象是否是完全相同的对象，需改用 identical()函数。==运算符的工作方式如下：

(1) 判断 x 或 y 是否为 null 的情况：如果两个均为 null，则返回 true；如果只有一个为 null，则返回 false。

(2) 其余情况下返回方法调用 x.==(y)的结果。值得注意的是：==之类的运算符是在第一个操作数上调用的方法。

示例代码如下：

```
//chapter4/bin/example_03.dart
void main(){
  var a = 10;
  var b = 10;
  var c = 30;
  var d = 40;
  //定义 e 和 f 且未赋值,默认值为 null
  var e,f;
  //相等运算
  print('a == b: ${a == b}');
  print('e == f: ${e == f}');
  print('a == e, ${a == e}');
  //不等运算
  print('a!= c, ${a!= c}');
  //大于运算
  print('d > c, ${d > c}');
  print('d > c, ${d > c}');
  //小于运算
  print('b < d, ${b < d}');
  //大于或等于运算
  print('a > = b, ${a > = b}');
  //小于或等于运算
  print('b < = c, ${b < = c}');
}
```

4.3 类型测试运算符

5min

表 4-4 中的类型测试运算符可以在运行时检查类型。

表 4-4　类型测试运算符

运算符	说明
as	类型转换
is	类型判断,如果对象是指定对象则返回 true
is!	类型判断,如果对象不是指定对象则返回 true

运算符 is 检查的不是对象是否属于某个类或其子类,而是检查对象所属的类是否实现了某个类的接口(直接或间接)。如果对象 obj 实现了类 T 定义的接口,obj is T 则返回 true。obj is Object 则始终返回 true,因为 Object 是所有类的父类。

运算符 as 也是用来检查对象是否属于某个类型或其子类,区别在于,如果测试成功则返回被检测的对象，如果测试失败则会抛出一个 CastError。示例代码如下：

```
//chapter4/bin/example_04.dart
//定义类 Car
class Car{
  String length;
  String color;
}
//定义类 Taxi 继承类 Car
class Taxi extends Car{
  double fee;
}
void main(){
  var t = Taxi();
  (t as Car).color = 'RED';
  print('变量 t 的引用对象运行时类型为 ${t.runtimeType}');
  print('t 的引用对象是类 Car 的子类：${t is Car}');
  print('t as Car 运行时类型为 ${(t as Car).runtimeType}');
}
```

使用 as 运算符，即类型转换的工作逻辑大致可以通过以下代码表述：

```
var t = someObject;
t is T ? t : throw CastError();
```

运算符 as 的主要使用场景是数据校验：

```
List ls = getTaxis() as List;
```

代码片段中的 getTaxis()方法用于获取一个泛型为 Taxi 的列表，如果返回值确实是列表，则继续其他代码的执行，如果返回的不是列表，则立即抛出错误。

```
(t as Car).color = 'RED';
```

上述代码片段实际上是对 as 运算符的滥用，因为类型转换是在运行时才执行的，因此会造成计算资源消耗。这里只是为了演示它的返回值是被检测对象 t，因为已经明确知道对象 t 所属的类就是 Car 类的子类，在实际开发中应当避免类似操作。

4.4 赋值运算符

4min

表 4-5 中列出了常用赋值运算符及含义，它们的作用是按相应规则为变量赋值。

表 4-5 常用赋值运算符及含义

运 算 符	含 义
=	a = b
??=	a ??= b 等价于 a 等于 null 时 a=b
-=	a -= b 等价于 a = a-b
/=	a /= b 等价于 a = a/b
%=	a %= b 等价于 a = a%b
>>=	a >>= b 等价于 a = a >> b
^=	a^= b 等价于 a = a^b
+=	a += b 等价于 a = a+b
*=	a *= b 等价于 a = a*b
~/=	a ~/= b 等价于 a = a~/b
<<=	a <<= b 等价于 a = a << b
&=	a &= b 等价于 a = a&b
|=	a |= b 等价于 a = a|b

可以使用符号=为变量赋值，使用符号?? =为值为 null 的变量赋值。示例代码如下：

```
//chapter4/bin/example_05.dart
void main(){
  //为变量 a 赋值
  var a = 12;
  print('a: $ a');
  var b1 = 10,b2;
  //b1 已有初始值所以将保持原值
  b1?? = 13;
  //b2 没有初始值所以将被赋值
  b2?? = 13;
  print('b1: $ b1, b2: $ b2');
}
```

复合赋值运算符的使用示例代码如下：

```
//chapter4/bin/example_06.dart
void main(){
  var a = 12,b = 5;
  print('a % = b: $ {a % = b}');
  print('a~/ = b: $ {a~/ = b}');
}
```

4.5　逻辑运算符

8min

表 4-6 中的逻辑运算符运算后的结果都为布尔值。逻辑与和逻辑或都是先计算左操作数,再计算右操作数。如果逻辑与左操作数计算结果是假值,则返回假值,无须计算右操作数。如果逻辑或左操作数计算结果是真值,则返回真值,无须计算右操作数。

表 4-6　逻辑运算符

运　算　符	描　　述
!expr	对表达式结果取反
||	逻辑或
&&	逻辑与

常见使用示例代码如下：

```
//chapter4/bin/example_07.dart
void main(){
  var c = true,d = false,e = true;
  //非运算
  print('!c: ${!c}');
  print('!d: ${!d}');
  //逻辑或运算
  print('c||d: ${c||d}');
  print('c||e: ${c||e}');
  //逻辑与运算
  print('c&&d: ${c&&d}');
  print('c&&e: ${c&&e}');
}
```

4.6　位运算符

6min

使用表 4-7 中的位运算符可以操作数字的各个二进制位。通常,可以对整数使用这些位运算符和移位运算符。

表 4-7　位运算符

运　算　符	描　　述
&	按位进行与操作
|	按位进行或操作
^	按位进行异或操作
~expr	按位进行取反操作
<<	按位进行左移操作
>>	按位进行右移操作

下例中使用了两个十六进制数，它们实际上是先转换为二进制数，然后进行位运算。示例代码如下：

```
//chapter4/bin/example_08.dart
void main(){
  final value = 0x22;
  final bitmask = 0x0f;
  //按位进行与操作
  print((value & bitmask) == 0x02);
  //取反后按位进行与操作
  print((value & ~bitmask) == 0x20);
  //按位进行或操作
  print((value | bitmask) == 0x2f);
  //按位进行异或操作
  print((value ^ bitmask) == 0x2d);
  //按位进行左移操作
  print((value << 4) == 0x220);
  //按位进行右移操作
  print((value >> 4) == 0x02);
}
```

4.7 条件表达式

5min

表 4-8 中列出的两种条件表达式常用于构建一个赋值语句，它们也用于替代简单的 if-else 语句。

表 4-8 条件表达式

运算符	描述
condition ? expr1 : expr2	如果条件为真则返回 expr1，否则返回 expr2
expr1 ?? expr2	如果 expr1 为非 null 则返回 expr1，否则返回 expr2

使用示例代码如下：

```
//chapter4/bin/example_09.dart
void main(){
  bool isPublic = true;
  //三元条件表达式
  print('${isPublic ? 'public' : 'private'}');
  var x,y = 10,z = 20;
  //二元条件表达式
  print('${x ?? y}');
  print('${y ?? z}');
}
```

4.8 其他运算符

7min

表 4-9 中列出的其他运算符涉及后续的知识，这里了解即可，可以在掌握相关知识后再理解。

表 4-9　其他运算符

运　算　符	描　　述
()	代表调用的是函数
[]	访问 List 特定位置的元素
.	访问对象的成员
?.	有条件地访问对象的成员，若对象为 null 则返回 null

使用示例代码如下：

```
//chapter4/bin/example_10.dart
//定义类 Point
class Point{
  int x,y;
  Point(this.x,this.y);
}
//定义顶层函数 printTime
void printTime() {
  print(DateTime.now());
}
void main() {
  //通过使用()运算符标识调用的是函数
  printTime();

  var ls = [1, 2, 3];
  //通过[]运算符访问 ls 数组的第一个元素
  var a = ls[0];

  var p1 = Point(3,6);
  //通过.运算符访问 p1 的成员 x
  print('${p1.x}');

  var p2;
  //通过?.运算符访问 p2 的成员 x
  //因为 p2 没有成员 x，所以返回 null
  print('${p2?.x}');
}
```

第 5 章

流程控制语句

在一个程序执行的过程中，各条语句的执行顺序对程序的结果是有直接影响的。控制语句用实现对程序流程的选择、循环、转向和返回等进行控制。只有在清楚每条语句的执行流程的前提下，才能通过控制语句的执行顺序实现要完成的任务。

Dart 中的控制语句分为以下几类：

（1）分支语句：if 和 switch。

（2）循环语句：while、do-while 和 for。

（3）跳转语句：break、continue、return 和 throw。

创建命令行应用程序，将项目命名为 chapter5。本章所有文件都在该项目的 bin 文件夹下创建与运行。

11min

5.1 分支语句

分支语句又称条件语句，条件语句使得部分程序块根据表达式的值有选择地执行。Dart 语言提供了 if 和 switch 两种分支语句。

5.1.1 if 语句

具体来说分支语句由选择结构组成，if 语句引导的选择结构包括 if 结构、if-else 结构和 else-if 结构。

1. if 结构

if 结构包含条件表达式和语句块，当条件表达式为 true 时执行语句块，否则执行 if 结构后面的语句。通常来说语句块由大括号（{}）包裹，如果 if 结构中的语句块只有一条语句，则可以省略大括号。

if 结构声明语法如下：

```
if(条件表达式){
  //语句块
}
```

示例代码如下：

```
//chapter5/bin/example_01.dart
void main(){
  var x = true,y = 10;
  if(x){
    //此处 x 为真,语句体得到执行
    print('条件 x 为 $ x');
  }
  if(!x){
    //此处!x 为假,语句体得不到执行
    print('条件!x 为 $ x,if 语句体得到执行.');
  }
  //此处省略大括号
  if(y<=10)
    //条件 y<=10 为真,语句体得到执行
    print('条件 y<=10 为 $ {y<=10}');
}
```

2. if-else 结构

在 if 结构中，只有条件表达式为 true 时提供相应的处理语句块。有时无论条件表达式为 true 还是 false 都需进行相应处理，此时可以使用 if-else 结构。声明格式如下：

```
if(条件表达式){
  //语句块 1
}else{
  //语句块 2
}
```

当程序执行到 if 语句时，先判断条件表达式，如果值为 true，则执行语句块 1，然后跳过 else 语句块，继续执行后续语句。如果条件表达式值为 false，则忽略语句块 1 直接执行语句块 2，然后继续执行后续语句。

if-else 结构使用示例代码如下：

```
//chapter5/bin/example_02.dart
void main(){
  var x = true;
  if(!x){
    //条件!x 为真时,if 语句块得到执行
    print('if !x: $ {!x}');
  }else{
    //条件!x 为假时,else 语句块得到执行
    print('else !x: $ {!x}');
  }
}
```

3. else-if 结构

else-if 结构是 if-else 结构的多层嵌套形式，它会在多个分支中执行一个语句块，其他分支将不会得到执行，所以这种结构常用于有多种判断结果的分支中。

else-if 结构如下：

```
if(条件表达式 1){
  //语句块 1
}else if(条件表达式 2){
  //语句块 2
}
...
}else if(条件表达式 n){
  //语句块 n
}else{
  //语句块 n+1
}
```

else-if 结构使用示例代码如下：

```
//chapter5/bin/example_03.dart
void main(){
  var x = true,y = 10;
  if(!x){
    //条件!x 为假, if 语句块不会执行
    print('条件 x 为 $x');
  }else if(y<=10){
    //如果条件 y<=10 为真,则 else if 语句块得到执行
    print('条件 y<=10 为 ${y<=10}');
  }else{
    //当所有条件都为假时,else 语句块才会执行
    print('else !x: ${!x}.');
  }
}
```

5.1.2 switch 语句

可以使用 switch 语句选择要执行的多个代码块中的一个代码块，语法如下：

```
switch(expr){
  case 值 1:
    //语句块 1
    break;
  case 值 2:
    //语句块 2
```

```
      break;
      ...
    case 值 n:
      //语句块 n
      break;
    default:
      //语句块 n + 1
}
```

首先设置表达式 expr，通常是一个变量。随后表达式的值会与结构中的每个 case 的值做比较。如果存在匹配，则与该 case 子句关联的代码块会被执行。如果 case 子句中没有与 expr 匹配的值，则会执行 default 子句的语句块，然后跳出 switch 结构。

switch 语句使用==来比较整数、字符串或编译时常量，比较的两个对象必须是同一个类的实例且没有覆写==运算符。

每个非空的 case 子句必须包含一个 break 语句。当没有匹配的子句时，可以使用 default 子句匹配这种情况。示例代码如下：

```
//chapter5/bin/example_04.dart
void main(){
  var color = 'Red';
  switch(color){
    case 'Green':
      print('color 值为 Green');
      break;
    case 'Orange':
      print('color 值为 Orange');
      break;
    case 'Red':
      print('color 值为 Red');
      break;
    default:
      print('color 值未匹配');
  }
}
```

下面的例子忽略了 break 子句，因此产生错误。

```
//chapter5/bin/example_05.dart
void main(){
  var color = 'Red';
  switch(color){
    case 'Green':
      print('color 值为 Green');
```

```
  //执行出错,没有 break 语句
    case 'Orange':
      print('color 值为 Orange');
      break;
    case 'Red':
      print('color 值为 Red');
      break;
    default:
      print('color 值未匹配');
  }
}
```

Dart 语言支持 case 子句的内容为空,这种情况称为贯穿。下例中值为 Orange 的 case 子句内容为空,它会造成贯穿。当传入变量值为 Orange 时,程序会继续执行与它相邻的值为 Red 的 case 子句。示例代码如下:

```
//chapter5/bin/example_06.dart
void main(){
  var color = 'Orange';
  switch(color){
    case 'Green':
      print('color 值为 Green');
      break;
    case 'Orange':
      //空语句贯穿
    case 'Red':
      //color 值为 Orange 和 Red 都会被执行
      print('color 值为 Red');
      break;
    default:
      print('color 值未匹配');
  }
}
```

非空 case 子句实现贯穿可以通过 continue 语句和标签来完成。下例中匹配到值为 Green 的 case 子句时,不仅会执行该子句的代码块,还会执行值为 Red 的 case 子句中的代码块。示例代码如下:

```
//chapter5/bin/example_07.dart
void main(){
  var color = 'Green';
  switch(color){
    case 'Green':
      print('color 值为 Green');
```

```
      //继续执行标签为 red 的 case 子句
      continue red;
    case 'Orange':
      print('color 值为 Orange');
      break;
    //标签 red,它代表了 case 值为 Red 的子句
    red:
    case 'Red':
      //color 值为 Green 和 Red 都会被执行
      print('color 值为 Red');
      break;
    default:
      print('color 值未匹配');
  }
}
```

每个 case 子句都可以有局部变量且仅在该 case 子句内有效。

5.2 循环语句

9min

循环语句能够根据循环条件使程序代码重复执行。Dart 支持 3 种循环结构：for、while 和 do-while。

5.2.1 for 语句

for 语句是一种应用广泛、功能最强的循环语句。声明格式如下：

```
for(表达式 1;表达式 2;表达式 3){
  //循环语句块
}
```

表达式 1 是初始化语句，用于初始化循环变量和其他变量。表达式 2 是循环条件，表达式 3 用于更新循环变量，使循环变量趋近于某个值，直到使循环条件变为 false。

开始 for 循环时，会执行表达式 1 初始化循环变量，表达式 1 只会执行一次。然后执行表达式 2，判断循环条件是否满足，如果满足，则继续执行循环体。执行完成后继续执行表达式 3，更新循环变量，然后接着再判断循环条件。如此反复，直到循环条件不满足时跳出循环。执行流程如图 5-1 所示。

for 循环使用示例代码如下：

```
//chapter5/bin/example_08.dart
void main(){
  for (var i = 0;i < 5;i++){
```

```
        //此处 i 的初始值为 0,所以使用 i + 1 保证其与实际循环次数一致
        print('第 ${i + 1}次循环.');
    }
}
```

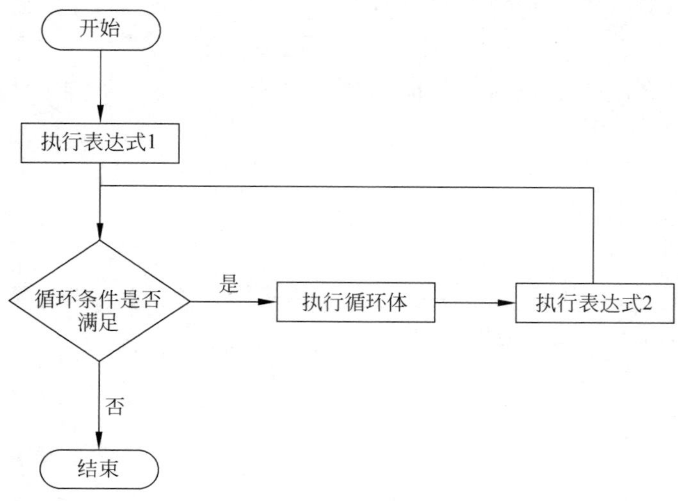

图 5-1 for 循环执行流程图

循环开始时,给循环变量 i 赋值为 0,每次执行循环体前都会判断 i 的值是否小于 5,如果为 true,则执行循环体,然后执行 i++,使得循环变量 i 的值加 1。然后继续判断循环条件,直到判断结果为 false 时跳出循环。

对于实现了 iterable 接口的类可以使用 forEach()函数进行迭代。例如 List 和 Set 类,它们还支持 for-in 形式的迭代。示例代码如下:

```
//chapter5/bin/example_09.dart
void main(){
  var collection = [0, 1, 2];
  //这里给 forEach 函数传入了一个匿名函数
  collection.forEach((i) => print(i));
  //for - in 形式的迭代
  for (var x in collection) {
    print(x);
  }
}
```

在 Dart 语言中,for 循环中的闭包会自动捕获索引值,这在某些情况下很有用。示例代码如下:

```
//chapter5/bin/example_10.dart
void main(){
  var callbacks = [];
  for (var i = 0; i < 2; i++) {
    callbacks.add(() => print(i));
  }
  callbacks.forEach((c) => c());
}
```

5.2.2　while 语句

while 循环在执行循环体前先判断循环条件，如果条件为真则执行循环体，如果条件为假则结束循环。

声明格式如下：

```
while(循环条件){
    //语句块
  }
```

在 while 循环体中一定要存在影响循环条件的语句，其作用是使循环条件最终变为 false 以避免无限循环。

示例代码如下：

```
//chapter5/bin/example_11.dart
void main(){
  var i = 0,x = 10;
  while(i < x){
    print('第 ${i + 1}次 while 循环.');
    //影响条件表达式并使其趋于假的语句
    i++;
  }
}
```

示例中 i＋＋是影响循环条件的因子，也是它使得循环能够在有限次循环后结束。

5.2.3　do-while 语句

do-while 循环与 while 循环类似。声明格式如下：

```
do
    //语句块
}while(循环条件){
```

不同之处在于 do-while 循环会先执行一次循环体，再判断循环条件，如果条件为真则执行循环体，如果条件为假则结束循环。

示例代码如下：

```
//chapter5/bin/example_12.dart
void main(){
  var i = 0,x = 10;
  do{
    print('第 ${i + 1}次 do while 循环.');
    //影响条件并使其趋于假的语句
    i++;
  }while(i < x);
}
```

5.3 跳转语句

6min

跳转语句可以改变代码的执行顺序，从而可以实现跳转。

5.3.1 break 语句

break 语句可用于 for、while 和 do-while 循环结构，它的作用是强行退出循环体，且不再执行循环体中剩余的语句。

在 while 循环中 break 语句使用示例代码如下：

```
//chapter5/bin/example_13.dart
void main(){
  var i = 0,x = 10;
  while(i < x){
    print('第 ${i + 1}次 while 循环.');
    if(i == 5){
      //当循环变量 i 等于 5 时跳出 for 循环
      break;
    }
    //影响条件表达式并使其趋于假的语句
    i++;
  }
}
```

5.3.2 continue 语句

continue 语句的作用是结束本次循环，跳过循环体中尚未执行的语句，接着继续判断循环条件，以决定是否继续循环。

在 for 循环中 continue 语句使用示例代码如下：

```
//chapter5/bin/example_14.dart
void main(){
  for (var i = 0; i < 10; i++) {
    if(i == 5) continue;
    //此处 i 的初始值为 0,所以使用 i + 1 保证其与实际循环次数一致
    print('第 ${i + 1}次循环.');
  }
}
```

5.3.3　assert

assert 函数又称断言，它的第一个参数是布尔表达式，在表达式的值为 false 时中断代码执行，断言在开发过程中用于检测对象是否符合期待。示例代码如下：

```
//chapter5/bin/example_15.dart
void main(){
  var x, y = 9;
  //确保变量 x 的值非空
  assert(x != null);
  //确保变量 x 的值小于 8
  assert(y < 8);
}
```

也可以向 assert 函数传递一个 String 类型的可选参数。示例代码如下：

```
//chapter5/bin/example_16.dart
void main(){
  var URLString = 'http://dartlang.org';
  //确保网址应该以 https 开始
  assert(URLString.startsWith('https'));
}
```

断言什么时候起作用取决于使用的工具和框架：

(1) Flutter 在调试模式下启用断言。

(2) 默认情况下，仅开发工具(例如 dartdevc)通常启用断言。

(3) 某些工具(例如 dart 和 dart2js)通过命令行标志---enable-asserts 支持断言。

(4) 在生产代码中，断言将被忽略，并且不会评估断言的参数。

第 6 章 类

Dart 是一种面向对象的编程语言，它支持基于 mixin 的继承机制和扩展方法。Dart 中每个对象都是一个类的实例，并且所有类都源自 Object 类。

使用 class 关键字声明类，类由类名和花括号包裹的类体构成，类体可以为空，格式如下：

```
class Point{}
```

可以通过以下格式创建该类的新实例。

```
Point  p = Point();
```

当然也可以使用 var、dynamic、Object 代替变量 p 前的类型名。

```
var  p = Point();
```

在这里该类不具有任何属性和方法，单纯地仅仅是一个符合规范的类而已，它不具有任何实用价值。

创建命令行应用程序，将项目命名为 chapter6。本章所有文件都在该项目的 bin 文件夹下创建与运行。

8min

6.1 属性

所谓的属性，就是在类中定义的变量。根据使用方式的不同，属性分为实例属性和类属性。

声明属性的格式如下：

```
[修饰符] 数据类型 属性名 [ = 值];
```

修饰符是可选的，修饰符可以是多个。数据类型可以是内置类型也可以是自定义类型，属性的命名应当符合规范，属性的赋值是可选的。

没有 static 修饰的属性就是实例属性，也可以称作实例变量。实例属性只能通过类的实例引用，声明实例属性的示例代码如下：

```
class Point{
  //声明实例属性 x,也可称作实例变量 x,默认初始值为 null
  num x;
  //声明实例属性 y,也可称作实例变量 y,设置初始值为 0
  num y = 0;
}
```

所有未初始化的属性的值默认为 null，如果在声明处就为属性初始化，其值在实例创建之前就被设置，甚至在构造函数初始化列表执行之前。所有的属性都会生成一个隐式的 getter 方法，非 final 修饰的属性还会生成一个 setter 方法。

实例对象通过使用点号(.)来引用属性。示例代码如下：

```
//chapter6/bin/example_01.dart
//声明 Point 类
class Point{
  num x;
  num y;
}
void main(){
  //创建 Point 类的实例
  var point = Point();
  //调用属性 x 的 setter 方法
  point.x = 4;
  //调用属性 x 的 getter 方法
  print(point.x);
  //属性 y 的值默认为 null
  print(point.y);
}
```

为了避免对象 p 为 null 而引起异常，可以使用?.代替.。

```
void main(){
  //如果对象 p 非 null,设置 y 的值为 4
  p?.y = 4;
}
```

类属性是由关键字 static 修饰的属性，也可以称作类变量。类属性对于表示类范围的状态变量和常量很有用，类属性在首次使用时会被初始化。类属性通过类直接访问和修改值，类属性不能通过类的实例访问和修改值，定义类属性 z 的示例代码如下：

```
//chapter6/bin/example_02.dart
class Point{
  num x;
  num y;
  //定义类属性 z,也可称作类变量 z
  static double z = 1;
}
void main(){
  //直接通过类访问类属性 z
  print(Point.z);
  //调用类属性 z 的 setter 方法来修改值
  Point.z = 10;
  //查看类属性 z 修改后的值
  print(Point.z);
}
```

14min

6.2 构造函数

通过创建与类名相同的方法声明构造函数，下例中 Point 方法就是 Point 类的同名构造函数。示例代码如下：

```
//chapter6/bin/example_03.dart
class Point{
  num x,y;
  Point(num x,num y){
    //构造函数体
    //通过传入的参数为属性赋值
    this.x = x;
    this.y = y;
  }
}
```

this 关键字引用的是当前实例，将构造函数参数的值分配给实例属性的模式非常普遍，Dart 拥有语法糖使其变得简单。示例代码如下：

```
class Point{
  num x,y;
  //使用语法糖设置 x 和 y,在构造函数体运行之前执行
  Point(this.x,this.y);
}
```

在 Dart 中与类名相同的构造函数只能有一个，这与其他语言中可以拥有多个同名构造函数不同，为了实现相同的功能，可以将所有与实例属性相关的参数放在可选参数列表中。

示例代码如下：

```
//chapter6/bin/example_04.dart
class Person{
  int age;
  double hight;
  String name;
  //将所有参数置于可选参数列表中
  Person({this.age,this.name,this.hight});
}
void main(){
  //创建只为 age 属性赋值的 Person 类的实例
  var p1 = Person(age:10);
  //创建为 name 和 age 属性赋值的 Person 类的实例
  var p2 = Person(name:'jobs',age:33);
  //创建为所有属性赋值的 Person 类的实例
  var p3 = Person(name:'jobs',age:33,hight:1.76);
  print(p1.age);
  print(p2.name);
  print(p3.hight);
}
```

如上例所示，向构造函数传入不同数量或类型的参数来创建新实例以满足实际需求，这将非常实用。

6.2.1 默认构造函数

如果在定义一个类时未声明构造函数，将会为此类提供一个默认构造函数，默认构造函数没有参数，并会调用父类中的无参数构造函数。

6.2.2 命名构造函数

通过附加的标识符声明命名构造函数，使用命名构造函数可为一个类实现多个构造函数以供满足明确的使用场景。定义命名构造函数 Point.origin 和 Point.fromJson 的示例代码如下：

```
//chapter6/bin/example_05.dart
class Point{
  num x, y;
  //同名构造函数
  Point(this.x,this.y);
  //命名构造函数 origin
  Point.origin(){
    x = 0;
    y = 0;
```

```
  }
  //命名构造函数 fromJson
  Point.fromJson(Map json){
    this.x = json['x'];
    this.y = json['y'];
  }
}
void main(){
  //使用命名构造函数 origin 来创建实例
  var p1 = Point.origin();
  print('p1:( ${p1.x}, ${p1.y})');
  //创建 json 数据,很多情况下 Map 类型等同于 json
  Map data = {'x':1,'y':2};
  //使用命名构造函数 fromJson 来创建实例
  var p2 = Point.fromJson(data);
  print('p2:( ${p2.x}, ${p2.y})');
}
```

构造函数不会被继承,这意味着父类的命名构造函数不会被子类继承。如果要使用父类中定义的命名构造函数创建子类,则必须在子类中实现该构造函数。

6.2.3 初始化列表

可以在构造函数主体运行之前初始化实例属性。在构造函数的括号后面使用冒号引导初始化列表,用逗号分隔初始化语句,初始化列表表达式等号右边不能使用 this 关键字。示例代码如下:

```
//chapter6/bin/example_06.dart
class Point{
  num x, y;
  //同名构造函数
  Point(this.x,this.y);
  //命名构造函数 origin
  Point.origin(){
    x = 0;
    y = 0;
  }
  //命名构造函数 fromJson 使用初始化列表为属性赋值
  Point.fromJson(Map json):x = json['x'],y = json['y'];
}
void main(){
  Map data = {'x':1,'y':2};
  var p1 = Point.fromJson(data);
  print('p1:( ${p1.x}, ${p1.y})');
}
```

在开发模式中,可以在初始化列表中使用断言验证输入。

```
Point.withAssert(this.x,this.y):assert(x >= 0){
  print('在 Point.withAssert()中:($x, $y)');
}
```

设置 final 修饰的属性时,初始化列表很方便。下面的示例在初始化列表中初始化 4 个 final 属性。示例代码如下:

```
//chapter6/bin/example_07.dart
class Cube{
  final num l;
  final num w;
  final num h;
  final num volume;
  //声明方法时可以省略参数前面的数据类型
  Cube(l,w,h):l = l,w = w,h = h,volume = l * w * h;
}
main(){
  var c = Cube(2,3,6);
  print('长方体的体积为:${c.volume}');
}
```

6.2.4 重定向构造函数

有时,构造函数的唯一目的是重定向到另一个构造函数。重定向构造函数的主体为空,构造函数调用出现在冒号后面。示例代码如下:

```
//chapter6/bin/example_08.dart
class Point{
  num x,y;
  //该类的主构造函数
  Point(this.x, this.y);
  //委托实现给主构造函数
  Point.alongXAxis(num x) : this(x,0);
}
main(){
  var p = Point.alongXAxis(6);
  print('p:(${p.x}, ${p.y})');
}
```

6.2.5 常量构造函数

如果类生成的对象不会改变,则可以在生成这些对象时将其定义为编译时常量。可以

在构造函数前使用 const 修饰,并确保所有实例属性都由 final 修饰。示例代码如下:

```
class ImmutablePoint{
  static final ImmutablePoint origin = const ImmutablePoint(0,0);
  final num x,y;
  const ImmutablePoint(this.x,this.y);
}
```

要使用常量构造函数创建编译时常量,需将 const 关键字放在构造函数名称之前。示例代码如下:

```
//chapter6/bin/example_09.dart
class ImmutablePoint{
  static final ImmutablePoint origin = const ImmutablePoint(0,0);
  final num x,y;
  const ImmutablePoint(this.x,this.y);
}
main(){
  //使用构造函数创建编译时常量
  var p = const ImmutablePoint(2,2);
  print('p:( ${p.x}, ${p.y})');
}
```

构造两个相同的编译时常量会产生一个规范的实例。示例代码如下:

```
//chapter6/bin/example_10.dart
class ImmutablePoint{
  static final ImmutablePoint origin = const ImmutablePoint(0,0);
  final num x,y;
  const ImmutablePoint(this.x,this.y);
}
main(){
  var a = const ImmutablePoint(1,1);
  var b = const ImmutablePoint(1,1);
  //它们是同一个实例
  print(identical(a,b));
}
```

6.2.6 工厂构造函数

当一个类不需要总是创建新实例时可以使用 factory 关键字来修饰构造函数。例如,工厂构造函数可以从缓存中返回实例,或者返回子类型的实例。

以下示例演示了工厂构造函数从缓存中返回对象的方法。工厂构造函数接收一个参数值,如果在缓存对象_cache 中存在与该值相等的键,则返回与键关联的值,即 Logger 类的实例。如果不存在,则使用该值作为键,并创建一个 Logger 类的实例作为值,并将此键值对存入缓存对象_cache 中。示例代码如下:

```
//chapter6/bin/example_11.dart
class Logger{
  final String name;
  bool mute = false;
  //在名字_cache 前有下画线代表库是私有的
  //用于存储键时字符串值是 Logger 对象的键值对集合
  static final Map<String, Logger> _cache =
  <String, Logger>{};
  //定义工厂构造函数
  //在工厂构造函数中无法访问 this
  factory Logger(String name){
    //判断_cache 的键中是否存在参数值
    //若存在则直接返回对应值
    //若不存在则创建与参数值关联的键值对并返回关联的值
    return _cache.putIfAbsent(
        name, () => Logger._internal(name));
  }
  //私有构造函数
  Logger._internal(this.name);
  //方法
  void log(String msg){
    if (!mute) print(msg);
  }
}
main(){
  //通过工厂构造函数创建实例
  //像普通构造函数一样调用工厂构造函数
  var logger1 = Logger('UI');
  var logger2 = Logger('UI');
  //判断是否是同一个实例
  print(identical(logger1,logger2));
  logger1.log('Button clicked');
}
```

6.3 方法

13min

方法是指在类中提供对象行为的函数,方法与函数是等价的,只是习惯将类中定义的函数称为方法。有 static 关键字修饰的方法就是类方法,无 static 关键字修饰的方法就是实例方法。

6.3.1 实例方法

声明实例方法同普通函数一样，只是将它放在类中。对象的实例方法可以访问实例变量和 this。示例代码如下：

```
//chapter6/bin/example_12.dart
import 'dart:math';
class Point{
  num x, y;
  Point(this.x, this.y);
  //实例方法
  num distanceTo(Point other){
    var dx = x - other.x;
    var dy = y - other.y;
    return sqrt(dx * dx + dy * dy);
  }
}
```

6.3.2 类方法

在方法前面使用 static 关键字修饰，该方法就是类方法或静态方法。类方法不能在实例上操作，因此实例无法访问它，只能通过类直接引用类方法。示例代码如下：

```
//chapter6/bin/example_13.dart
import 'dart:math';
class Point{
  num x, y;
  Point(this.x, this.y);
  //类方法又称静态方法
  static num distanceBetween(Point a, Point b) {
    var dx = a.x - b.x;
    var dy = a.y - b.y;
    return sqrt(dx * dx + dy * dy);
  }
}

void main(){
  var a = Point(2, 2);
  var b = Point(4, 4);
  //直接通过类调用类方法
  var distance = Point.distanceBetween(a, b);
  print(distance);
}
```

6.3.3 方法 getter 和 setter

getter 和 setter 是一组特殊的方法，它们提供了对对象属性读和写的能力。在 Dart 中对属性的访问，实际上都是调用 getter 方法，对属性的赋值实际上都是调用 setter 方法。每个属性都有一个与之关联的 getter 方法，如果是非 final 修饰的属性还有一个关联的 setter 方法。

```
class Circle{
  final pi = 3.14;
  num r;
}
```

上例中，Circle 类的 pi 属性由 final 修饰，因此只有一个 setter 方法与之关联。r 属性则有一对 setter 和 getter 方法与之关联。

属性的 getter 和 setter 方法可以通过使用 get 和 set 关键字修饰来显式声明，方法名是需要定义的属性名。getter 方法不需要参数列表，getter 的调用语法与变量的访问没有区别。setter 方法只接收一个参数，setter 的调用语法与变量赋值一致。示例代码如下：

```
//chapter6/bin/example_14.dart
class Circle{
  final pi = 3.14;
  num r;
  //定义 getter 方法
  num get area => pi * r * r;
  //定义 setter 方法
  set area(num area) => r = area/(2 * pi);
  Circle(this.r);
}

void main(){
  //创建 Circle 类的实例
  var c = Circle(9);
  //调用属性 area 的 getter 方法
  print('面积: ${c.area}');
  //调用属性 area 的 setter 方法
  c.area = 30;
  print('半径: ${c.r}');
}
```

上例中对 Circle 类通过显式定义 getter 和 setter 方法添加 area 属性。从 getter 方法的返回值可以看出 area 的值是由 pi 和 r 的属性决定的，setter 方法则是更改 r 属性的值。应当仔细分析此例中的 area 属性与 r 属性和 pi 属性取值与赋值的区别，这是理解与使用

getter 和 setter 方法的关键。通常来说，如果一个属性的值与其他属性有关，则可为其显式定义 getter 方法。如果修改一个属性的值会影响其他属性的值，则可为其显式定义 setter 方法。

6.4 继承

8min

继承是重用代码的一种途径，继承可以使子类获得父类的实例成员，即实例属性和实例方法。继承也会使子类继承父类的接口，后续会详细说明。在 Dart 语言中采用单继承的方式，除 Object 类之外的所有类都只有一个父类，因为 Object 类没有父类，因此在 Dart 中类层次结构是以 Object 类为根的树。

先定义一个类 Person：

```
class Person{
  String name;
  int age;
}
```

再定义一个类 Employee：

```
class Employee{
  String name;
  int age;
  num salary;
}
```

可以从上面两个类看出来它们的属性有很多相同之处，Employee 只比 Person 多一个 salary 属性。这样重复性的工作显然是极其没有效率的，因此采用继承机制完成代码的重用。

在 Employee 类的基础上附加 extends 关键字和需要继承的 Person 类，这样子类 Employee 就拥有了父类 Person 的实例属性和实例方法。这里我们没有定义构造函数，因此两个类都只有默认构造函数。示例代码如下：

```
//chapter6/bin/example_15.dart
class Person{
  String name;
  int age;
}
//继承 Person 类
class Employee extends Person{
  num salary;
}
```

```
void main(){
  var emp = Employee();
  //判断 emp 是否是类 Person 子类的实例
  print(emp is Person);
}
```

构造函数、类属性和类方法不会被子类继承，即子类不会从其父类继承构造函数，声明没有构造函数的子类仅具有默认构造函数。

默认情况下，子类会调用父类的默认构造函数，父类的构造函数在子类构造函数体执行之前调用。如果存在初始化列表，它将在父类默认构造函数被执行之前执行，执行顺序如下：

（1）初始化列表。

（2）父类的默认构造函数。

（3）子类的默认构造函数。

6.4.1　调用父类的非默认构造函数

如果在父类中提供了构造函数，那么必须主动调用父类中所有构造函数中的一个构造函数。在构造函数主体之前，在冒号(:)之后指定父类构造函数。

在 Employee 类的命名构造函数的初始化列表中调用父类 Person 的命名构造函数 fromJson。示例代码如下：

```
//chapter6/bin/example_16.dart
class Person{
  String name;
  int age;
  //命名构造函数 fromJson
  Person.fromJson(Map json){
    this.name = json['name'];
    this.age = json['age'];
    print('Person.fromJson 构造函数');
  }
}

class Employee extends Person{
  num salary;
  //Person 提供了构造函数
  //因此必须调用父类中所有构造函数中的一个构造函数
  Employee.fromJson(Map json) : super.fromJson(json){
    print('Employee.fromJson 构造函数');
  }
}
```

```
main(){
  Map json = {'name':'张三','age':23};
  //通过命名构造函数创建实例
  var e = Employee.fromJson(json);
  e.salary = 10968;
  print('名字:${e.name},年龄:${e.age},薪资:${e.salary}');
}
```

因为父类构造函数的参数是在调用构造函数之前求值的，所以参数可以是表达式。示例代码如下：

```
//chapter6/bin/example_17.dart
class Person{
  String name;
  int age;
  Person.fromJson(Map json){
    this.name = json['name'];
    this.age = json['age'];
  }
}
class Employee extends Person{
  static Map defaultData = {'name':'张三','age':23};
  num salary;
  //向父类构造函数提供表达式
  Employee() : super.fromJson(defaultData);
}
main(){
  var e = Employee();
  e.salary = 10968;
  print('名字:${e.name},年龄:${e.age},薪资:${e.salary}');
}
```

传递给父类构造函数的参数不能使用 this 关键字，在此之前子类构造函数尚未被执行，子类的实例对象也就还未被初始化，因此实例成员都无法访问，但子类的类成员可以被访问。

6.4.2 覆写类成员

子类可以覆盖实例方法、getter 方法和 setter 方法。可以使用@override 注解指示有意覆盖父类的实例成员，在子类中不使用@override 注解也不影响运行结果，其主要作用是为了说明该成员在父类中定义过。示例代码如下：

```
//chapter6/bin/example_18.dart
class Person{
```

```
  String name;
  int age;
  void say(String msg){
    print('Person: $ msg');
  }
}
class Employee extends Person{
  num salary;
  //覆写父类 say 方法
  @override
  void say(String msg){
    print('Employee: $ msg');
  }
}
main(){
  var e = Employee();
  e.say("hello");
}
```

6.4.3 覆写操作符

可以覆盖如表 6-1 所示的运算符。例如,如果定义一个 Vector 类,则可以定义一个+方法来添加两个向量。

表 6-1 可覆写操作符

<	+	|	[]	>	/	^	[]=	<=
~/	&	~	>=	*	<<	==	-	%
>>								

类覆盖+和-运算符的示例代码如下:

```
//chapter6/bin/example_19.dart
class Vector{
  final int x, y;
  Vector(this.x, this.y);
  //覆写 + 操作符
  Vector operator + (Vector v) => Vector(x + v.x, y + v.y);
  //覆写 - 操作符
  Vector operator - (Vector v) => Vector(x - v.x, y - v.y);
  toString(){
    return 'x: $ x, y: $ y';
  }
}
```

```
void main(){
  final v = Vector(2, 3);
  final w = Vector(2, 2);
  //调用 Vector 类的 + 操作符
  print(v + w);
  //调用 Vector 类的 - 操作符
  print(v - w);
}
```

如果覆写==操作符，则还应该覆写对象的属性 hashCode 的 getter 方法。覆写==和 hashCode 的示例代码如下：

```
//chapter6/bin/example_20.dart
class Person{
  final String firstName, lastName;
  Person(this.firstName, this.lastName);
  //覆写类的 hashCode 的 getter 方法
  @override
  int get hashCode{
    int result = 17;
    result = 37 * result + firstName.hashCode;
    result = 37 * result + lastName.hashCode;
    return result;
  }
  //覆写类的 == 操作符
  @override
  bool operator == (dynamic other){
    if (other is! Person) return false;
    Person person = other;
    return (person.firstName == firstName &&
        person.lastName == lastName);
  }
}

void main(){
  var p1 = Person('Bob', 'Smith');
  var p2 = Person('Bob', 'Smith');
  var p3 = 'not a person';
  //比较实例 p1 和 p2 的 hashCode 是否相等
  print(p1.hashCode == p2.hashCode);
  //比较实例 p1 和 p2 是否相等
  print(p1 == p2);
  print(p1 != p3);
}
```

6.4.4 未定义函数

需要在代码尝试使用不存在的方法或实例变量时进行检测或做出反应，可以重写 noSuchMethod 方法。示例代码如下：

```
//chapter6/bin/example_21.dart
class A{
  //覆写 noSuchMethod 方法
  //否则当尝试调用一个不存在的成员时将抛出异常 NoSuchMethodError
  @override
  noSuchMethod(Invocation invocation){
     print('你尝试使用一个不存在的成员：${invocation.memberName}');
  }
}
void main(){
  //这里必须使用 dynamic
  dynamic a = A();
  //调用类中不存在的成员
  a.w;
}
```

6.5 抽象类和接口

11min

6.5.1 抽象类

抽象类是使用 abstract 关键字修饰的类，抽象类是无法实例化的类。抽象类通常具有抽象方法，直接使用分号(;)替代方法体即可声明一个抽象方法，抽象方法只能存在于抽象类中。抽象类对于定义接口很有用且可以带有一些方法实现。

声明具有抽象方法的抽象类的示例代码如下：

```
//chapter6/bin/example_22.dart
//定义一个抽象类
abstract class Person{
  //定义实例属性
  String name;
  int age;
  //定义实例方法
  void say(String msg){
    print('Person: $msg');
  }
  //声明一个抽象方法
  void doSomething(String job);
```

```
}
//继承抽象类
class Employee extends Person{
  //在子类中提供抽象方法的具体实现
  void doSomething(String job){
    print('Employee $ job');
  }
}
main(){
  var e = Employee();
  e.doSomething('打扫卫生');
}
```

实例方法、getter 方法和 setter 方法都可以是抽象的，只需将它们定义为一个接口方法，并将其实现留给其他类。

6.5.2 隐式接口

在 Dart 语言中每个类都隐式定义了一个接口，该接口包含该类所有实现的接口和实例成员。如果想创建一个类 A，它支持类 B 的所有实例属性和实例方法，但又不继承类 B 的实现，那么类 A 应该实现类 B 的接口，然后类 A 再提供自己的实现。

类通过在 implements 子句中提供一个或多个接口，然后提供接口所需的实例属性和实例方法实现一个或多个接口。示例代码如下：

```
//chapter6/bin/example_23.dart
class Person {
  //在接口中，但仅对此库可见，因为它是私有的
  final _name;
  //不在接口中，因为它是构造函数
  Person(this._name);
  //在接口中
  String greet(String who) => '你好 $ who. 我是 $ _name.';
}

//Person 接口的一种实现
class Impostor implements Person {
  get _name => '';
  //提供自己的实现
  String greet(String who) => '你好 $ who.';
}
//定义顶层方法
String greetBob(Person person) => person.greet('Bob');

void main() {
```

```
  print(greetBob(Person('Kathy')));
  print(greetBob(Impostor()));
}
```

上述代码中Impostor类不是Person类的子类，它没有继承Person类的任何成员。implements子句的作用是在接口间建立关联，而不是共享实现。

下面是指定一个类实现多个接口的示例：

```
class Point implements Comparable,Location {...}
```

因为每个类都提供了隐式接口，因此在Dart语言中没有正式的接口声明，但是在Dart中可以定义一个纯抽象类来描述传统意义上的接口。

6.6　向类添加特征

5min

mixin是一种在多个类层次结构中重用类代码的方式，定义类时使用关键字mixin代替class，且不为该类提供构造函数，mixin类属于抽象类，因此不能被实例化。示例代码如下：

```
mixin Fly{
  bool canFly = true;
  void flying(){
    if(canFly)
      print('飞行中');
  }
}
```

要使用mixin，需使用with关键字，后跟一个或多个mixin类名称。该类会获得mixin类定义的实例属性和实例方法，使用mixin的示例代码如下：

```
//chapter6/bin/example_24.dart
mixin Fly{
  bool canFly = true;
  void flying(){
    if(canFly)
      print('飞行中');
  }
}
mixin Walk{
  bool canWalk = true;
  void walking(){
    if(canWalk)
```

```
      print('行走中');
    }
}
//使用 mixin
class Dove with Fly,Walk{
}
void main(){
  var d = Dove();
  //使用从 mixin 类 Fly 获得的方法
  d.flying();
  //使用从 mixin 类 Walk 获得的方法
  d.walking();
}
```

有时希望指定类型才能使用 mixin 类，可以使用 on 子句指定父类约束。也就是说 on 子句后面的类是使用该 mixin 类的父类，并且 on 子句后面的类也是该 mixin 类的父类，在 mixin 中可以使用 super 调用父类中的实例成员。

定义一个类 Bird，让 Dove 类继承 Bird 类，然后通过 on 子句向 mixin 类 Fly 添加父类约束。示例代码如下：

```
//chapter6/bin/example_25.dart
//用 on 关键字指定使用此 mixin 类的父类约束 Bird
//Bird 类同时是此 mixin 类的父类
mixin Fly on Bird{
  bool canFly = true;
  void flying(){
    if(canFly)
      print('飞行中');
  }
}
//此 mixin 类的使用没有任何限制
mixin Walk{
  bool canWalk = true;
  void walking(){
    if(canWalk)
      print('行走中');
  }
}
//定义 Bird 类
class Bird{
}
//让 Dove 类继承 Bird 类,使 Dove 类可以使用 mixin 类 Fly
class Dove extends Bird with Fly,Walk{
}
```

```
void main() {
  var d = Dove();
  d.flying();
}
```

使用 mixin 的类是其父类的子类，也是 mixin 表示的类型的子类。

mixin 的工作原理是生成一个新类，该新类将 mixin 的实现置于父类之上，这意味着在新类中 mixin 类的实现会覆盖父类中的同名实例属性和实例方法，这里的父类是指 extends 关键字修饰的类。如果有多个 mixin，则依次创建新类。

当前类没有除 Object 之外的父类时，示例代码如下：

```
mixin Walk{
  void walking(){
    print('行走中');
  }
}
class Dog with Walk{
}
```

在语义上等价于如下代码：

```
mixin Walk{
  void walking(){
    print('行走中');
  }
}
class Dog extends Object with Walk{
}
```

它实际上是 Object 与 mixin 类 Walk 产生一个新类，然后 Dog 再继承这个新类。

如果当前类存在除 Object 之外的父类，示例代码如下：

```
class Dove extends Bird with Fly,Walk{
}
```

它实际上分为三步完成。

第 1 步：Bird 类和 mixin 类 Fly 产生新类，我们把它看作类 BirdFly。

```
class Bird with Fly{
}
```

第 2 步：产生的新类 BirdFly 再和 mixin 类 Walk 产生新类，我们把它看作类

BirdFlyWalk。

```
class BirdFly with Walk{
}
```

第 3 步：产生的新类 BirdFlyWalk 与 Dove 完成继承关系。

```
class Dove extends BirdFlyWalk{
}
```

继承顺序依次是 Bird 类、Fly 类、Walk 类、Dove 类。这就是它们生成新类的顺序和逻辑。如果多个类有相同的属性或方法，排在后面的类会覆盖前面的类。最终继承关系如图 6-1 所示。

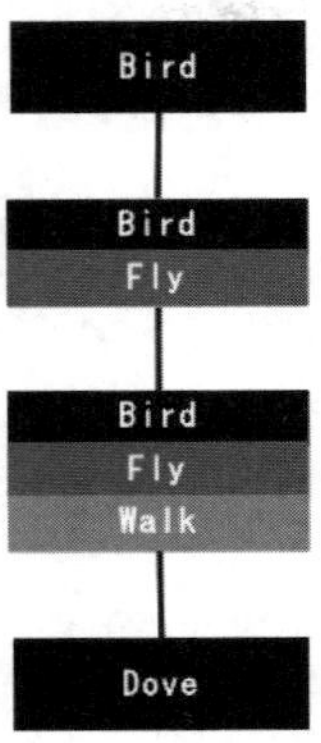

图 6-1 继承关系图

7min

6.7 枚举类

枚举类是一种特殊的类，用于表示固定数量的常量值。

使用 enum 关键字声明枚举类型的结构如下：

```
enum 类名{
  值 1,
  值 2,
  值 3
}
```

每个枚举值都有一个名为 index 的 getter 方法，该方法将会返回以 0 为基准索引的位置值。例如：第一个值的索引为 0，第二个值的索引为 1。示例代码如下：

```
//chapter6/bin/example_26.dart
//定义枚举类
enum Color{
  //枚举值列表
  red,
  green,
  blue
}
void main(){
  //打印枚举值在枚举类中的索引
  print(Color.red.index);
  print(Color.green.index);
  print(Color.blue.index)
}
```

要获取枚举中所有值的列表,需使用枚举的 values 常量。

```
var values = Color.values
```

可以在 switch 语句中使用枚举,如果不处理所有枚举值,则会收到警告,示例代码如下:

```
//chapter6/bin/example_27.dart
//定义枚举类
enum Color{
  //枚举值列表
  red,
  green,
  blue
}
void main(){
  //打印枚举类中的所有枚举值
  print(Color.values);
  //为变量赋值一个枚举值
  var aColor = Color.blue;
  //switch 语句是枚举常见的使用场景
  switch (aColor) {
    case Color.red:
      print('红色');
      break;
    case Color.green:
      print('绿色');
      break;
    default:
```

```
      print('蓝色');
  }
}
```

枚举类有以下限制：

(1) 不能作为子类、mixin 或实现枚举类。

(2) 无法显式实例化枚举。

第7章 异　常

Dart代码可以抛出和捕获异常，异常表示发生了某些意外的错误。如果异常未被捕获，则引起异常的isolate将被暂停，并且isolate及其程序将被中止。

在Dart语言中所有的异常都是非必检异常，方法不声明它们可能会引发哪些异常，并且不需要捕获任何异常。

Dart语言提供了Exception和Error类型，以及许多预定义的子类型，也可以自定义异常。Error是程序无法恢复的严重错误，表示程序出现较严重问题，而又无法通过编程处理，只能终止程序。大多数错误与代码编写者执行的操作无关。Exception是程序可以恢复的异常，它可以通过编程解决。Exception可能是由参数错误、网络连接中断或字符串解析失败等引起的。本章所关注的是对Exception及其子类的异常处理。

7.1 抛出异常

通过throw语句来主动抛出异常。示例代码如下：

```
//chapter7/bin/example_01.dart
void main(){
  throw FormatException('格式转换异常');
}
```

异常不仅仅是Exception和Error对象，还可以将任何非空对象作为异常抛出，例如：字符串。示例代码如下：

```
//chapter7/bin/example_02.dart
void main(){
  throw '这是一个字符串';
}
```

因为throw语句是一个表达式，因此可以在胖箭头语句或其他使用表达式的地方出现。示例代码如下：

```
//chapter7/bin/example_03.dart
void main(){
  void distanceTo() => throw UnimplementedError();
  distanceTo();
}
```

6min

7.2 捕获异常

捕获异常会阻止异常传播,除非重新抛出异常,捕获一个异常以便能够对其做进一步处理。

使用 try 语句来捕获异常,使用 catch 语句处理异常,catch 语句有两个参数,第一个是必选的异常对象,第二个是可选的堆栈对象。

try-catch 语句声明格式如下:

```
try{
  //可能引起异常的语句
}catch(e,s){
  //处理异常
}
```

示例代码如下:

```
//chapter7/bin/example_04.dart
void main(){
  try {
    //抛出异常
    throw Exception;
  }catch (e,s) {
    //捕获异常
    print('异常详情:\n $e');
    print('堆栈跟踪:\n $s');
  }
}
```

可以通过 on 加异常名来捕获指定类型的异常。示例代码如下:

```
//chapter7/bin/example_05.dart
void main(){
  try {
    throw FormatException();
  }on FormatException{
```

```
    print('格式转换异常');
  }
}
```

try 子句中的代码块可能引发多个异常,此时可以使用 on 子句捕获指定异常。如果需要异常对象,则可以同时使用 on 和 catch 子句。与抛出的异常对象类型相匹配的第一个 on 或 catch 子句会处理异常,如果 catch 子句未指定异常类型,则该子句可以处理引发的任何类型的异常。示例代码如下:

```
//chapter7/bin/example_06.dart
void main(){
  try{
    //抛出异常
    throw IntegerDivisionByZeroException();
  }on FormatException{
    //处理 FormatException 类型的异常
    print('格式转换异常');
  }on Exception catch(e){
    //处理 Exception 类型的异常
    print('异常:$e');
  }catch(e){
    //未指定类型,可以处理所有异常
    print('其他异常:$e');
  }
}
```

对异常进行部分处理后,如果需要异常继续传播,则可以使用 rethrow 关键字。示例代码如下:

```
//chapter7/bin/example_07.dart
void action(){
  try {
    throw FormatException();
  }catch(e){
    //部分处理异常
    print('action()部分处理 ${e.runtimeType}');
    //重新抛出异常
    rethrow;
  }
}
void main(){
  try {
    action();
```

```
  } catch (e) {
    print('main()完成处理 ${e.runtimeType}.');
  }
}
```

7.3 最终操作

1min

finally 语句在 try 和 catch 语句之后，无论是否触发异常，该语句都会被执行。如果提供该语句，则其语法格式如下：

```
try{
  //可能引起异常的语句
}catch(e,s){
  //处理异常
}finally{
  //无论是否引发异常都需要执行的代码块
}
```

示例代码如下：

```
//chapter7/bin/example_08.dart
void main(){
  try{
    //此处不会有异常抛出
    var i = 1;
  }catch(e){
    //有异常时才会执行
    print('Error: $e');
  }finally{
    //始终会得到执行
    print('finally 语句块');
  }
}
```

如果没有 catch 子句，则在 finally 子句执行后异常继续传播。示例代码如下：

```
//chapter7/bin/example_09.dart
void main(){
  try {
    //此处抛出异常
    throw Exception();
  }finally{
    //始终会得到执行
```

```
    print('finally 语句块');
  }
}
```

7.4 自定义异常

可以通过实现 Exception 接口来定义自定义异常。示例代码如下：

```
//chapter7/bin/example_10.dart
//实现 Exception 接口来自定义异常
class MyException implements Exception{
  //接收消息的变量
  final String msg;
  //常量构造函数
  const MyException([this.msg]);
  //覆写 toString 方法
  @override
  String toString() => msg ?? 'MyException';
}
void main(){
  //抛出自定义异常
  throw MyException('自定义异常');
}
```

第8章

泛型

查看集合类型 List 的 API 文档时会发现其类型实际上是 List<E>。符号<>将 List 标记为可泛型化的类，即类型可参数化。通常使用一个字母来表示类型参数，例如 E、T、S、K 和 V 等。

泛型常用于要求类型一致的情况，它还可以减少代码重复。例如现在声明一个只包含 String 类型的数组，可以声明为 List<String>，读作字符串类型的 List。这样声明可以避免因在该数组中放入非 String 类型的值而出现异常，同时编译器也可以及时发现并定位问题。导致异常的示例代码如下：

```
var names = List<String>();
names.addAll(['Seth', 'Kathy', 'Lars']);
//错误
names.add(42);
```

创建命令行应用程序，将项目命名为 chapter8。本章所有文件都在该项目的 bin 文件夹下创建与运行。

7min

8.1 使用泛型

List、Set 和 Map 字面量可以被参数化。参数化字面量与已经介绍的字面量一样，只不过在左括号之前添加了<type>(用于 List 和 Set)或<keyType,valueType>(用于 Map)。使用类型字面量的示例代码如下：

```
//chapter8/bin/example_01.dart
//创建 List<String>的实例
var names = <String>['Seth','Kathy','Lars'];
//创建 Set<String>的实例
var uniqueNames = <String>{'Seth','Kathy','Lars'};
//创建 Map<String,String>的实例
var website = <String, String>{
```

```
    'qq.com': '腾讯 QQ',
    'aliyun.com': '阿里云',
    'toutiao.com': '头条新闻'
};
```

可以参数化集合的构造函数以便使用泛型。在使用构造函数时指定一种或多种类型，将类型放在类名之后的尖括号(< type >或< keyType, valueType >)中。示例代码如下：

```
//chapter8/bin/example_02.dart
//创建 List < String >的实例
var values = List < String >();
//创建 Set < String >的实例
var setValues = Set < String >();
//创建 Map < String,String >的实例
var mapValues = Map < String,String >();
```

Dart 泛型类型已经过规范化，这意味着它们会在运行时携带其类型信息。因此可以测试集合的类型，示例代码如下：

```
//chapter8/bin/example_03.dart
print(names is List < String >);
print(uniqueNames is Set < String >);
print(website is Map < String,String >);
```

8.2 自定义泛型

8min

使用泛型的另一个原因是减少代码重复。泛型可以在多种类型之间共享单个接口和实现，同时可以利用静态分析。

泛型类的定义结构如下：

```
//chapter8/bin/example_04.dart
//定义泛型类，泛型参数 T
class ClassName < T >{
  //使用泛型参数定义实例变量
  T t;
  //定义实例方法，方法的返回值和参数都可以使用泛型参数 T
  T method(List < T > ts){
    //泛型参数 T 定义局部变量
    T ts1;
    ts1 = ts[0];
```

```
    return ts1;
  }
}
```

类名后面加上泛型参数 T,这样在类的实例变量和实例方法中也可以使用泛型参数 T。

8.2.1 泛型类

创建一个用于缓存对象的接口,示例代码如下:

```
//chapter8/bin/example_05.dart
//缓存 Object 的接口
abstract class ObjectCache{
  Object getByKey(String key);
  void setByKey(String key, Object value);
}
```

需要字符串版本的相同接口,因此创建了另一个接口,示例代码如下:

```
//chapter8/bin/example_06.dart
//缓存字符串的接口
abstract class StringCache{
  String getByKey(String key);
  void setByKey(String key, String value);
}
```

当需要更多其他类型的类似接口时,会发现非常烦琐。泛型就可以省去创建这些相似接口的麻烦,此时只需创建一个带有类型参数的接口。示例代码如下:

```
//chapter8/bin/example_07.dart
//缓存任意对象的泛型接口
abstract class Cache<T>{
  T getByKey(String key);
  void setByKey(String key,T value);
}
```

在此代码中,T 是替代类型。可以将其视为占位符,作为开发人员以后定义的类型。

提供接口的具体实现并创建实例。示例代码如下:

```
//chapter8/bin/example_08.dart
//实现 Cache 接口
class CachePool<U> implements Cache<U>{
  final Map pool = Map();
  //实现 getByKey 方法,返回类型为泛型参数 U
```

```
  @override
  U getByKey(String key){
    return pool[key];
  }
  //实现 setByKey 方法,返回类型为泛型参数 U
  @override
  void setByKey(String key,U value){
    pool[key] = value;
  }
}
void main(){
  //创建实例,泛型参数为 int
  var intMap = CachePool<int>();
  intMap.setByKey('first',1);
  print(intMap.getByKey('first'));

  //创建实例,泛型参数为 String
  var stringMap = CachePool<String>();
  stringMap.setByKey('hi','hello');
  print(stringMap.getByKey('hi'));
}
```

在上例中 CachePool 类的实例等价于 Map<String,dynamic>的实例,这里主要是为了说明泛型接口的定义与实现。

8.2.2 泛型方法

最初 Dart 的泛型支持仅限于类。一种称为泛型方法的较新语法,允许在方法上使用类型参数。示例代码如下:

```
//chapter8/bin/example_09.dart
//定义泛型方法,可为方法提供泛型参数
T first<T>(List<T> ts) {
  //用泛型参数定义局部变量
  T tmp;
  tmp = ts[0];
  return tmp;
}
void main(){
  //使用泛型方法,泛型参数为 int
  var intList = [1,2,3];
  print(first<int>(intList));

  //使用泛型方法,泛型参数为 String
```

```
  var strList = ['one','two','three'];
  print(first<String>(strList));
}
```

这是使用泛型方法的一个比较极端的例子，其主要意图在于演示各个可以使用泛型参数的位置，在实际使用中只会在少量位置使用泛型参数。

从 first 方法可以看出可以在多个地方使用类型参数 T：方法的泛型(first<T>)、函数的返回值类型(T)、参数的类型(List<T>)、局部变量(T tmp)。

8.2.3 限制类型

在实现泛型类型时，可以限制其参数的类型，可以使用 extends 做到这一点。示例代码如下：

```
//chapter8/bin/example_10.dart
//定义基础类
class BaseClass{}
//定义子类
class Extender extends BaseClass{}
//限制泛型参数的类型
class Generic<T extends BaseClass>{
  //覆写 toString 方法
  @override
  String toString() {
    return 'Generic<$T>的实例';
  }
}
void main(){
  //使用基础类作为泛型参数
  var baseGen = Generic<BaseClass>();
  print(baseGen);

  //使用子类作为泛型参数
  var exteGen = Generic<Extender>();
  print(exteGen);
}
```

在为 Generic 类指定泛型参数时，可以将 BaseClass 或其任何子类用作通用参数。指定任何非 BaseClass 类型都会导致错误。

第 9 章 库

Dart 程序是由被称为库的模块化单元组成的，一个库由多个顶层声明组成，这些声明可能包含函数、变量及类。

9.1 声明与使用

27min

使用 library 关键字显式声明库。示例代码如下：

```
//chapter9/bin/example_01.dart
//声明库
library stack;

//声明顶层变量，带下画线代表此变量是库私有的，不对外公开
final _contents = [];

//声明顶级函数
//判断堆栈是否为空
get isEmpty => _contents.isEmpty;
//获取堆栈最上面的元素
get top => isEmpty ? throw '堆栈为空，不能获取元素' : _contents.last;
//弹出元素
get pop => isEmpty ? throw '堆栈为空，不能弹出元素' : _contents.removeLast();
//推入元素
dynamic push(ele){
  _contents.add(ele);
  return ele;
}

//声明库范围的类
class StackTool{
  //清空堆栈
  void clear(){
    _contents.clear();
```

```
    _contents.length;
  }
  //获取堆栈的长度
  int length(){
    return _contents.length;
  }
}
```

由关键字 library 修饰的单词 stack 就是库名。

代码中_contents 是顶层变量，它的初始值是空列表。顶层变量是延迟初始化的，在它们的 getter 第一次被调用时才初始化。因此变量_contents 在某个访问它的方法被调用时才被设置成[]。

顶层变量属于静态变量，因此也被称为库变量。顶层变量的作用域覆盖了声明它们的整个库，库的作用域通常由多个类与顶层函数构成。顶层变量也可以由 final 修饰，这意味着它们没有定义 setter 并且必须在声明时就初始化。

代码中 isEmpty、top、pop 和 push 都是顶层函数，顶层函数的作用域覆盖整个库，它们可以是普通方法、setter 和 getter 方法。

在库中还可以声明顶层类，如代码中的 StackTool 类。在 Dart 中类都是顶层的，因为 Dart 不支持嵌套类。

以下画线(_)开头的变量都是库私有的，顶层变量_contents 便是库私有的。它只对库可见，因此只能在库内部访问变量_contents。

下面是一个简单的程序，与 stack 库不一样，这里没有显式的库声明且功能单一，但它也是一个库。示例代码如下：

```
//chapter9/bin/example_02.dart
void main(){
  print("Hello World");
}
```

上述程序同时也是一个脚本，脚本是从 main()函数开始执行的。意味着这是一个可被直接执行的库，对于快速、简单的任务，通过脚本可以更方便地编写实验性代码。

9.1.1 导入库

使用 import 指令指定一个库的命名空间，唯一必须指定的参数是库的 URI。对于内置库使用 dart 前缀，再加上库名。

应用程序需要使用内置 math 库来生成随机数，可以在文件中导入并使用。示例代码如下：

```
//chapter9/bin/example_03.dart
//导入内置库 math,dart:前缀代表内置库
import 'dart:math';

void main() {
  //Random 是 math 库中的类
  var random = Random();
  //Random 类中的方法 nextInt(int max)有一个参数,用于指定能随机生成的最大整数
  var randomInt = random.nextInt(100);
  print("0～100 的随机数：$ randomInt");
}
```

Dart 软件生态采用软件包来管理和共享软件，例如库和工具。获取软件包，需要采用 pub 包管理器。对于第三方库，导入包前需要添加依赖，Dart 应用程序和库的根目录都有 pubspec.yaml 文件，其中列出了软件包的依赖关系，并包括其他元数据，例如版本号等。

pubspec.yaml 文件支持的依赖如表 9-1 所示。

表 9-1　依赖文件字段

字　段	选　填	说　明
name	必选	包名，全部小写且以下画线分割单词，应当是有效的 Dart 标识符，建议在 pub.dev 上搜索软件包，避免重名
version	可选	版本号，由点分割的 3 个数字，例如 1.24.3 它可以拥有构建版本(＋1,＋2)或预发行版(-dev.1、-alpha.1、-beta.1、-rc.1)后缀。默认 0.0.0，若发布软件包则是必选
environment	必选	运行环境，配合 sdk 约束或 flutter sdk 约束指定运行环境版本
description	可选	描述，说明软件包的作用，应为纯英文，60～180 个字符。若发布软件包则是必选
homepage	可选	主页，指向软件包的网址，对于托管的软件包指向软件包页面的链接
repository	可选	软件包源代码的存储库链接，例如：https://GitHub.com/< user >/< repository >
issue tracker	可选	软件包问题跟踪链接，若存在存储库并指向 GitHub，则 pub.dev 站将使用默认的问题跟踪器 https://GitHub.com/< user >/< repository >/issues
documentation	可选	文档链接
dependencies	可选	常规依赖项，存在依赖项时必选
dev_dependencies	可选	开发时，所需依赖项，存在依赖项时必选
dependency_overrides	可选	在开发过程中，可能需要临时覆盖依赖项，存在依赖项时必选
executables	可选	软件包可以将一个或多个脚本公开为可执行文件
publish_to	可选	默认使用 pub.dev 网站。不指定任何内容以防止发布程序包。此设置可用于指定要发布的自定义发布包服务器

最简单的 pubspec. yaml 文件仅列出项目名和平台依赖信息。

```
name: my_app
environment:
  sdk: '>= 2.7.0 < 3.0.0'
```

在依赖文件中添加第三方包依赖。

```
name: my_app
environment:
  sdk: '>= 2.7.0 < 3.0.0'
dependencies:
  json_string: ^2.0.1
```

当修改过 pubspec. yaml 文件后需要在应用程序根目录执行 pub get 命令获取新增软件包，pub 工具将在应用根目录下创建 pubspec. lock 文件，表明依赖信息已经更新，可以在项目中正常使用新增包。同时还会更新. packages 文件，该文件会将项目所依赖的每个包名称映射到系统缓存中的相应包目录。

在文件中引入第三方包，需要使用 package:scheme 指定库，指定库后可以使用库中提供的功能。示例代码如下：

```
//chapter9/bin/example_04.dart
//导入 pub 包管理器提供的第三方包
import 'package:json_string/json_string.dart';
//使用 json_string 库中定义的 mixin 类 Jsonable
class Person with Jsonable{
  String name;
  int age;
  Person(this.name,this.age);
  //提供 toJson 方法的实现
  @override
  Map < String, dynamic > toJson() {
    return {'name':name,'age':age};
  }
}
void main(){
  var p = Person('jobs',66);
  //调用 toJson 方法
  print(p.toJson());
}
```

库的组成至少包含 lib 目录和 pubspec. yaml 文件。库代码位于 lib 目录下，可以根据需要在 lib 下创建任何层次结构。按照规约，实现代码放在 lib/src 目录下，该目录是库私有的，如果要公开 lib/src 下的 API，则可以在 lib 目录下创建主库文件<package-name>. dart，在该文件导出该库的所有公共 API。

对于本地库，当两个文件都在 lib 目录内或两个文件都在 lib 目录外时，可以使用相对路径导入。但是当 lib 目录中的文件被外部的文件引用时必须使用 package:前缀。

在本项目的 bin 目录下的 main. dart 文件中，可以发现已经使用 package:前缀导入了 lib 目录下的 chapter9. dart 文件。现在以相对路径的形式导入当前目录下的 example_01. dart 文件。示例代码如下：

```
//chapter9/bin/main.dart
//使用 package:前缀导入本项目 lib 目录下的文件
import 'package:chapter9/chapter9.dart' as chapter9;
//使用相对路径导入当前目录下的文件
import 'example_01.dart';
void main(List<String> arguments) {
  print('Hello world: ${chapter9.calculate()}!');
}
```

9.1.2 指定库前缀

如果导入两个标识符冲突的库，则可以使用 as 为一个或两个库指定前缀。使用冲突的库成员时，则需要在库成员前指定前缀。例如：如果 library1 和 library2 都具有 Element 类，则需要为其中一个或两者指定库前缀。示例代码如下：

```
import 'package:lib1/lib1.dart';
import 'package:lib2/lib2.dart' as lib2;

//使用来自库 lib1 的 Element 类
Element element1 = Element();

//使用来自库 lib2 的 Element 类
lib2.Element element2 = lib2.Element();
```

实际使用中并不一定在库冲突时才使用，可以为任何库指定前缀，只要它们的名字不冲突就可以。示例代码如下：

```
//chapter9/bin/example_05.dart

//指定库前缀，也可以叫别名
//使用该库的成员必须指定前缀
import 'dart:collection' as co;

void main(){
```

```
    //通过库前缀 co 使用库成员 Queue 类
    co.Queue<int> queue;
}
```

9.1.3 导入库的一部分

库中通常定义了大量可用的库成员，而使用时只需一小部分就能满足要求，显然导入整个库会影响应用程序的性能。如果只想使用库的一部分，则可以有选择地导入该库。与库相关的关键字中可以使用 show 导入库的一部分成员，使用 hide 以表明导入某些成员。示例代码如下：

```
//chapter9/bin/example_06.dart
//只导入 collection 库中的成员 Queue 和 LinkedList
import 'dart:collection' show Queue,LinkedList;
//导入 math 库中除 sin 和 cos 之外的成员
import 'dart:math' hide sin,cos;
```

9.1.4 导出库

当定义了很多功能且位于多个不同的文件中时，如果要使用它们就需要导入所有的文件。这非常麻烦且容易出错，因此可以通过在主库文件中使用 export 关键字导出所有API。示例代码如下：

```
//chapter9/bin/example_07.dart
library example;
//导出库
export 'example_01.dart';
export 'example_02.dart';
```

9.2 核心库

2min

核心库 dart:core 包含内置类型、集合和其他核心功能，该库会自动被导入每个 Dart 程序中。

控制台打印：顶层 print()方法接收一个任意对象的参数，并在控制台输出该对象的字符串值，实际上是调用该对象的 toString()方法。示例代码如下：

```
print(anObject);
print('I drink $tea.');
```

9.2.1 数字

dart:core 库定义了 num、int 和 double 类，它们具有一些用于处理数字的基本实用方法。

1. int 常用属性

(1) sign：返回此整数的符号，对于 0 返回 0，对于小于 0 的数返回－1，对于大于 0 的数返回＋1。

(2) bitLength：返回存储此整数所需的最大位数，位数是指二进制位的个数。位数不包括符号位，带符号的数需要加 1，即 x. bitLength＋1。

(3) isEven：判断此整数是不是偶数，当此整数为偶数时，返回 true。

(4) isOdd：判断此整数是不是奇数，当此整数为奇数时，返回 true。

属性使用示例代码如下：

```
//chapter9/bin/example_08.dart
void main(){
  //查看符号
  print('4 的符号:${4.sign}');
  print('-3 的符号:${(-3).sign}');

  //查看存储所需长度
  //00000100
  print('存储 4 所需的位数:${4.bitLength}');
  //00000001
  print('存储 1 所需的位数:${1.bitLength}');
  //11111111
  print('存储 -1 所需的位数:${(-1).bitLength+1}');
  //11111100
  print('存储 -4 所需的位数:${(-4).bitLength+1}');

  //判断奇偶
  print('4 是偶数吗:${4.isEven}');
  print('3 是奇数吗:${3.isOdd}');
}
```

2. int 常用方法

(1) toRadixString(int radix)：将此整数根据参数 radix 指定的进制转化为字符串的表示形式，参数 radix 的范围在整数 2～36。在字符串表示中，小写字母用于表示 9 以上的数字，a 表示 10，依次类推，z 表示 35。

(2) int tryParse(String source，{int radix})：将字符串解析为可能带符号的整数文字，然后返回其值。传入的字符串不能为 null，当解析出现异常时返回 null。可通过可选参数 radix 指定采用的进制，参数 radix 的范围在整数 2～36。

(3) parse():将字符串转换为整数。

方法使用示例代码如下:

```
//chapter9/bin/example_09.dart
void main(){
  //64 根据十六进制转换为字符串
  print('64 转换为十六进制: $ {64.toRadixString(16)}');
  //64 根据二进制转换为字符串
  print('64 转换为二进制: $ {64.toRadixString(2)}');

  //解析八进制的字符串'64'解析为十进制的整数
  print('解析 64 根据二进制转换的字符串: $ {int.tryParse('64',radix:8)}');

  //将字符串解析为整数
  print(int.parse('42'));
  print(int.parse('0x42'));
  //将字符串解析为浮点数
  print(double.parse('0.50'));
}
```

9.2.2 字符串

Dart 中的字符串是 UTF-16 代码单元的不变序列,可以使用正则表达式(RegExp 对象)在字符串中搜索并替换部分字符串。

String 类将此类方法定义为 split()、contains()、startsWith()、endsWith()等。

1. 字符串查找

(1) endsWith(String other):判断此字符串是否以 other 结尾,如果此字符串以 other 结尾,则返回 true。

(2) contains(Pattern other, [int startIndex = 0]):判断此字符串是否包含其他匹配项,如果包含则返回 true。如果提供了可选参数 startIndex,则此方法仅在该索引处或之后匹配。

(3) indexOf(Pattern pattern, [int start]):返回此字符串中模式的第一个匹配项的位置,如果提供可选参数 start,则从该索引处或之后开始匹配。如果没有匹配到,则返回−1。

(4) lastIndexOf(Pattern pattern, [int start]):返回此字符串中模式的最后一个匹配项的位置,如果没有匹配到,则返回−1。

(5) startsWith(Pattern pattern, [int index = 0]):判断此字符串是否以特定模式开头,如果此字符串以模式匹配开头,则返回 true。如果提供了可选参数 index,则此方法仅在该索引处或之后匹配。

字符串查找示例代码如下:

```
//chapter9/bin/example_10.dart
void main(){
  'Dart'.endsWith('t');

  var str1 = 'Dart strings';
  str1.contains('D');
  str1.contains(new RegExp(r'[A-Z]'));
  str1.contains('X', 1);
  str1.contains(new RegExp(r'[A-Z]'), 1);

  var str2 = 'Dartisans';
  str2.indexOf('art');
  str2.indexOf(new RegExp(r'[A-Z][a-z]'));

  var str3 = 'Dartisans';
  str3.lastIndexOf('a');
  str3.lastIndexOf(new RegExp(r'a(r|n)'));

  var str4 = 'Dart';
  str4.startsWith('D');
  str4.startsWith(new RegExp(r'[A-Z][a-z]'));
}
```

2. 字符串截取

(1) substring(int startIndex, [int endIndex]): 返回此字符串的子字符串,该子字符串从 startIndex(包含)开始到 endIndex(不包含)结束。如果不提供 endIndex 参数,则从索引 startIndex 处开始直到结束。

(2) split(Pattern pattern): 在 pattern 匹配项处拆分字符串,并返回子字符串列表。

可以从字符串中提取单个字符,将其作为字符串或整数。确切地说,实际上获得了单独的 UTF-16 代码单元。例如高音谱号符号(\ u {1D11E})是两个代码单元。

字符串截取示例代码如下:

```
//chapter9/bin/example_11.dart
void main(){
  //提取子字符串
  print('Never odd or even'.substring(6,9));
  var string = 'dartlang';
  print(string.substring(1));
  print(string.substring(1,4));

  //根据提供的模式分隔字符串
  var parts = 'structured web apps'.split(' ');
  print(parts.length);
```

```
  print(parts[0]);

  //通过索引获取单个字符
  print('Never odd or even'[0]);

  //将 split()与空字符串参数一起使用以获取所有字符的列表，有利于迭代
  for (var char in 'hello'.split('')) {
    print(char);
  }

  //获取字符串中的所有 UTF-16 代码单元
  var codeUnitList =
  'Never odd or even'.codeUnits.toList();
  print(codeUnitList[0]);
}
```

3. **大小写转换**

(1) toLowerCase()：将此字符串中的所有字符转换为小写。如果字符串已经全部小写，则此方法返回此字符串。

(2) toUpperCase()：将此字符串中的所有字符转换为大写。如果字符串已经全部大写，则此方法返回此字符串。

大小写转换示例代码如下：

```
//chapter9/bin/example_12.dart
void main(){
  //转化为大写
  print('structured web apps'.toUpperCase());
  //转化为小写
  print('STRUCTURED WEB APPS'.toLowerCase());
}
```

4. **裁剪和空字符串**

(1) trim()：返回没有任何前导和尾随空格的字符串。

(2) trimLeft()：返回没有任何前导空格的字符串。

(3) trimRight()：返回没有任何尾随空格的字符串。

使用 trim()方法移除字符串头部和尾部的所有空白，使用属性 isEmpty 判断字符串是否为空。示例代码如下：

```
//chapter9/bin/example_13.dart
void main(){
  //裁剪字符串
  print(' hello '.trim());
  //检查字符串是否为空
```

```
    print(''.isEmpty);
    //字符串中只含有空白而不是空字符串
    print(' '.isNotEmpty);
  }
```

5. 字符串替换

replaceAll(Pattern from, String replace)：使用 replace 替换所有匹配的子字符串 from。

字符串是不可变的对象，这意味着可以创建它们，但不能更改它们。仔细查看 String API 参考，会注意到，没有任何一种方法可以实际更改 String 的状态。例如，方法 replaceAll()返回一个新的 String 而不更改原始的 String。示例代码如下：

```
//chapter9/bin/example_14.dart
void main(){
  var greetingTemplate = 'Hello, NAME!';
  var greeting =
  greetingTemplate.replaceAll(RegExp('NAME'), 'Bob');
  //greetingTemplate 没有改变
  print(greetingTemplate);
  //greeting 是替换过后的结果
  print(greeting);
}
```

6. 构建字符串

StringBuffer 是一个有效的串联字符串的类，允许使用 write*()方法增量构建字符串，在调用 toString()方法时，StringBuffer 才会创建新的 String 对象。

(1) clear()：清除字符串缓冲区。

(2) write(Object obj)：将已转换为字符串的 obj 的内容添加到缓冲区。

(3) writeAll(Iterable objects, [String separator = ""])：遍历给定对象并按顺序写入它们，可选参数 separator 用于在写入时提供分隔符。

writeAll()方法的第二个参数是可选的，可用于指定分隔符，在本例中为空格。示例代码如下：

```
//chapter9/bin/example_15.dart
void main(){
  var sb = StringBuffer();
  //向字符串缓冲区写入数据
  sb..write('Use a StringBuffer for ')
    ..writeAll(['efficient', 'string', 'creation'], ' ')
    ..write('.');
```

```
  var fullString = sb.toString();
  //打印最终字符串
  print(fullString);
  //清空字符串缓冲区
  sb.clear();
}
```

7. 正则表达式

(1) Pattern 类有两个实现，一个是 String，另一个是 RegExp。

(2) RegExp 类提供与 JavaScript 正则表达式相同的功能。使用正则表达式进行有效字符串搜索和模式匹配。

使用正则表达式的示例代码如下：

```
//chapter9/bin/example_16.dart
void main(){
  //这是一个或多个数字的正则表达式
  var numbers = RegExp(r'\d+');

  var allCharacters = 'llamas live fifteen to twenty years';
  var someDigits = 'llamas live 15 to 20 years';

  //contains()可以使用正则表达式
  print(allCharacters.contains(numbers));
  print(someDigits.contains(numbers));

  //将每个匹配项替换为另一个字符串
  var exedOut = someDigits.replaceAll(numbers, 'XX');
  print(exedOut);
}
```

也可以直接使用 RegExp 和 Match 类提供的功能，对正则表达式的匹配项进行访问和控制。示例代码如下：

```
//chapter9/bin/example_17.dart
void main(){
  var numbers = RegExp(r'\d+');
  var someDigits = 'llamas live 15 to 20 years';
  //检查一个正则表达式在字符串中是否有匹配项
  print(numbers.hasMatch(someDigits));
  //循环所有匹配项
  for (var match in numbers.allMatches(someDigits)) {
    print(match.group(0));
  }
}
```

9.2.3　URIs

Uri 类提供了用于对 URI 中使用的字符串进行编码和解码的函数。这些函数处理 URI 专用的字符，例如“&”和“=”。Uri 类还解析并公开 URI 的组件，例如主机、端口、方案等。

1. 编码和解码标准 URI

(1) decodeFull(String uri)：uri 使用百分比编码对字符串进行编码，以使其可以安全地用作完整的 URI。除大写和小写字母、数字和字符外的所有字符！# $ &'() * +,-./:;=? @_~均按百分比编码。这是 ECMA-262 版本 5.1 中为 encodeURI 函数指定的字符集。

(2) encodeFull(String uri)：解码 uri 中的百分比编码。

这些方法非常适合编码或解码完全标准的 URI，而保留完整的特殊 URI 字符。示例代码如下：

```
//chapter9/bin/example_18.dart
void main(){
  var uri = 'https://example.org/api?foo = some message';
  //编码 URI
  var encoded = Uri.encodeFull(uri);
  print(encoded);

  //解码 URI
  var decoded = Uri.decodeFull(encoded);
  print(decoded);
}
```

注意，只有 some 和 message 之间的空格被编码。

2. 编码和解码 URI 组件

(1) encodeComponent(String component)：使用百分比编码对字符串 component 进行编码，使其可以安全地用作 URI 组件。

(2) decodeComponent(String encodedComponent)：解码 encodeComponent 中的百分比编码。

注意，对 URI 组件进行解码可能会更改其含义，因为某些解码的字符可能具有给定 URI 组件类型的分隔符的字符。在解码各个部分之前，需始终使用该组件的分隔符来拆分 URI 组件。

编码与解码 URL 组件的示例代码如下：

```
//chapter9/bin/example_19.dart
void main(){
  var uri = 'https://example.org/api?foo = some message';
```

```
  //编码 URI 及组件
  var encoded = Uri.encodeComponent(uri);
  print(encoded);

  //解码 URI 及组件
  var decoded = Uri.decodeComponent(encoded);
  print(uri);
}
```

3. 解析 URIs

(1) parse(String uri, [int start = 0, int end]): 通过解析 URI 字符串创建一个 Uri 对象。如果提供了 start 和 end,则它们必须指定 uri 的有效子字符串,并且只有从 start 到 end 的子字符串才被解析为 URI。

(2) Uri({String scheme, String userInfo, String host, int port, String path, Iterable < String > pathSegments, String query, Map < String, dynamic > queryParameters, String fragment}): 使用 Uri 组件构建 Uri。

如果有一个 Uri 对象或 URI 字符串,可以使用 Uri 属性获取其组成部分,例如: path。要从字符串创建 Uri,需使用 parse()静态方法。示例代码如下:

```
//chapter9/bin/example_20.dart
void main(){
  //解析 URIs
  var uri = Uri.parse('https://example.org:8080/foo/bar#frag');
  //访问各个组件
  print(uri.scheme);
  print(uri.host);
  print(uri.path);
  print(uri.fragment);
  print(uri.origin);
}
```

可以使用 Uri()构造函数从各个部分构建 URI。示例代码如下:

```
//chapter9/bin/example_21.dart
void main(){
  //构建 URI
  var uri = Uri(
      scheme: 'https',
      host: 'example.org',
      path: '/foo/bar',
      fragment: 'frag');
  print(uri);
}
```

9.2.4 时间和日期

时间和日期(DateTime)对象是一个时间点,可以采用UTC(世界统一时间)或本地时区来生成时间。可以使用几种构造函数创建DateTime对象:

(1) DateTime(int year, [int month = 1, int day = 1, int hour = 0, int minute = 0, int second = 0, int millisecond = 0, int microsecond = 0]):根据本地时区来创建DateTime实例。

(2) DateTime.now():使用本地时区中的当前日期和时间构造一个DateTime实例。

(3) DateTime.utc(int year, [int month = 1, int day = 1, int hour = 0, int minute = 0, int second = 0, int millisecond = 0, int microsecond = 0]):使用UTC构造一个DateTime实例。

(4) tryParse(String formattedString):解析字符串formattedString来构造一个DateTime实例。如果解析出错,则返回null。

日期创建与解析示例代码如下:

```
//chapter9/bin/example_22.dart
void main(){
  //获取当前日期和时间
  var now = DateTime.now();
  print(now);
  //使用本地时区创建一个新的DateTime
  var t1 = DateTime(2000);
  print(t1);
  //指定月份和日期
  var t2 = DateTime(2000, 1, 2);
  print(t2);
  //指定日期使用UTC时间
  var t3 = DateTime.utc(2000);
  print(t3);
  //解析ISO 8601日期
  var t4 = DateTime.tryParse('2000-01-01T00:00:00Z');
  print(t4);
}
```

9.3 数学库

1min

数学库(dart:math)提供了常用功能,例如正弦和余弦、最大值和最小值,以及常数,例如pi和e。数学库中的大多数功能都作为顶层方法使用。

要在应用中使用此库,需导入dart:math。示例代码如下:

```
import 'dart:math';
```

1. 常用函数

(1) max<T extends num>(T a, T b): 返回两个数字中的较大者。

(2) min<T extends num>(T a, T b): 返回两个数字中的较小者。

(3) sqrt(num x): 将 x 转换为 double 类型并返回值的正平方根。如果 x 为-0.0,则返回-0.0; 否则,如果 x 为负或 NaN,则返回 NaN。

示例代码如下:

```
//chapter9/bin/example_23.dart
import 'dart:math';

void main(){
  //取最大或最小值的函数
  print(max(1,1000));
  print(min(1, -1000));
}
```

2. 数学常量

此库中还定义了常用的数学常量,包括 e、pi、sqrt2 等。示例代码如下:

```
//2.718281828459045
print(e);
//3.141592653589793
print(pi);
//1.4142135623730951
print(sqrt2);
```

3. 随机数

(1) Random([int seed]): 创建一个随机数生成器,可选的 seed 参数用于初始化生成器的内部状态。

(2) nextBool(): 随机生成一个布尔值。

(3) nextDouble(): 生成非负的随机浮点值,该值均匀地分布在从 0.0(含)到 1.0(不含)。

(4) nextInt(int max): 生成一个非负的随机整数,其范围为 0(含)到 max(不包括)均匀分布。

使用 Random 类生成随机数时,可以选择提供一个种子给 Random 构造函数。示例代码如下:

```
var random = Random();
//0.0～1.0
random.nextDouble();
//0～9
random.nextInt(10);
```

甚至可以生成随机布尔值。示例代码如下：

```
var random = Random();
//true 或者 false
random.nextBool();
```

9.4　转换库

dart:convert 库具有编码器和解码器，用于在不同的数据表示形式之间进行转换。包括常见的数据形式 JSON 和 UTF-8，并支持创建其他转换器。转换器是指对数据进行编码和解码。JSON 是一种简单的文本格式，用于表示结构化对象和集合。UTF-8 是一种常见的可变宽度编码，可以表示 Unicode 字符集中的每个字符。

dart:convert 库可在 Web 应用程序和命令行应用程序中使用。要使用它，需导入 dart:convert。示例代码如下：

```
import 'dart:convert';
```

9.4.1　编码和解码 JSON

(1) jsonEncode(Object value, {Object toEncodable(Object nonEncodable)})：将值转换为 JSON 字符串，如果 value 包含无法直接编码为 JSON 字符串的对象，则可使用 toEncodable 函数将其转换为直接可编码的对象。如果省略 toEncodable，则默认调用该不可调用对象的 toJson()方法。

(2) jsonDecode(String source, {Object reviver(Object key, Object value)})：解析字符串并返回生成的 Json 对象。对于在解码过程中已解析的每个对象或列表属性，都会调用一次可选的 reviver 函数。key 参数可以是列表属性的整数列表索引、对象属性的字符串映射键，或者最终结果为 null。

使用 jsonDecode()将 JSON 编码的字符串解码为 Dart 对象。示例代码如下：

```
//chapter9/bin/example_24.dart
import 'dart:convert';
void main(){
  //确保在 JSON 字符串中使用双引号 ("),不能使用单引号(')
  var jsonString = '''[{"score": 40},{"score": 80}]''';
  //将 JSON 字符串解码为列表
  var scores1 = jsonDecode(jsonString);
  print(scores1 is List);

  var firstScore = scores1[0];
```

```
  print(firstScore is Map);
  print(firstScore['score']);
}
```

使用 jsonEncode()将受支持的 Dart 对象编码为 JSON 格式的字符串。示例代码如下：

```
var scores2 = [
  {'score': 40},
  {'score': 80},
  {'score': 100, 'overtime': true, 'special_guest': null}
];
//将列表 scores2 编码为 JSON 格式字符串
var jsonText = jsonEncode(scores2);
print(jsonText);
```

只有类型为 int、double、String、bool、null、List 或 Map(带有字符串键)的对象可以直接编码为 JSON,而 List 和 Map 对象是递归编码的。

9.4.2 解码和编码 UTF-8 字符

(1) encode(String input)：将字符串 input 转换为 UTF-8 编码的字节列表。

(2) decode(List<int> codeUnits, {bool allowMalformed})：将 UTF-8 代码单元解码为相应的字符串。如果 allowMalformed 为 true,解码器使用 Unicode 替换字符 U + FFFD 替换无效(或未终止)的字符序列。否则,它将引发 FormatException。

使用 utf8.decode()方法将 UTF-8 编码的字节解码为 Dart 字符串。示例代码如下：

```
//chapter9/bin/example_25.dart
import 'dart:convert';
void main(){
  //ignore: omit_local_variable_types
  List<int> utf8Bytes = [
    0xc3, 0x8e, 0xc3, 0xb1, 0xc5, 0xa3, 0xc3, 0xa9,
    0x72, 0xc3, 0xb1, 0xc3, 0xa5, 0xc5, 0xa3, 0xc3,
    0xae, 0xc3, 0xb6, 0xc3, 0xb1, 0xc3, 0xa5, 0xc4,
    0xbc, 0xc3, 0xae, 0xc5, 0xbe, 0xc3, 0xa5, 0xc5,
    0xa3, 0xc3, 0xae, 0xe1, 0xbb, 0x9d, 0xc3, 0xb1
  ];
  //对字节数据进行解码
  var funnyWord = utf8.decode(utf8Bytes);
  print(funnyWord);
}
```

使用 utf8.encode()将 Dart 字符串编码为 UTF-8 编码的字节列表。示例代码如下：

```
List<int> encoded = utf8.encode('Îñţérñåţîöñåļîžåţîờñ');
print(encoded.length == utf8Bytes.length);
for (int i = 0; i < encoded.length; i++) {
  //比较字节是否相等
  print(encoded[i] == utf8Bytes[i]);
}
```

9.5 输入和输出库

1min

dart:io 库提供用于处理文件、目录、进程、sockets、WebSocket、HTTP 客户端和服务器的 API。

通常,dart:io 库实现并提升了异步 API,同步方法很容易阻塞应用程序,从而难以扩展。因此,大多数操作都是通过 Future 或 Stream 对象返回结果的,Future 或 Stream 对象是现代服务器平台常见的模式。

要使用 dart:io 库,必须将其导入。示例代码如下:

```
import 'dart:io';
```

通过 I/O 库,命令行应用程序可以读取和写入文件及浏览目录。有两种选择来读取文件的内容:一次全部读取和流式传输。一次读取一个文件需要足够的内存来存储文件的所有内容。如果文件很大,或者想在读取文件时对其进行处理,则应使用流式传输。

1. 文件类常用构造函数和方法

(1) File(String path):创建 File(文件)对象,如果 path 是相对路径,则在使用时相对当前工作目录。如果 path 是绝对路径,则与当前工作目录无关。

(2) File.fromUri(Uri uri):从 URI 创建一个 File 对象。

(3) openRead([int start,int end]):为此文件的内容创建一个新的独立 Stream。如果存在 start,则从字节偏移量开始读取文件。否则从头开始(索引 0)。如果存在 end,则仅读取直到字节索引的 end。否则,直到文件结束。为了确保释放系统资源,必须将流读取完整,否则必须取消对流的订阅。

(4) openWrite({FileMode mode: FileMode.write,Encoding encoding: utf8}):为该文件创建一个新的独立 IOSink。当该 IOSink 不再使用时,必须关闭,以释放系统资源。IOSink 支持两种 FileMode 值:FileMode.write,将初始写入位置设置为文件的开头;FileMode.append,将初始写入位置设置为文件的末尾。

(5) readAsString({Encoding encoding: utf8}):使用给定的 Encoding 以字符串形式读取整个文件内容。读取文件内容后,返回一个以字符串结尾的 Future <String>。

(6) readAsLines({Encoding encoding: utf8}):使用给定的 Encoding 以行读取整个文

件的内容。返回一个 Future < List < String >>。

(7) readAsBytes():以字节列表的形式读取整个文件的内容。返回 Future < Uint8List >。

2. 以文本形式读取文件

读取使用 UTF-8 编码的文本文件时,可以使用 readAsString()读取整个文件内容。当各行很重要时,可以使用 readAsLines()。在这两种情况下,都将返回一个 Future 对象,该对象以一个或多个字符串的形式提供文件的内容。示例代码如下:

```
//chapter9/bin/example_26.dart
import 'dart:io';
Future main() async{
  var config = File('config.txt');
  var contents;
  //将整个文件的内容放在单个字符串中
  contents = await config.readAsString();
  print('文件内容的字符长度: ${contents.length}');
  //将文件内容以行作为分割,拆分成多个字符串
  contents = await config.readAsLines();
  print('文件内容的行数 ${contents.length}');
}
```

3. 以二进制形式读取文件

以下代码将整个文件作为字节读取到字节列表中,对 readAsBytes()的调用返回一个 Future 对象,当可用时会提供结果。示例代码如下:

```
//chapter9/bin/example_27.dart
import 'dart:io';

Future main() async{
  var config = File('config.txt');
  try {
    var contents = await config.readAsBytes();
    print('文件内容的字节长度为 ${contents.length}');
  } catch (e) {
    print(e);
  }
}
```

为了捕获错误,以免导致未捕获的异常,可以在 Future 上注册 catchError 处理程序,或在异步函数中使用 try-catch。示例代码如下:

```
Future main() async{
  var config = File('config.txt');
  try {
```

```
    var contents = await config.readAsString();
    print(contents);
  } catch (e) {
    print(e);
  }
}
```

4. 以流读取文件内容

可以使用 Stream 读文件，每次读取一点。可以使用 Stream API 或者 await for。示例代码如下：

```
//chapter9/bin/example_28.dart
import 'dart:io';
import 'dart:convert';

Future main() async{
  var config = File('config.txt');
  //ignore: omit_local_variable_types
  Stream<List<int>> inputStream = config.openRead();

  var lines =
  utf8.decoder.bind(inputStream).transform(LineSplitter());
  try {
    await for (var line in lines) {
      print('从流中获取的字符长度：${line.length}');
    }
    print('文件关闭');
  } catch (e) {
    print(e);
  }
}
```

5. 写文件内容

可以使用 IOSink 将数据写入文件。使用 File 的 openWrite()方法获取可以写入的 IOSink 对象。IOSink 类有以下常用函数：

(1) IOSink(StreamConsumer<List<int>> target,{Encoding encoding: utf8})：构造函数，创建一个 IOSink 对象，StreamConsumer 是接收多个完整流的“接收器”的抽象接口。使用者可以使用 addStream 接收多个连续的流，并且当不需要添加更多数据时，close 方法将告诉使用者完成其工作并关闭。

(2) add(List<int> data)：将字节数据添加到目标使用者。

(3) close()：关闭目标使用者。

(4) flush()：返回一个 Future，一旦基础 StreamConsumer 接收了所有缓冲的数据，该

Future 将完成。

(5) write(Object obj)：将 obj 转换为 String，并将结果的编码添加到目标使用者。

(6) writeAll(Iterable objects，[String separator = ""])：遍历给定 objects 并按顺序写入它们。可选的 separator 是指遍历的对象间写入时的分隔符。

默认模式 FileMode. write 完全覆盖文件中的现有数据。示例代码如下：

```
//chapter9/bin/example_29.dart
import 'dart:io';
void main()async{
  var logFile = File('log.txt');
  var sink = logFile.openWrite();
  sink.write('文件访问时间 ${DateTime.now()}\n');
  await sink.flush();
  await sink.close();
}
```

要添加到文件的末尾，需使用可选的 mode 参数指定 FileMode. append。示例代码如下：

```
var sink = logFile.openWrite(mode: FileMode.append);
```

6. 列出目录中的文件

查找目录的所有文件和子目录是异步操作，list()方法返回一个 Stream，当遇到文件或目录时该 Stream 发出一个对象。Directory 类有以下常用函数：

(1) Directory(String path)：构造函数，创建一个 Directory(目录)对象。如果 path 是相对路径，则在使用时对应当前工作目录。如果 path 是绝对路径，则与当前工作目录无关。

(2) list({bool recursive：false，bool followLinks：true})：列出此目录的子目录和文件。可选参数 recursive 表示递归到子目录。返回用于目录、文件和链接的 FileSystemEntity 对象流。

目录遍历示例代码如下：

```
//chapter9/bin/example_30.dart
import 'dart:io';
Future main() async {
  //当前项目的根目录
  var dir = Directory('');
  try {
    var dirList = dir.list();
    await for (FileSystemEntity f in dirList) {
      if (f is File) {
        print('发现文件 ${f.path}');
```

```
        } else if (f is Directory) {
          print('发现目录${f.path}');
        }
      }
    } catch (e) {
      print(e.toString());
    }
  }
```

第 10 章 异　　步

Dart 库有许多返回 Future 或 Stream 对象的函数。这些函数是异步的：它们在设置可能耗时的操作(例如 I/O)之后返回，而无须等待该操作完成。

async 和 await 关键字支持异步编程，可以编写出看起来类似于同步代码的异步代码。

12min

10.1 Future

Future 表示异步操作的结果，它有两种状态：

(1) 未完成状态：当调用异步函数时，它返回未完成的 Future，并且持续到异步函数操作完成。

(2) 完成状态：如果异步函数操作成功，则返回一个值；如果异步函数操作失败，则返回一个错误。

10.1.1 创建 Future

可以通过构造函数创建 Future，构造函数的参数是一个函数，该函数的返回值类型为 Future<T>或 T，其中 T 代表的是任何类型。下例中 getInt 函数的返回值是 int 类型。示例代码如下：

```
//chapter10/bin/example_01.dart
import 'dart:math';
import 'dart:async';
//同步函数，返回 int 类型的值
int getInt(){
  print('执行 getInt 方法');
  //创建随机对象实例
  Random rng = Random();
  //返回 0～100 的随机数
  return rng.nextInt(100);
}
void main(){
```

```
  //通过 Future 类的构造函数创建 Future
  //此处调用的 getInt 函数返回值是 int 类型
  Future<int> future = Future(getInt);
  //验证是否是 Future<int>的实例
  print(Future(getInt));
}
```

也可以使用 async 标记函数，使函数成为异步函数，异步函数会自动将返回值包装成一个 Future。此时 getInt 函数的返回值是 Future<int>类型。示例代码如下：

```
//chapter10/bin/example_02.dart
import 'dart:math';
import 'dart:async';
//向函数添加 async 标记以表明是异步函数
//返回值使用 Future 包装并提供泛型参数 int
Future<int> getInt() async{
  print('执行 getInt 方法');
  Random rng = Random();
  return rng.nextInt(100);
}
void main(){
  //getInt 方法的返回值是 Future<int>类型
  Future<int> future = getInt();
  //验证是否是 Future<int>类型
  print(getInt());
}
```

10.1.2　使用 Future

可以使用 Future 类提供的 then、catchError 和 whenComplete 方法对 Future 对象进行进一步处理，当异步操作成功时，执行 then 方法，then 方法接收一个参数为异步操作返回值的回调函数。当异步操作失败时，执行 catchError 方法，catchError 方法接收一个参数为错误对象的回调函数。当异步操作完成时，无论执行失败还是成功都会执行 whenComplete 方法，whenComplete 方法接收一个无参的自定义回调函数。示例代码如下：

```
//chapter10/bin/example_03.dart
import 'dart:math';
import 'dart:async';
Future<int> getInt() async{
  print('执行 getInt 方法');
  Random rng = Random();
  return rng.nextInt(100);
}
```

```
void main(){
  //通过 Future 类的构造函数创建 Future
  Future<int> future = getInt();
  future.then((Object onValue){
    //异步调用成功
    print('异步操作成功,值为 $onValue');
  }).catchError((Object onError){
    //异步调用失败,用于捕获异常
    print('异步操作失败: $onError');
  }).whenComplete((){
    //异步调用完成时调用,与是否成功无关
    print('异步操作完成');
  });
}
```

也可以使用 await 关键字等待异步操作完成，使用 await 关键字的函数必须使用 async 标记，并且使用 try 语句捕获异常。示例代码如下：

```
//chapter10/bin/example_04.dart
import 'dart:math';
import 'dart:async';
Future<int> getInt() async{
  print('执行 getInt 方法');
  Random rng = Random();
  return rng.nextInt(100);
}
void main() async{
  //通过 Future 类的构造函数创建 Future
  Future<int> future = getInt();
  try{
    //使用 await 等待异步调用完成,使其等同于同步代码
    var onValue = await future;
    print('异步操作成功,值为 $onValue');
  }on Exception catch(onError){
    print('异步操作失败: $onError');
  }finally{
    print('异步操作完成');
  }
}
```

尽管异步功能可能会执行耗时的操作，但它不会等待这些操作。相反，异步函数会照常执行直到遇见第一个 await 表达式。然后，它返回 Future 对象，仅在 await 表达式完成后才恢复执行。

在 await 表达式中，表达式的值通常是 Future，如果不是，则该值将自动包装在 Future

中。此 Future 对象表示承诺返回一个对象。await 表达式的值就是返回的对象。await 表达式使执行暂停，直到该对象可用为止。

如果异步函数不需要返回值，则需将返回类型修改为 Future<void>。

10.2 Stream

14min

Stream 是一系列异步事件的源。Stream 提供了一种接收事件序列的方式，每个事件要么是数据事件（又被称为 Stream 的元素），要么是错误事件（发生故障时的通知），当 Stream 发出所有事件后，单个 done 事件将通知侦听器已完成。

Stream 和 Future 的很多特性类似，但也有一些区别：

（1）Future 在异步操作完成时提供单个结果、错误或者值。Stream 可以提供多个结果。

（2）Future 使用 then、catchError、whenComplete 方法获取或处理结果，Stream 则只需通过 listen（监听）即可处理所有值。

（3）Future 发送和接收相同的值，而 Stream 可以使用辅助方法在值到达前进行处理。

10.2.1 创建 Stream

首先创建一个 StreamController 对象，然后使用 StreamController 对象的 stream 属性返回一个 Stream 对象。示例代码如下：

```
StreamController<int> controller = StreamController<int>(
    onListen: startTimer,
    onPause: stopTimer,
    onResume: startTimer,
    onCancel: stopTimer);

Stream stream = controller.stream;
```

StreamController 构造函数支持泛型，这里使用 int 类型。构造函数提供了多个可选参数：

（1）onListen：监听 stream 时调用的回调函数。

（2）onPause：stream 暂停时调用的回调函数。

（3）onResume：stream 恢复时调用的回调函数。

（4）onCancel：取消 stream 时调用的回调函数。

（5）sync：布尔值，默认值为 false，同步 stream 标记。

10.2.2 使用 Stream

使用 Stream 类的 listen 方法监听 stream，该方法提供以下参数：

（1）onData：必选参数，回调函数，该函数的参数是 Stream 事件发出的值。

（2）onError：可选参数，回调函数，来自 Stream 的错误。该回调函数类型必须是 void onError(error)或 void onError(error,StackTrace stackTrace)，该函数的两个参数一个是错误对象，另一个是可选的堆栈跟踪信息。如果省略此函数且 stream 发生错误，则会将错误信息向外传递。

（3）onDone：可选参数，回调函数，当此 stream 关闭并发送完成事件时，将调用此回调函数。

（4）cancelOnError：可选参数，布尔值，默认值为 false。如果值为 true，则在 stream 传递第一个错误事件时自动取消订阅。

listen 方法返回用于订阅 Stream 中事件的对象 StreamSubscription，该对象保留上述处理事件的回调函数，还可以发出取消订阅事件和临时暂停 stream 中的事件。完整示例代码如下：

```
//chapter10/bin/example_05.dart
import 'dart:async';
//用于返回 stream 对象,stream 又称为流
Stream<int> createStream(Duration interval,int maxCount){
  //定义流控制器
  StreamController<int> controller;
  //定义定时器
  Timer timer;
  //计数变量
  int counter = 0;

  void tick(_){
    counter++;
    //将 counter 的值作为事件发送给 stream
    controller.add(counter);
    //判断计数变量是否达到最大值
    if (counter == maxCount) {
      //终止计时器
      timer.cancel();
      //关闭 stream 并通知监听器
      controller.close();
    }
  }
  //启动计时器
  void startTimer(){
    print('createStream 开始执行');
    //interval 是调用 tick 函数的时间间隔
    timer = Timer.periodic(interval,tick);
  }
  //终止计时器
  void stopTimer(){
```

```
    if (timer != null){
      timer.cancel();
      timer = null;
    }
    print('createStream 结束执行');
  }
  //创建流控制器
  controller = StreamController< int >(
      onListen: startTimer,
      onPause: stopTimer,
      onResume: startTimer,
      onCancel: stopTimer);
  //返回流
  return controller.stream;
}
void main() async{
  //接收返回的 stream 对象
  //时间间隔为 1s,最大值为 10s
  Stream< int > stream = createStream(const Duration(seconds:1),10);
  //监听 stream
  stream.listen((int value) {
    print('来自 createStream 的值: $ value');
  });
}
```

函数 createStream 返回 Stream 对象，接收两个参数，Duration 类型的参数表示持续时间，参数 maxCount 表示发起事件的最大次数。当对 stream 执行 listen 方法时，回调函数 startTimer 开始执行。

Timer 是计时器类，Timer.periodic 命名构造函数创建一个重复的计时器，该构造函数有两个参数，第一个参数表示持续时间，第二个参数是回调函数，该回调函数接收一个 Timer 类型的参数。计时器从指定的持续时间倒计时到 0。当计时器达到 0 时，计时器将调用指定的回调函数。

因为在回调函数 tick 内部可以直接使用对象 timer，所以在传入参数处使用“_”做占位符。StreamController 对象使用 add 方法向 stream 发送事件，使用 close 方法关闭 stream。

10.3　生成器函数

10min

传统函数只会返回单个值，生成器函数生成值的序列。生成器函数可以采用同步返回带有值的 Iterable 对象，在异步中返回 Stream 对象。

关键字 yield 返回单个值到序列，但是不会停止生成器函数。

生成器函数按需生成值，当开始迭代 iterator 或者开始监听 stream 时才生成值。

10.3.1 同步生成器

要实现同步生成器功能，需将函数主体标记为 sync *，并使用 yield 语句传递值到序列。示例代码如下：

```
//chapter10/bin/example_06.dart
Iterable< int > getNumbers(int number) sync * {
  print('生成器开始执行');
  int k = 0;
  while (k < number) yield k++;
  print('生成器执行结束');
}
void main() {
  print('创建 iterator');
  Iterable< int > numbers = getNumbers(10);
  print('开始迭代');
  for (int val in numbers) {
    print(' $ val');
  }
}
```

10.3.2 异步生成器

要实现异步生成器函数，需将函数主体标记为 async *，并使用 yield 语句传递值到 stream。示例代码如下：

```
//chapter10/bin/example_07.dart
Stream< int > getStream(int number) async * {
  print('异步生成器开始执行');
  int k = 0;
  while (k < number) yield k++;
  print('异步生成器执行结束');
}
void main() {
  //创建 stream
  Stream< int > stream = getStream(10);
  //监听 stream
  stream.listen((int value) {
    print('异步生成器生成的值: $ value');
  });
}
```

10.3.3　递归生成器

递归是指在函数内部调用函数本身，如果生成器是递归的，则可以使用 yield * 来提高其性能。示例代码如下：

```
//chapter10/bin/example_08.dart
Iterable<int> getNumbersRecursive(int number) sync* {
  print('生成 $number 开始');
  if (number > 0) {
    yield* getNumbersRecursive(number - 1);
  }
  yield number;
  print('生成 $number 结束');
}
void main() {
  print('创建 iterator');
  Iterable<int> numbers = getNumbersRecursive(3);
  print('开始迭代');
  for (int val in numbers) {
    print('$val');
  }
  print('main 函数结束');
}
```

第 11 章 Isolate

大多数计算机,甚至在移动平台上,都具有多核 CPU。为了利用所有这些核心,开发人员通常使用并发运行的共享内存线程。但是,共享状态并发易于出错,并且可能导致复杂的代码。

2min

11.1 什么是 Isolate

所有 Dart 代码始终在 Isolate 中执行,一个 Isolate 只有一个线程,一个专用的内存区域和它自己的事件循环,完全独立于其他 Isolate。如图 11-1 所示的是单个 Isolate 的示意图,长方形区块代表内存块,由循环箭头图标代表事件循环管理。

图 11-1 Isolate

Dart 应用具有 main()方法,该方法具有标准的 Isolate,并且从 main()运行的代码都在该 Isolate 上运行,其内部的事件由事件循环管理执行。绝大多数 Dart 应用程序都在此单一标准 Isolate 上运行,如果需要在后台执行耗时任务,就需要通过 Isolate.spawn()创建新的 Isolate。不同的 Isolate 彼此完全独立,并且如果需要在它们之间访问数据或交换信息,则唯一的方法是通过消息,如图 11-2 所示。

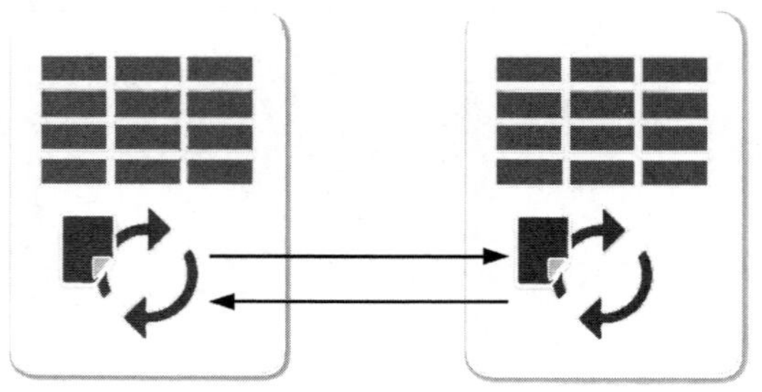

图 11-2 Isolate 间传递消息

11.2 事件循环

10min

每个 Isolate 都用一个自己的事件循环，Dart 中的事件循环包含两个队列：事件队列和微任务队列。

事件队列是 Dart 处理此 Isolate 外部事件的唯一途径，例如：I/O 操作、Stream、用户的单击、网络请求的回调、来自其他 Isolate 的消息或由用户交互触发的任何操作。这些被称为事件，并且会被添加到事件队列，事件循环将根据队列顺序处理事件。

微任务是在使用顶层函数 scheduleMicrotask()时传入的函数所包含的任务，该函数通常执行耗时工作，常用于异步任务，并且该微任务将被添加到微任务队列。scheduleMicrotask()会异步执行一个函数，通过此函数注册的回调始终按顺序执行，并保证在其他异步事件（如 Timer 事件或 DOM 事件）之前运行。微任务队列不处理外部事件。

启动应用程序时，将创建并启动一个新线程。创建主线程后，Dart 会将微任务和事件队列初始化为空，同时执行 main()方法，执行过程中将微任务放入微任务队列，事件放入事件队列，一直到执行完 main()方法中的最后一条代码，然后启动事件循环。值得注意的是，微任务队列的执行优先级高于事件队列，只有当微任务队列中的任务全部执行完成，并且微任务队列为空时，事件循环才开始处理事件队列中的事件，执行流程如图 11-3 所示。

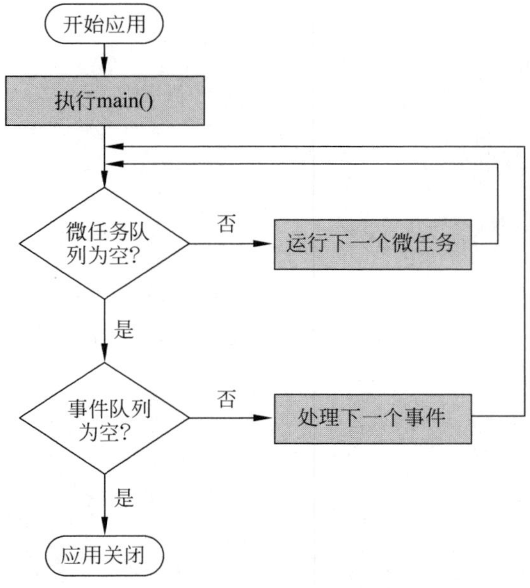

图 11-3 应用执行流程

Dart 采用单线程，事件循环通过并发访问线程来允许并行任务执行。线程负责执行代码，一旦代码进入该线程，它的执行顺序就是线性的，并且直到所有代码执行完毕后才会停止。事件循环控制线程将按顺序执行内容，它不会更改或执行 Dart 代码，仅用来控制其执行。事件循环会将事件队列和任务队列中的条目放入线程的调用堆栈中，由于外部事件可以不断扩充事件队列，进而使得 Dart 中的异步和响应成为可能。

为了更加直观地理解事件循环控制代码执行的顺序，示例代码如下：

```
//chapter11/bin/example_01.dart
import 'dart:async';
main(){
  print('main 方法 #1');
  scheduleMicrotask(() => print('微任务 #1'));

  Future.delayed(Duration(seconds:1),
        () => print('future 发起的事件 #1 (delayed)'));
  Future(() => print('future 发起的事件 #2'));
  Future(() => print('future 发起的事件 #3'));

  scheduleMicrotask(() => print('微任务 #2 '));
  print('main 方法 #2');
}
```

这里使用了 Future()和 Future.delayed()在事件队列中安排任务，可以查看运行结果：

```
main 方法 #1
main 方法 #2
微任务 #1
微任务 #2
future 发起的事件 #2
future 发起的事件 #3
future 发起的事件 #1 (delayed)
```

当在执行事件队列中的某一个事件时，如果事件中的某个任务必须被执行完成后才能执行下一个事件，但是该事件中某个任务又比较消耗时间，则可以使用 scheduleMicrotask()将其添加到微任务队列。示例代码如下：

```
//chapter11/bin/example_02.dart
import 'dart:async';
main() {
  print('main #1');
  scheduleMicrotask(() => print('微任务 #1'));

  Future.delayed(Duration(seconds:1),
          () => print('事件 #1 (延迟)'));

  Future(() => print('事件 #2 '))
```

```
        .then((_) => print('事件 #2a'))
        .then((_) {
      print('事件 #2b');
      scheduleMicrotask(() => print('微任务 #0 来自 事件 #2b'));
    }).then((_) => print('事件 #2c'));

  scheduleMicrotask(() => print('微任务 #2'));

  Future(() => print('事件 #3'))
      .then((_) => Future(
          () => print('事件 #3a 新的 Future')))
      .then((_) => print('事件 #3b'));

  Future(() => print('事件 #4'));
  scheduleMicrotask(() => print('微任务 #3'));
  print('main #2');
}
```

运行结果如下：

```
main #1
main #2
微任务 #1
微任务 #2
微任务 #3
事件 #2
事件 #2a
事件 #2b
事件 #2c
微任务 #0 来自 事件 #2b
事件 #3
事件 #4
事件 #3a 新的 Future
事件 #3b
事件 #1 (延迟)
```

11.3 创建 Isolate

10min

与传统多线程语言不同，Dart 采用 Isolate 实现多线程。可以使用 Isolate 运行异步、后台、耗时的任务，可以在单个应用程序中创建和运行多个 Isolate。

所有 Dart 代码都运行在 Isolate 中，当调用 main 函数时，其实该函数就位于一个 Isolate 中，被称作 main Isolate。可以从以下几方面来深入理解 Isolate：

(1) 用于实现并发编程。

(2) 拥有自己独立的内存块。

(3) 每个 Isolate 都有独立的事件循环,该循环始终运行并捕获事件(例如:从网络获取数据)且加以处理。

(4) 与线程相似但不共享内存。

(5) 只能通过消息与其他 Isolate 通信。

使用 Isolate 类的 spawn 方法创建新的 Isolate,它包含两个参数,第一个参数是函数名,第二个参数可以是任何类型,它是当前 Isolate 传递给新 Isolate 的消息,该参数会被第一个参数提供的函数捕获。示例代码如下:

```
Isolate.spawn(newIsolate, '执行耗时任务');
```

这里传递名为 newIsolate 的函数,并且传递了一个字符串类型的消息。

可以使用 Isolate 的 kill 方法关闭创建的 Isolate。示例代码如下:

```
isolate.kill(priority: Isolate.immediate);
```

完整示例代码如下:

```
//chapter11/bin/example_03.dart
import 'dart:io';
import 'dart:isolate';

void newIsolate(String arg) {
  print('新的 Isolate 被创建并带有消息: $arg');
  for(int progress = 0; progress < 500 ;progress++){
    print('新的 Isolate 的任务进度: $progress');
  }
  print('新的 Isolate 任务完成');
}

void main() async{
  print('创建新的 Isolate');
  Isolate isolate = await Isolate.spawn(newIsolate, '执行耗时任务');
  print('新的 Isolate 被创建,main Isolate 开始执行自己的任务');
  for(int progress = 0;progress < 500;progress++){
    print('main Isolate 任务进度 : $progress');
  }
  //stdin 允许在命令行中同步或异步读取输入流
  //stdin.first 表示在收到输入流中的第一个元素后,停止监听此流,继续执行后续代码
  await stdin.first;
  //关闭新的 Isolate
  isolate.kill(priority: Isolate.immediate);
```

```
  isolate = null;
  print('新的 Isolate 被关闭');
}
```

11.4 获取消息

13min

Isolate 之间通信需要使用 ReceivePort 类，该类有两个重要方法：

（1）receivePort.sendPort：返回发送到此接收端口的 SendPort 对象，SendPort 对象用于将消息发送到 ReceivePort 对象。

（2）receivePort.listen((dynamic receivedData){})：用于接收 sendPort 发送的消息。

示例代码如下：

```
//chapter11/bin/example_04.dart
import 'dart:isolate';

void newIsolate(SendPort sendPort) {
  print('新的 Isolate 被创建并带有消息: $ sendPort');
  for(int progress = 0; progress < 500 ;progress++){
    print('新的 Isolate 的任务进度: $ progress');
  }
  sendPort.send("任务完成");
  print('新的 Isolate 任务完成');
}

void main() async{
  ReceivePort receivePort =  ReceivePort();
  print('创建新的 Isolate');
  Isolate isolate = await Isolate.spawn(newIsolate, receivePort.sendPort);
  print('新的 Isolate 被创建,main Isolate 开始执行自己的任务');
  for(int progress = 0;progress < 500;progress++){
    print('main Isolate 任务进度 : $ progress');
  }

  receivePort.listen((dynamic receivedData) {
    print('数据来源于新的 Isolate : $ receivedData');

    //有条件地关闭新的 Isolate
    if(receivedData is String && receivedData.toString().contains('任务完成')){
      //关闭新的 Isolate
      isolate.kill(priority: Isolate.immediate);
      isolate = null;
    }
  });
}
```

11.5 相互通信

15min

在 Isolates 间传递消息有两种方式：第一种，通过使用 ReceivePort 类，使用方式同获取消息一致；第二种，使用 Dart 团队开发的 stream_channel 包。

11.5.1 使用 ReceivePort

与获取消息一样，只是将 ReceivePort 对象在两个 Isolates 间相互传递。示例代码如下：

```
//chapter11/bin/example_05.dart
import 'dart:isolate';

void newIsolate(SendPort sendPortOfMainIsolate) {
  ReceivePort receivePort = ReceivePort();
  print('新的 Isolate 被创建并带有消息: $ sendPortOfMainIsolate');
  //监听 main Isolate 发送的消息
  receivePort.listen((dynamic receivedData) {
    print('新的 Isolate 任务命令: $ receivedData');
    sendPortOfMainIsolate.send(receivedData);
  });
  //将新 Isolate 的 sendPort 发送到 main Isolate,使其可以发送消息到此新 Isolate
  sendPortOfMainIsolate.send(receivePort.sendPort);
}

void main() async{
  ReceivePort receivePort =  ReceivePort();
  print('创建新的 Isolate');
  Isolate isolate = await Isolate.spawn(newIsolate, receivePort.sendPort);
  print('新的 Isolate 被创建,main Isolate 开始执行自己的任务');
  for(int progress = 0;progress < 100;progress++){
    print('main Isolate 任务进度 : $ progress');
  }
  //监听新的 Isolate 发送的消息
  receivePort.listen((dynamic receivedData) {
    print('数据来源于新的 Isolate : $ receivedData');
    if (receivedData is SendPort) {
      SendPort sendPortOfNewIsolate  =  receivedData;
      for(int commond = 0;commond < 50;commond++){
        print('main Isolate 命令: $ commond');
        sendPortOfNewIsolate.send(commond);
      }
    }else{
      print('数据源于新的 Isolate, 命令 : $ receivedData');
```

```
    }
  });
}
```

11.5.2　使用 stream_channel

在 pubspec. yaml 文件中添加 stream_channel 依赖项。

```
dependencies:
  stream_channel: ^2.0.0
```

将 receivePort 对象设置为 IsolateChannel. connectReceive(),并将同一个 receivePort 对象的 sendPort 设置为 IsolateChannel. connectSend()。

使用 channel. sink. add('message')发送消息,使用 channel. stream. listen((dynamic receiveData){});接收消息。示例代码如下:

```
//chapter11/bin/example_06.dart
import 'dart:isolate';
import 'package:stream_channel/isolate_channel.dart';

void elIsolate(SendPort sPort) {
  print("新的 Isolate 被创建");
  IsolateChannel channel = IsolateChannel.connectSend(sPort);
  //监听根 Isolate 发送过来的消息
  channel.stream.listen((data) {
    print('新的 Isolate 收到消息: $data');
  });
  //向根 Isolate 发送消息
  channel.sink.add("hi");
}
void main() async {
  ReceivePort rPort = ReceivePort();
  IsolateChannel channel = IsolateChannel.connectReceive(rPort);
//监听新的 Isolate 发送过来的消息
  channel.stream.listen((data) {
    print('根 Isolate 收到消息: $data');
    //向新的 Isolate 发送消息
    channel.sink.add('How are you');
  });

  await Isolate.spawn(elIsolate, rPort.sendPort);
}
```

第 12 章 拓展阅读

5min

12.1 可调用类

为了允许像函数一样调用 Dart 类的实例，需实现 call()方法。

在 WannabeFunction 类定义了一个 call()方法，该函数接收 3 个字符串并将它们串联起来，每个字符串之间用空格分隔，并附加一个感叹号。示例代码如下：

```
//chapter12/bin/example_01.dart
//声明类
class WannabeFunction{
  //实现 call 方法
  String call(String a, String b, String c) => '$a $b $c!';
}
main(){
  //创建类的实例
  var wf = WannabeFunction();
  //像方法一样调用类
  var out = wf('Hi', 'there,', 'gang');
  print(out);
}
```

8min

12.2 扩展方法

当使用其他人的 API 或被广泛使用的库时，更改 API 通常是不切实际或不可能的。但是，如果仍想添加一些功能，那么扩展方法将是一种途径。扩展方法允许对已有 API 进行覆盖或拓展。扩展的成员可以是方法、getters、setters、运算符，扩展还可以具有静态字段和静态帮助方法。

扩展方法的声明语法如下：

```
extension < extension name > on < type >{
  (< member definition >) *
}
```

例如：可以对 String 类进行扩展，覆盖其原 API 中的 parseInt()方法，并拓展一个返回重复一次值的方法 doubleWrite()。

扩展方法声明的示例代码如下：

```
//对内置类型 String 进行扩展
extension parsingString on String {
  int parseInt() {
    //在解析的值的基础上加 1
    return int.parse(this) + 1;
  }
  String doubleIt(){
    //复写原值
    return this + this;
  }
}
```

将上述代码保持在 string_apis.dart 文件中，假设现在使用它的代码与其在同一目录下，导入并使用。示例代码如下：

```
//chapter12/bin/example_02.dart
//导入扩展库
import 'string_apis.dart';
main(){
  //使用扩展方法 parseInt(),该方法覆盖原有 API
  print('${'12'.parseInt()}');
  //使用扩展方法 doubleIt(),该方法覆盖原有 API
  print('${'go'.doubleIt()}');
}
```

在使用扩展方法时，不能对类型为 dynamic 的变量调用控制方法，从而导致异常的示例代码如下：

```
import 'string_apis.dart';
main(){
  dynamic d = '2';
  //运行时异常:NoSuchMethodError
  print(d.parseInt());
}
```

扩展方法可以和类型推断一起使用。示例代码如下：

```
//chapter12/bin/example_03.dart
import 'string_apis.dart';
main(){
  var v = '12';
  //类型推断
  print(v.parseInt());
}
```

使用时也可以使用扩展名显式调用。示例代码如下：

```
//chapter12/bin/example_04.dart
import 'string_apis.dart';
main(){
  //通过扩展名显式调用
  print(parsingString('Go').doubleIt());
}
```

扩展方法还可以对支持泛型的类进行扩展。

11min

12.3 类型定义

在 Dart 中函数是对象，就像字符串和数字是对象一样。typedef 用于声明函数类型，它为函数类型提供一个名称，可以在声明字段和返回类型时使用该名称。当将 typedef 定义的函数类型分配给变量时，会保留类型信息。

不使用 typedef 声明的函数类型，使用示例代码如下：

```
//chapter12/bin/example_05.dart
class SortedCollection{
  Function compare;

  SortedCollection(int f(Object a, Object b)) {
    compare = f;
  }
}
int sort(Object a, Object b) => 0;
void main(){
  SortedCollection coll = SortedCollection(sort);
  //我们知道 compare 是一个函数
  //但是它的函数类型是声明
  print(coll.compare is Function);
}
```

分配 f 给 compare 时，类型信息会丢失。参数 f 的类型是（Object，Object）→int，而

compare 的类型是 Function。如果为函数提供类型定义，并保留类型信息，则开发人员和工具都可以使用该信息。示例代码如下：

```
//chapter12/bin/example_06.dart
//定义类型别名
typedef Compare = int Function(Object a, Object b);
class SortedCollection {
  Compare compare;
  SortedCollection(this.compare);
}
int sort(Object a, Object b) => 0;

void main() {
  SortedCollection coll = SortedCollection(sort);
  print(coll.compare is Function);
  //判断是不是 Compare 类型
  print(coll.compare is Compare);
}
```

因为 typedef 只是定义类型别名，所以它们提供了一种检查任何函数类型的方法。示例代码如下：

```
//chapter12/bin/example_07.dart
//声明函数别名且可提供泛型参数
typedef Compare<T> = int Function(T a, T b);

int sort(int a, int b) => a - b;

void main() {
  //判断 sort 是否是 Compare<int>类型的函数
  print(sort is Compare<int>);
}
```

12.4 元数据

使用元数据提供有关代码的附加信息。元数据注解以字符@开头，跟着字符@的是对编译时常量的引用(例如 deprecated)或对常量构造函数的调用。

所有 Dart 代码都有两个注解：@deprecated 和@override。有关使用@override 的示例，参见扩展一个类。

注解@deprecated 表示弃用，使用@deprecated 注解的示例代码如下：

```
//chapter12/bin/example_08.dart
class Television {
  //弃用:使用[turnOn]代替
  @deprecated
  void activate() {
    turnOn();
  }
  //打开电视电源
  void turnOn() {
    //实现细节
  }
}
```

可以定义自己的元数据注解。定义带有两个参数的@todo注解的示例代码如下：

```
//chapter12/bin/todo.dart
library todo;
//定义注解
class Todo{
  final String who;
  final String what;
  const Todo(this.who, this.what);
}
```

使用@todo注解的示例代码如下：

```
//chapter12/bin/example_09.dart
//导入注解库
import 'todo.dart';
//使用注解
@Todo('jobs', '做家务')
void doSomething() {
  print('做一些事');
}
```

元数据可以出现在库、类、类型别名、类型参数、构造函数、工厂、函数、字段、参数或变量声明之前，也可以出现在import或export指令之前。可以在运行时使用反射来检索元数据。

3min

12.5 注释

Dart支持单行注释、多行注释、文档注释。

单行注释以“//”开头，Dart编译器会忽略“//”和行尾之间的所有内容。示例代码如下：

```
//chapter12/bin/example_10.dart
void main(){
  //打印当前时间
  print(DateTime.now());
}
```

多行注释以“/ * ”开头，以“ * /”结尾。Dart 编译器将忽略“/ * ”和“ * /”之间的所有内容，多行注释可以嵌套。示例代码如下：

```
//chapter12/bin/example_11.dart
void main(){
  / *
   * 打印当前时间
   * 使用 UTC 时间
   *  * /
  print(DateTime.now().toUtc());
}
```

文档注释是以“///”或“/ ** ”开头的多行或单行注释。在连续的行上使用“///”与多行文档注释具有相同的效果。

在文档注释中，Dart 编译器将忽略所有文本，除非将其括在方括号中。使用方括号，可以引用类、方法、字段、顶级变量、函数和参数。方括号中的名称在文档化的程序元素的词法范围内解析。

参考其他类和参数的文档注释示例代码如下：

```
//chapter12/bin/example_12.dart
///连接 MySQL 数据库配置类
class ConnectionSettings{
  //...
}
///连接 MySQL 数据库的类
class MySQLConnection{
  ///关闭数据库连接
  Future close(){
    //...
  }
  ///连接给定端口上的 MySQL 服务器
  ///[ConnectionSettings]类包含用户名、密码、数据库等信息
  static Future < MySQLConnection > connect(ConnectionSettings c) async{
    //...
  }
  ///为[values]中每组 sql 参数执行多次[sql]语句
```

```
  Future<List<Object>> queryMultiString(String sql,Iterable<List<Object>> values){
    //...
  }
}
```

在生成的文档中，[ConnectionSettings]成为指向 ConnectionSettings 类的 API 文档的链接。

要解析 Dart 代码并生成 HTML 文档，可以使用 SDK 的文档生成工具。

第二部分

第 13 章

服务端开发

13.1 HTTP 请求与响应

13.1.1 服务端

HttpServer 是使用 HTTP 传输内容的服务端对象。使用 HttpServer 类的静态方法 bind 创建 HttpServer 对象，使用时需要向该方法传递主机和端口两个必选参数，该 HttpServer 对象就绑定到了相应的主机和端口。示例代码如下：

```
import 'dart:io';
main()async{
  //创建 HttpServer 实例,并绑定主机地址和端口
  var server = await HttpServer.bind('localhost',80);
}
```

方法 bind 的第一个参数是主机名，可以是特定的主机名或字符串。也可以使用 InternetAddress 类提供的预定义值指定主机：

(1) loopbackIPv6：使用地址 localhost，IP 版本为 6。

(2) loopbackIPv4：使用地址 localhost，IP 版本为 4。

(3) anyIPv6：将自动分配 IP 地址，IP 版本为 6。

(4) anyIPv4：将自动分配 IP 地址，IP 版本为 4。

方法 bind 的第二个参数是 int 类型的端口名，该端口是该服务在计算机上的唯一标识。1024 以下是为系统标准服务保留的端口，HTTP 的服务端口为 80，也可以指定为 1024 及以上的端口号。

HttpServer 是提供 HttpRequest 对象的 Stream，因此可以对其进行监听。监听函数会接收一个回调函数，回调函数的参数是 HttpRequest 对象。可以通过 HttpRequest 对象获取请求相关的信息。示例代码如下：

```
//监听请求流
server.listen((HttpRequest request){
  //打印请求地址
  print('请求地址 ${request.requestedUri}');
  //写入响应信息
  request.response.write('访问成功!');
  //关闭当前请求的响应对象
  request.response.close();
});
```

HttpRequest 对象包含一些有用的属性：

（1）method：请求方法。可能的值：GET、POST、PUT、DELETE 等。

（2）uri：请求地址。一个 Uri 对象，包含请求相关的主机、端口、查询字符串等信息。

（3）headers：请求的标头。HttpHeaders 对象包含内容类型、内容长度、日期等信息。

（4）response：返回与 HttpRequest 关联的 HttpResponse 对象，服务器利用该对象响应请求。

使用 method 属性获取请求方法名，并针对相应方法处理相关事务。示例代码如下：

```
if(request.method == 'GET'){
  //处理 GET 请求
}else if(request.method == 'POST'){
  //处理 POST 请求
}
```

属性 uri 的最大作用是从请求地址中获取查询参数及少量数据，常用于处理 GET 请求。示例代码如下：

```
//获取所有查询参数及值组成的 Map 对象
//常在处理 GET 请求中使用
var queryMap = request.uri.queryParameters;
//获取指定参数名对应的值
var name = queryMap['name'];
```

使用 headers 属性获取 HttpHeaders 对象，常用该对象获取请求内容类型。示例代码如下：

```
//请求的内容类型
var contentType = request.headers.contentType;
```

HttpResponse 对象包含针对响应的一些属性和方法：

（1）encoding：响应数据采用的编码，常用的编码方式是 JSON 和 UTF-8。

（2）headers：响应的标头，它返回一个 HttpHeaders 对象。

(3) statusCode：响应状态码，用于指示请求成功或失败。

(4) write(Object obj)：将 obj 转化为字符串，并写入响应数据中。

(5) writeln([Object obj = ""])：将 obj 转化为字符串，并在响应数据新的一行中写入该字符串。

(6) writeAll(Iterable objects, [String separator = ""])：遍历可迭代对象 objects 并按顺序写入响应数据中，可以为 objects 中的元素间指定分隔符 separator。

(7) addStream(Stream<List<int>> stream)：将流添加到响应数据中，返回 Future 对象。

响应状态码是由 HttpStatus 对象提供的，它预定义了一部分状态码。常用的有以下常量：

(1) HttpStatus. ok：表示请求成功。

(2) HttpStatus. notFound：表示不存在该页面。

(3) HttpStatus. methodNotAllowed：表示不响应该方法。

设置响应码和数据的示例代码如下：

```
//响应 GET 请求
//设置响应状态码
request.response.statusCode = HttpStatus.ok;
//向响应对象 HttpResponse 主体写入数据
request.response.write('GET 请求成功!');
```

除了处理请求地址上的查询参数，有时还需要处理请求主体中的数据。HttpRequest 的主体数据是 List<int>类型的 Stream，可以使用 Utf8Decoder 的 bind 方法将其转换为 String 类型的 Stream。POST 请求对于可以发送的数据量没有限制，并且可以分块传输，Stream 的 join 方法可以将这些块拼凑在一起。示例代码如下：

```
//处理 POST 请求
//对请求解码
var content = await utf8.decoder.bind(request).join();
//将字符串解析为 json 数据
var data = jsonDecode(content) as Map;
//迭代 json 数据
data.forEach((k,v){
  print('键: $ k,值: $ v');
});
```

在请求使用 UTF-8 编码格式时，才可以使用 Utf8Decoder 解码请求主体数据。

如果当前是在本地运行服务端和客户端代码，则它们的访问与响应不受影响。在实际部署应用时，可能需要接收不同地址或不同端口的请求，此时就需要配置跨域参数。示例代码如下：

```
//跨域请求配置
void addCorsHeaders(HttpResponse response){
  //允许的源,即主机地址
  response.headers.add('Access-Control-Allow-Origin','*');
  //允许请求使用的方法
  response.headers.add('Access-Control-Allow-Methods',
      'GET,POST,OPTIONS');
  //允许请求使用的标头
  response.headers.add('Access-Control-Allow-Headers',
      'Origin,X-Requested-With,Content-Type,Accept');
}
```

服务端代码需要与浏览器或客户端配合使用,此处只提供了接收和响应请求的服务,发出请求和处理响应将使用客户端来完成。服务端完整代码如下:

```
//chapter13/bin/http_server.dart
import 'dart:io';
import 'dart:convert';

main()async{
  //创建 HttpServer 实例并绑定主机地址和端口
  var server = await HttpServer.bind('localhost',80);
  //监听请求流
  server.listen((HttpRequest request)async{
    //缓存请求的内容类型 contentType
    var contentType = request.headers.contentType;
    //根据请求方法名进入不同的处理代码块
    if(request.method == 'GET'){
      await handleGet(request);
    }else if(request.method == 'POST'
        && contentType?.mimeType == 'application/json'){
      await handlePost(request);
    }else{
      request.response.statusCode = HttpStatus.methodNotAllowed;
      request.response.write('${request.method}请求不被服务端支持!');
    }
    //关闭当前请求的响应对象
    await request.response.close();
  });
}
//处理 GET 请求的实现代码
void handleGet(HttpRequest request)async{
  //应用跨域请求配置
  addCorsHeaders(request.response);
  try{
```

```
    //处理 GET 请求
    //获取所有查询参数及值组成的 Map 对象
    //常在处理 GET 请求中使用
    var queryMap = request.uri.queryParameters;
    //获取指定参数名对应的值
    var name = queryMap['name'];
    //打印参数 name 的值
    if(name != null) print('参数 name 的值:$name');
    //响应 GET 请求
    request.response.statusCode = HttpStatus.ok;
    request.response.write('GET 请求成功!');
  }catch(e){
    request.response.statusCode = HttpStatus.internalServerError;
    request.response.write('GET 请求异常:$e');
  }
}
//处理 POST 请求的实现代码
void handlePost(HttpRequest request)async{
  //应用跨域请求配置
  addCorsHeaders(request.response);
  try{
    //处理 POST 请求
    //对请求解码
    var content = await utf8.decoder.bind(request).join();
    //将字符串解析为 json 数据
    var data = jsonDecode(content) as Map;
    //迭代 json 数据
    data.forEach((k,v){
      print('键:$k,值:$v');
    });
    //响应 POST 请求
    //使用级联操作符设置对象的多个值
    request.response
      ..statusCode = HttpStatus.ok
      ..write('POST 请求成功!')
      ..write('这是新的一行');
  }catch(e){
    request.response.statusCode = HttpStatus.internalServerError;
    request.response.write('POST 请求异常:$e');
  }
}

//跨域请求配置
void addCorsHeaders(HttpResponse response){
  //允许的源,即主机地址
  response.headers.add('Access-Control-Allow-Origin','*');
```

```
  //允许的方法
  response.headers.add('Access-Control-Allow-Methods',
      'GET,POST,OPTIONS');
  //允许的标头
  response.headers.add('Access-Control-Allow-Headers',
      'Origin,X-Requested-With,Content-Type,Accept');
}
```

13.1.2 客户端

HttpClient 是采用 HTTP 向服务器发出请求并接收响应数据的客户端。它可以通过一系列方法将 HttpClientRequest 对象发送到服务器,并从服务器接收 HttpClientResponse 对象。

创建客户端实例:

```
//创建 HttpClient 实例
var client = HttpClient();
```

当需要关闭客户端实例时可以调用 close 方法,它有一个可选参数 force,该参数表示是否强制关闭客户端。其默认值为 false,表示在完成所有连接后关闭客户端。当值为 true 时,将关闭所有连接并立即释放资源。这些被关闭的连接将收到一个错误事件,以指示客户端已关闭。调用示例代码如下:

```
//完成所有与服务器的连接后关闭客户端
client.close();
//立即关闭客户端并释放资源
client.close(force:true);
```

HttpClient 定义了打开 HTTP 连接的常用方法,它们都返回一个 HttpClientRequest 并包装在 Future 对象中。HttpClientRequest 对象可通过 headers 属性返回的 HttpHeaders 设置请求标头信息,也可以通过继承的 write、writeln、writeAll 方法向请求主体写入数据。

常用方法如下:

(1) get(String host, int port, String path):使用 GET 方法打开 HTTP 连接。

(2) post(String host, int port, String path):使用 POST 方法打开 HTTP 连接。

(3) put(String host, int port, String path):使用 PUT 方法打开 HTTP 连接。

(4) delete(String host, int port, String path):使用 DELETE 方法打开 HTTP 连接。

通过 get 方法发出 GET 请求,需要向该方法指定服务器主机、端口号和路径,路径中可能包含查询字符串。该方法返回 HttpClientRequest 对象。示例代码如下:

```
//GET 请求
var reqGet = await client.get(_host,80,'?name=jobs');
```

调用 HttpClientRequest 对象的 close 方法会返回一个 HttpClientResponse 对象，因为它是包装在 Future 中的，所以可以调用 then 方法并提供回调函数。HttpClientResponse 的主体来自服务器的数据流，因此可以使用解码器解析响应主体信息。示例代码如下：

```
//调用 close 方法以在完成连接后获取 HttpClientResponse 对象
await reqGet.close().then((response){
  //使用 utf8.decoder 转换数据,将 List<int>转换为 String
  //因为它们是流,所以可以通过监听获取转换后的数据
  response.transform(utf8.decoder).listen((contents) {
    //处理数据
    print(contents);
  });
});
```

使用 post 方法发出 POST 请求，其参数与 get 方法一致。指定标头的 contentType 为 json 格式，并通过 write 方法写入 json 数据。示例代码如下：

```
//POST 请求
var reqPost = await client.post(_host,80,'')
    //指定内容类型为 json
    ..headers.contentType = ContentType.json
    //写入数据
    ..write(jsonEncode(jsonData));
```

使用 delete 发起一个 DELETE 请求，因为该请求不被服务器支持，所以请求会失败。可以通过 HttpClientResponse 的 statusCode 得到请求状态码以判断是否成功。示例代码如下：

```
//发出 DELETE 请求
var reqDelete = await client.delete(_host,80,'');
await reqDelete.close().then((response){
  //判断请求是否成功
  if(response.statusCode != HttpStatus.ok){
    print('DELETE 请求失败');
  }
});
```

客户端完整代码如下：

```
//chapter13/bin/http_client.dart
import 'dart:io';
```

```
import 'dart:convert';

//主机名等价于 localhost
String _host = InternetAddress.loopbackIPv6.host;

//json 数据
Map jsonData = {
  'city': '上海',
  'area': '16800 平方千米',
  'population': '1600 万'
};

Future main() async{
  //创建 HttpClient 实例
  var client = HttpClient();

  //发出 GET 请求
  var reqGet = await client.get(_host,80,'?name = jobs');
  //调用 close 方法以在完成连接后获得 HttpClientResponse 对象
  await reqGet.close().then((response){
    //使用 utf8.decoder 转换数据,将 List < int >转换为 String
    //因为它们是流,所以可以通过监听获取转换后的数据
    response.transform(utf8.decoder).listen((contents) {
      //处理数据
      print(contents);
    });
  });

  //发出 POST 请求
  var reqPost = await client.post(_host,80,'')
    //指定内容类型为 json
    ..headers.contentType = ContentType.json
    //写入数据
    ..write(jsonEncode(jsonData));
  await reqPost.close().then((response){
    response.transform(utf8.decoder).listen((contents) {
      print(contents);
    });
  });

  //发出 DELETE 请求
  var reqDelete = await client.delete(_host,80,'');
  await reqDelete.close().then((response){
    //判断请求是否成功
    if(response.statusCode != HttpStatus.ok){
      print('DELETE 请求失败');
```

```
      }
      response.transform(utf8.decoder).listen((contents) {
        print(contents);
      });
    });
  }
```

到此已经编写了服务端和客户端的代码,首先运行服务端代码以启动服务,然后运行客户端代码,并观察控制台打印信息。

13.2　shelf 框架

20min

shelf 框架使得创建和编写 Web 服务的各个部分变得容易。

(1) 公开一小组简单类型。

(2) 将服务逻辑映射为一个简单函数:拥有单个参数 request 和返回值 response。

(3) 轻松混合并匹配同步和异步处理。

(4) 灵活地返回具有相同模型的简单字符串或字节流。

简单示例代码如下:

```
//chapter13/bin/shelf_basic_server.dart
import 'package:shelf/shelf.dart';
import 'package:shelf/shelf_io.dart' as io;
void main(){
  //创建处理程序
  Response echoRequest(Request request) {
    //返回请求 URL 和 method 作为响应
    return Response.ok('请求 URL: ${request.URL},请求方法: ${request.method}');
  }
  //创建适配器实例
  var server = io.serve(echoRequest,'localhost',1024);
  //成功得到实例后打印服务主机地址与端口
  server.then((server){
    print('服务地址 http://${server.address.host}:${server.port}');
  });
}
```

在上述代码中涉及许多重要概念:处理程序、中间件、适配器。

13.2.1　处理程序

处理程序 handler 是可以处理 Request 对象并返回 Response 对象的函数。示例代码如下:

```
//创建处理程序
Response echoRequest(Request request) {
  //返回请求地址作为响应
  return Response.ok('请求地址: ${request.URL}');
}
```

该处理程序将请求的 URL 字符串作为响应并返回到请求的发出者。

Request 对象常用属性和方法：

(1) encoding：消息正文的编码。

(2) headers：HTTP 标头。

(3) method：HTTP 请求方法，例如：GET、POST 等。

(4) requestedUri：请求的原始 Uri。

(5) handlerPath：当前处理程序的 URL 路径，与 URL 构成 requestedUri。

(6) URL：当前处理程序到请求资源的 URL 路径及查询参数，相对于 handlerPath。

(7) change({Map<String, Object> headers, Map<String, Object> context, String path, dynamic body})：通过复制当前 Request 并应用指定的更改来创建新的 Request。

(8) read ()：返回表示正文的 Stream，类型为 Stream<List<int>>。

(9) readAsString([Encoding encoding])：将正文作为字符串的 Future。

Response 对象常用属性和方法：

(1) statusCode：响应的 HTTP 状态代码。

(2) headers：HTTP 标头。

(3) encoding：消息正文的编码。

(4) read()：返回表示消息正文的 Stream，类型为 Stream<List<int>>。

(5) readAsString([Encoding encoding])：将正文作为字符串的 Future。

(6) Response(int statusCode, {dynamic body, Map<String, Object> headers, Encoding encoding, Map<String, Object> context})：使用给定的 statusCode 创建一个 HTTP 响应。

(7) Response.ok(dynamic body, {Map<String, Object> headers, Encoding encoding, Map<String, Object> context})：创建一个状态码为 200 的成功响应。

(8) Response.notFound(dynamic body, {Map<String, Object> headers, Encoding encoding, Map<String, Object> context})：创建一个状态码为 404 的未找到响应。

(9) Response.internalServerError({dynamic body, Map<String, Object> headers, Encoding encoding, Map<String, Object> context})：创建一个状态码为 500 的内部服务器错误响应。

(10) change({Map<String, Object> headers, Map<String, Object> context, dynamic body})：通过复制当前 Response 并应用指定的更改来创建新的 Response。

13.2.2 适配器

适配器可以创建 Request 对象，将它们传递给处理程序，并处理由处理程序生成的 Response 对象。在大多数情况下，适配器转发来自底层 HTTP 服务器的请求，并且转发及响应到底层 HTTP 服务器。shelf_io. serve 就是这种适配器。适配器也可能在浏览器中使用 window. location 和 window. history 合成 HTTP 请求，或者可以直接从 HTTP 客户端将请求传递到 Shelf 处理程序。

serve(Handler handler, dynamic address, int port, {SecurityContext securityContext, int backlog, bool shared: false})→ Future < HttpServer >: 启动一个 HttpServer，它侦听指定的地址和端口，并将请求发送到处理程序。

在使用 serve 方法时处理程序可以是单独的处理程序，也可以是通过 Pipeline 类的实例返回的处理程序，在这里使用的是单独的处理程序。

示例代码如下：

```
io.serve(echoRequest,'localhost',1024);
```

运行代码，当服务启动成功后会在控制台打印服务地址和端口。

```
服务地址 http://localhost:1024
```

编写客户端代码，代码使用 GET 请求访问指定主机地址和端口的服务。示例代码如下：

```
//chapter13/bin/shelf_basic_client.dart
import 'dart:io';
import 'dart:convert';

Future main() async{
  //创建 HttpClient 实例
  var client = HttpClient();
  //发出 GET 请求
  var reqGet = await client.get('localhost',1024, '?msg = hello');
  //调用 close 方法以在完成连接后获得 HttpClientResponse 对象
  await reqGet.close().then((response){
    //使用 utf8.decoder 转换数据,将 List < int >转换为 String
    //因为它们是流,所以可以通过监听获取转换后的数据
    response.transform(utf8.decoder).listen((contents){
      //处理数据
      print(contents);
    });
  });
}
```

运行客户端代码，可以观察到服务端返回的响应。

```
请求 URL:?msg = hello,请求方法: GET
```

13.2.3 中间件

处理程序也可以进行部分处理，然后将请求转发给另一个处理程序，这种处理程序被称为中间件（middleware），因为它位于服务堆栈的中间。可以把中间件视作传入一个handler，然后返回一个新 handler 的函数。示意如下：

```
Handler Middleware(Handler innerHandler);
```

一个 shelf 应用通常由多层中间件组成，每个中间件中又有一个或多个处理程序。shelf. Pipeline 类使这类应用程序易于构建。

Pipeline 是一个帮助程序，可以将一组中间件和一个处理程序很容易地组合在一起。

（1）addMiddleware(Middleware middleware)：将中间件添加到中间件组中，并返回一个新的 Pipeline 对象。中间件组中的最后一个中间件将是处理请求的最后一个中间件，且是处理响应的第一个中间件。

（2）addHandler(Handler handler)：如果 Pipeline 中的所有中间件均已通过请求，将传入的 handler 作为请求的最终处理程序，并返回一个的新的处理程序。一个 shelf 应用程序只能有一个处理程序，因此此方法只能调用一次。

使用示例代码如下：

```
var handler = Pipeline()
      //添加中间件
      .addMiddleware(logRequests())
      //添加处理程序
      .addHandler(echoRequest);
//创建适配器实例
var server = io.serve(handler,'localhost',1024);
```

logRequests()方法是 shelf 框架内置的方法，它的作用是打印请求的时间、内部处理程序所用时间、请求方法、响应状态码和请求 Uri。它将返回一个 Middleware 对象，因此可以作为中间件在应用程序中使用。使用 addHandler 方法添加处理程序 echoRequest，该方法会返回一个新的处理程序，可以直接将返回值用作参数传递到适配器。

运行此程序并运行客户端代码将可以在控制台看到 logRequests 方法打印的日志信息。

```
服务地址 http://localhost:1024
2020 - 07 - 26T15:23:43.901771   0:00:00.014704 GET     [200] /?msg = hello
```

可以使用 createMiddleware 方法创建中间件，该方法包含 3 个可选参数：

（1）requestHandler：请求处理程序，接收 Request 对象，返回 Response 对象的函数。该处理程序用于对 Request 对象进行进一步处理，如果需要继续传递 Request 对象，则返回 null。否则，正常返回 Response 对象。

（2）responseHandler：响应处理程序，接收 Response 对象，返回 Response 对象的函数。它的作用是对 Response 对象进行进一步处理。

（3）errorHandler：错误处理程序，接收错误参数 error 和堆栈跟踪参数 StackTrace，返回 Response 对象的函数。在程序运行出错时对错误进行处理。

通常来说只会向中间件传递一个参数，因为部分中间件只对请求进行处理，部分只对响应进行处理。这里为了演示创建和使用的中间件提供了所有参数，在实际使用中需要传递所有参数的中间件极少。示例代码如下：

```
//创建中间件
var middleware = createMiddleware(requestHandler: reqHandler,
    responseHandler: resHandler,errorHandler: errHandler);
//请求处理程序

//如果希望直接响应请求则通过响应信息
Response reqHandler(Request request){
  print('请求处理程序 请求 Uri: ${request.requestedUri}');
  //返回 null 表示使 request 继续传递到下一个中间件或处理程序
  //否则将直接响应请求
  return null;
}
//响应处理程序
Response resHandler(Response response){
  print('响应处理程序 响应码: ${response.statusCode}');
  return response;
}
//错误处理程序
Response errHandler(error,StackTrace st){
  if (error is HijackException) throw error;
  print('错误处理程序: $ error');
  throw error;
}
```

完整示例代码如下：

```
//chapter13/bin/shelf_pipeline_server.dart
import 'package:shelf/shelf.dart';
import 'package:shelf/shelf_io.dart' as io;
void main(){
```

```
  var handler = Pipeline()
      //添加中间件
      .addMiddleware(logRequests())
      //添加自定义中间件
      .addMiddleware(middleware)
      //添加处理程序
      .addHandler(echoRequest);

  //创建适配器实例
  var server = io.serve(handler,'localhost',1024);
  //成功得到实例后打印服务主机地址与端口
  server.then((server){
    print('服务地址 http://${server.address.host}:${server.port}');
  });
}
//创建处理程序
Response echoRequest(Request request){
  //返回请求 URL 和 method 作为响应
  return Response.ok('请求 URL:${request.URL},请求方法:${request.method}');
}
//创建中间件
var middleware = createMiddleware(requestHandler: reqHandler,
    responseHandler: resHandler,errorHandler: errHandler);
//请求处理程序

//如果希望直接响应请求则通过响应信息
Response reqHandler(Request request){
  print('请求处理程序,请求 Uri:${request.requestedUri}');
  //返回 null 表示使 request 继续传递到下一个中间件或处理程序
  //否则将直接响应请求
  return null;
}
//响应处理程序
Response resHandler(Response response){
  print('响应处理程序,响应码:${response.statusCode}');
  return response;
}
//错误处理程序
Response errHandler(error,StackTrace st){
  if (error is HijackException) throw error;
  print('错误处理程序:$error');
  throw error;
}
```

运行此服务,并运行客户端应用。查看服务在控制台的输出信息:

```
服务地址 http://localhost:1024
请求处理程序 请求 Uri:http://localhost:1024/?msg = hello
响应处理程序 响应码:200
2020 - 07 - 26T15:23:43.901771   0:00:00.014704 GET   [200] /?msg = hello
```

13.3 路由包

42min

35min

一些中间件还可以采用多个处理程序，并为每个请求调用一个或多个处理程序。例如：路由器中间件可能会根据请求的 Uri 和 HTTP 方法选择要调用的处理程序。shelf_router 包就提供了路由的功能。

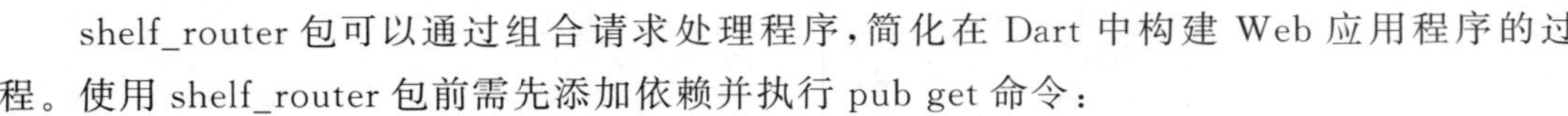

shelf_router 包可以通过组合请求处理程序，简化在 Dart 中构建 Web 应用程序的过程。使用 shelf_router 包前需先添加依赖并执行 pub get 命令：

```
dependencies:
  shelf: ^0.7.7
  shelf_router: ^0.7.2
```

13.3.1 定义路由

该包有一个 Router 类，基于 HTTP 方法和路由模式将请求匹配到对应的处理程序。它包含以下方法和属性：

（1）Router()：无参构造函数，创建一个 Router 对象。

（2）handler → Handler：获取一个处理程序，该处理程序会将传入的请求路由传到已注册的处理程序。

（3）get(String route,Function handler)：使用 handler 处理到路径的 GET 请求。

（4）post(String route,Function handler)：使用 handler 处理到路径的 POST 请求。

（5）put(String route,Function handler)：使用 handler 处理到路径的 PUT 请求。

（6）delete(String route, Function handler)：使用 handler 处理到路径的 DELETE 请求。

（7）all(String route, Function handler)：使用 handler 处理到路径的所有请求。

（8）mount(String prefix, Router router)：在前缀 prefix 下附加一个路由器，通过这样的形式可以将多个路由器对象组合在一起形成完整的 API。

该包还包含一个顶层函数：

params(Request request, String name)：通过路径捕获 URL 中参数的值。

这里先创建一个 Router 对象，然后使用 get 方法，示例代码如下：

```
//chapter13/bin/shelf_router_basic_server.dart
import 'package:shelf/shelf.dart';
import 'package:shelf/shelf_io.dart' as io;
import 'package:shelf_router/shelf_router.dart';

main(){
  //创建 Router 对象
  var app = Router();
  //使用 Router 类的 get 方法创建可以捕获路径/hello 上的 GET 请求
  //这里的处理程序是一个匿名函数
  app.get('/hello',(Request request){
    return Response.ok('Hello World');
  });
  //通过 shelf_io.serve 方法启动一个 HttpServer
  var server = io.serve(app.handler, 'localhost', 1024);
  //成功得到实例后打印服务主机地址与端口
  server.then((server){
    print('服务地址 http://${server.address.host}:${server.port}');
  });
}
```

启动服务，并创建客户端请求到路径/hello 下的 GET 方法。示例代码如下：

```
//chapter13/bin/shelf_router_basic_client.dart
import 'dart:io';
import 'dart:convert';

Future main() async{
  //创建 HttpClient 实例
  var client = HttpClient();
  //发出 GET 请求
  var reqGet = await client.get('localhost',1024,'/hello');
  //调用 close 方法以在完成连接后获得 HttpClientResponse 对象
  await reqGet.close().then((response){
    //使用 utf8.decoder 转换数据，将 List<int>转换为 String
    //因为它们是流，所以可以通过监听获取转换后的数据
    response.transform(utf8.decoder).listen((contents){
      //处理数据
      print(contents);
    });
  });
}
```

运行客户端代码结果如下：

```
Hello World
```

此时如果请求到除/hello以外的路径将导致错误。

13.3.2　路由参数

在路径中可以嵌入URL参数，其方法是使用尖括号将参数名包裹起来。参数不仅作为路由匹配的一部分，还可以在处理程序中使用。示例代码如下：

```
//可以将参数名 paramName 放在<>中以在路径中指定参数
app.get('/users/<userName>/whoami',(Request request) async{
  //可以使用 params(request,param)读取匹配参数的值
  var userName = params(request, 'userName');
  return Response.ok('你是 ${userName}');
});
```

也可以将URL参数传入处理程序的参数列表中。示例代码如下：

```
//可以将 URL 参数传入处理程序
app.get('/users/<userName>/sayhello', (Request request, String userName) async{
  return Response.ok('你好 ${userName}');
});
```

可以同时传递多个参数，此时处理程序必须接收所有参数或者一个也不接收，且必须按照URL参数顺序接收。示例代码如下：

```
//可以在 URL 中指定多个参数
//处理程序必须按 URL 中出现的顺序接收所有参数或者一个也不接收
app.get('/users/<userName>/<userId>',(Request request, String userName,String userId) async
{
  return Response.ok('你好 ${userName}, ${userId}');
});
```

可以通过<paramName|REGEXP>的形式指定满足正则表达式REGEXP的参数paramName。如果未指定正则表达式，则默认使用正则表达式'[^/]+'，它表示匹配除/之外的所有字符串。示例代码如下：

```
//可以使用<paramName|REGEXP>指定自定义正则表达式,其中 REGEXP 是正则表达式(省略^和$)
//如果未指定正则表达式,将使用'[^/]'
app.get('/users/<userName>/messages/<msgId|[0-9]+>',(Request request)
async{
  var userName = params(request, 'userName');
  var msgId = int.parse(params(request, 'msgId'));
  return Response.ok('userName: ${userName},msgId: ${msgId}');
});
```

正则表达式“[0－9]＋”表示匹配一串数字。

13.3.3 组合路由

定义一个 Router 对象，然后使用 mount 方法将其安装到现有 Router 对象中。示例代码如下：

```
//创建新路由器
var art = Router();
art.get('/<articleId>', (Request request) async {
  var articleId = params(request, 'articleId');
  return Response.ok('article ID: ${articleId}');
});
//将 art 路由器安装到现有路由器 app 上
//前缀可以使用根路径/
app.mount('/article/',art);
```

在实际使用中需要将每个路由器对象封装在类中。示例代码如下：

```
//chapter13/bin/shelf_router_server.dart
class App{
  //返回处理程序
  Handler get handler{
    //创建 Router 对象
    final app = Router();

    //可以将参数名 paramName 放在<>中以在路径中指定参数
    app.get('/users/<userName>/whoami',(Request request) async{
      //可以使用 params(request,param)读取匹配参数的值
      var userName = params(request, 'userName');
      return Response.ok('你是 ${userName}');
    });

    //可以将 URL 参数传入处理程序
    app.get('/users/<userName>/sayhello', (Request request, String userName) async{
      return Response.ok('你好 ${userName}');
    });

    //可以在 URL 中指定多个参数
    //处理程序必须按 URL 中出现的顺序接收所有参数或者一个也不接收
    app.get('/users/<userName>/<userId>',(Request request, String userName,String userId)
async{
      return Response.ok('你好 ${userName}, ${userId}');
    });
```

```
        //可以使用<paramName|REGEXP>指定自定义正则表达式,其中 REGEXP 是正则表达式(省略^和$)
        //如果未指定正则表达式,将使用'[^/]'
        app.get('/users/<userName>/messages/<msgId|[0-9]+>',(Request request) async{
          var userName = params(request, 'userName');
          var msgId = int.parse(params(request, 'msgId'));
          return Response.ok('userName: ${userName},msgId: ${msgId}');
        });

        //将 art 路由器安装到现有路由器上
        //前缀可以使用根路径/
        app.mount('/article/', Art().router);
        //捕获所有请求并过滤掉已定义的路由
        app.all('/<ignored|.*>', (Request request) {
          return Response.notFound('页面未发现');
        });

        //返回处理程序,它会将请求路由匹配到已注册的处理程序上
        return app.handler;
      }
    }
    class Art{
      //返回路由器
      Router get router{
        //创建新路由
        var art = Router();
        //捕获/article/String 下所有请求
        art.get('/<articleId>', (Request request) async{
          var articleId = params(request, 'articleId');
          return Response.ok('文字 ID: ${articleId}');
        });

        //捕获/article/下所有请求并过滤掉已定义的路由
        art.all('/<ignored|.*>',(Request request){
          return Response.notFound('页面未发现');
        });
        return art;
      }
    }
```

然后通过 Pipeline 组合中间件和 Router 对象返回的 handler。示例代码如下：

```
//chapter13/bin/shelf_router_server.dart
import 'package:shelf/shelf.dart';
import 'package:shelf/shelf_io.dart' as io;
import 'package:shelf_router/shelf_router.dart';
```

```
main(){
  //创建 Router 对象
  var app = App();
  //通过 shelf_io.serve 方法启动一个 HttpServer
  var server = io.serve(app.handler, 'localhost', 1024);
  //成功得到实例后打印服务主机地址与端口
  server.then((server){
    print('服务地址 http://${server.address.host}:${server.port}');
  });
}
```

启动服务后就可以访问路由器中定义的路由了。

13.3.4 路由注解

还可以通过注解使用路由，首先添加开发依赖项并执行 pub get 命令：

```
dev_dependencies:
  shelf_router_generator: ^0.7.0+1
```

每个方法都有对应的注解，这些注解在 shelf 包中定义，使用时需要依赖开发依赖包 shelf_router_generator。以下列出常用注解：

(1) Route.get(String route)：将路径 route 的 GET 请求与注解方法匹配。

(2) Route.post(String route)：将路径 route 的 POST 请求与注解方法匹配。

(3) Route.put(String route)：将路径 route 的 PUT 请求与注解方法匹配。

(4) Route.delete(String route)：将路径 route 的 DELETE 请求与注解方法匹配。

(5) Route.mount(String prefix)：将前缀 prefix 的请求路由与注解方法匹配。

(6) Route.all(String route)：将路径 route 的使用请求与注解方法匹配。

使用注解时将一组相关的处理程序封装在类中，处理程序需要显式声明，函数名以下画线开头。在处理程序上使用合适的注解，并配置路由参数。

使用 part 指令指示路由器生成程序生成的文件名并作为当前库的一部分：

```
part 'shelf_router_generator.g.dart';
```

路由器生成程序会按类为单位生成返回 Router 对象的函数，可以通过生成的函数获取当前类的 Router 对象和处理程序。生成函数的名字遵循一定规律，以_$开头，中间是封装处理程序的类名，然后以 Router 结尾。

完整示例代码如下：

```
//chapter13/bin/shelf_router_generator.dart
import 'dart:async' show Future;
import 'package:shelf/shelf.dart';
import 'package:shelf/shelf_io.dart' as io;
import 'package:shelf_router/shelf_router.dart';

//由 pub run build_runner build 生成的文件
part 'shelf_router_generator.g.dart';

main() async{
  final service = App();
  //通过 Pipeline 对象组合中间件和单个处理程序
  var handler = Pipeline()
      .addMiddleware(logRequests())
      .addHandler(service.handler);
  //通过 shelf_io.serve 方法启动一个 HttpServer
  var server = await io.serve(handler, 'localhost', 1024);
  print('Serving at http://${server.address.host}:${server.port}');
}

class App{
  @Route.get('/users/<userName>/whoami')
  Future<Response> _users(Request request) async{
    var userName = params(request, 'userName');
    return Response.ok('You are ${userName}');
  }

  @Route.get('/users/<userName>/<userId>')
  Future<Response> _usersId(Request request, String userName,String userId) async {
    return Response.ok('Hello ${userName}, ${userId}');
  }

  @Route.get('/users/<userName>/messages/<msgId|[0-9]+>')
  Future<Response> _msgId(Request request) async {
    var userName = params(request, 'userName');
    var msgId = int.parse(params(request, 'msgId'));
    return Response.ok('userName:${userName},msgId:${msgId}');
  }

  @Route.mount('/article/')
  Router get _art => Art().router;

  @Route.all('/<ignored|.*>')
  Future<Response> _notFound(Request request) async{
    return Response.notFound('Page not found');
  }
```

```
  //生成的_$AppRouter函数可用于获取此对象的handler
  Handler get handler => _$AppRouter(this).handler;
}

class Art{
  @Route.get('/<articleId>')
  Future<Response> _articleId(Request request) async {
    var articleId = params(request, 'articleId');
    return Response.ok('article ID: ${articleId}');
  }
  @Route.all('/<ignored|.*>')
  Future<Response> _notFound(Request request)async => Response.notFound('null');
  //生成的_$ArtRouter函数用于公开此对象的Router
  Router get router => _$ArtRouter(this);
}
```

因为此时函数还没有生成，所以需要根据函数名生成规律拼凑函数名，并通过函数获得当前类的 Router 对象或处理程序。

然后在编辑器的命令行中输入以下命令：

```
pub run build_runner build
```

生成的文件内容如下：

```
//chapter13/bin/shelf_router_generator.g.dart
//GENERATED CODE - DO NOT MODIFY BY HAND

part of 'shelf_router_generator.dart';

//****************************************************************************
//ShelfRouterGenerator
//****************************************************************************
//与路由定义类App关联的函数
Router _$AppRouter(App service) {
  final router = Router();
  router.add('GET', r'/users/<userName>/whoami', service._users);
  router.add('GET', r'/users/<userName>/<userId>', service._usersId);
  router.add(
      'GET', r'/users/<userName>/messages/<msgId|[0-9]+>', service._msgId);
  router.mount(r'/article/', service._art);
  router.all(r'/<ignored|.*>', service._notFound);
  return router;
}
```

```
//与路由定义类 Art 关联的函数
Router _$ArtRouter(Art service) {
  final router = Router();
  router.add('GET', r'/<articleId>', service._articleId);
  router.all(r'/<ignored|.*>', service._notFound);
  return router;
}
```

可以从代码中看出这与手动编写的代码非常相似。

第三部分

第 14 章

Angular 基础

AngularDart 是前端框架 Angular 的 Dart 版本，本文简称 Angular。它采用 HTML 和 Dart 构建 Web 端应用程序，并且以包的形式提供，包名叫 angular。如果需要成体系的 UI，可以使用材质化组件库，包名叫 angular_components。

应用程序启动后，Angular 框架会接管所有工作，控制显示视图和交互。

14.1 初始项目

初始项目是为了快速了解项目重要文件和一些概念，本节无须掌握，重在概览项目结构。

创建新项目时，在可选的模板列表中选择 AngularDartWebApp 创建 Web 客户端应用程序样例，该模板采用 angular 包和材质化组件库 angular_components，如图 14-1 所示。

应用程序模板是由 stagehand 包提供的，主要包括以下模板：

(1) Console Application：命令行应用程序。

(2) Dart Package：Dart 库或应用程序。

(3) Web Server：采用 shelf 框架的 Web 服务端应用程序。

(4) AngularDart Web App：采用 Angular 框架的 Web 客户端应用程序。

如果在创建项目时可选的模板没有完全显示，可以在命令行中执行如下命令：

```
pub global activate stagehand
```

该命令不仅可以获取还可以更新 stagehand 包。当命令执行成功后，重启编辑器就可以得到完整的模板列表。

将项目命名为 angulardart，如图 14-2 所示。该项目主要用于介绍 Angular 应用程序的基本构成要素。

创建好的项目目录结构如图 14-3 所示。

选中 web 目录下的 index.html 文件并单击，文件内容如图 14-4 所示。

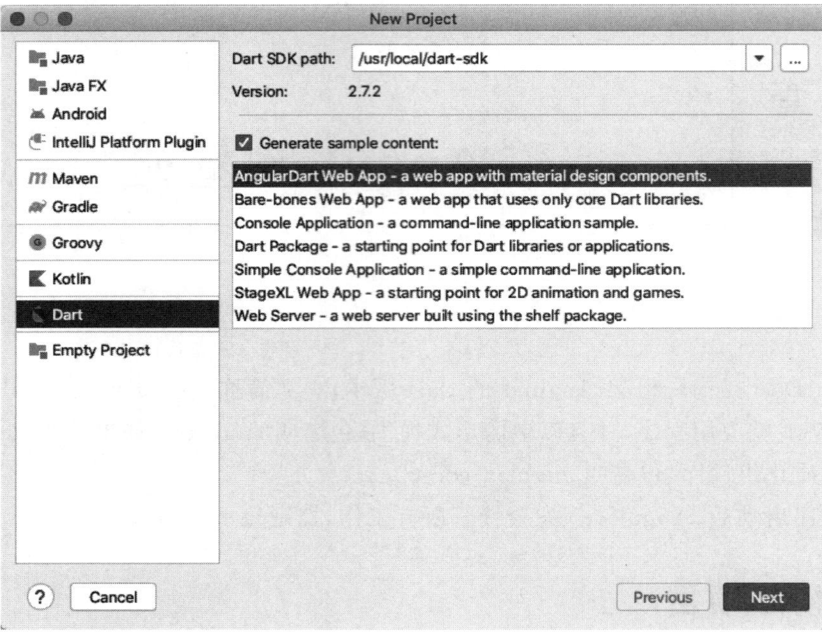

图 14-1　应用程序模板

New Project

Project name: angulardart

Project location: ~/IdeaProjects/angulardart

More Settings

Cancel　Previous　Finish

图 14-2　项目命名

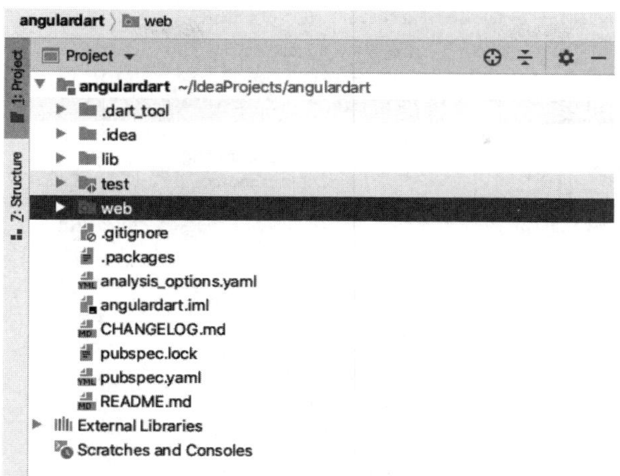

图 14-3　项目结构

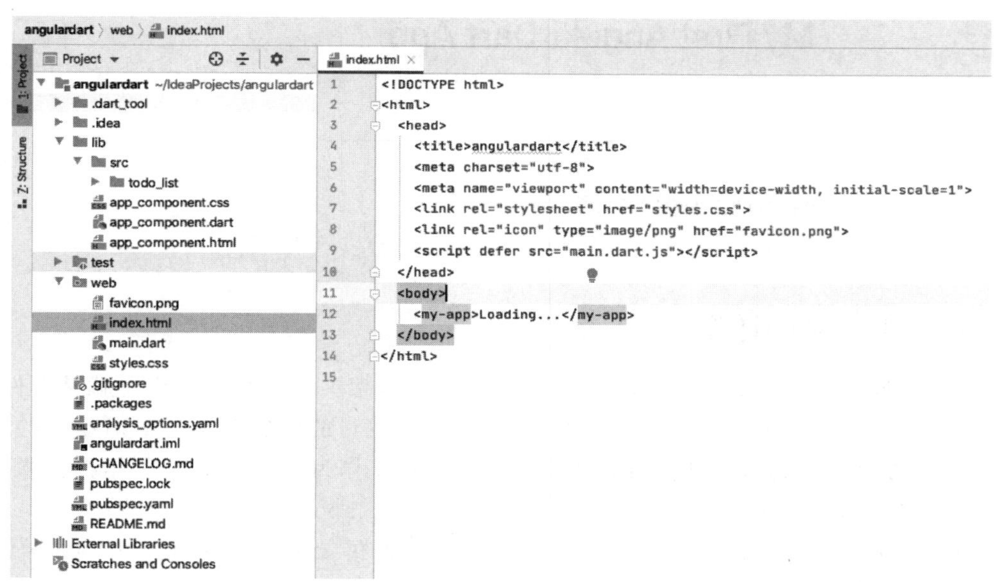

图 14-4　首页

在右边的视图中单击右键，选择 Run 'web/index.html'选项，项目将自动编译并运行，如图 14-5 所示。

Web 项目编译和运行需要使用工具 webdev，它采用了两种编译器：dartdevc 和 dart2js。在开发时 webdev 会选择 dartdevc，因为它支持增量编译。第一次编译需要较多时间，之后在修改代码时编译时间将会大大缩短。在部署应用程序时 webdev 会选择 dart2js，它做了大量针对性优化，并且可以保证在所有现代浏览器中正常运行。

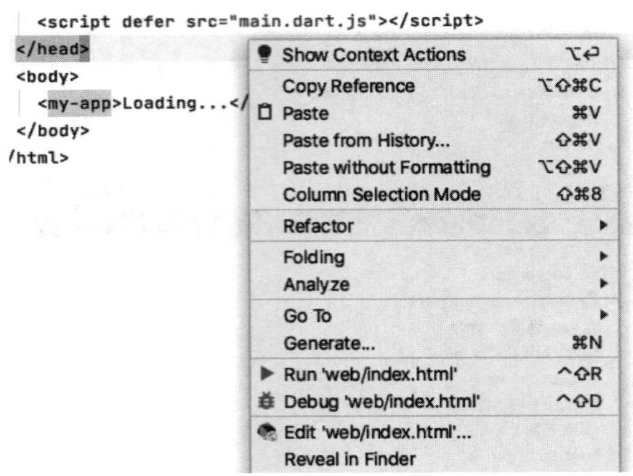

图 14-5 运行项目

初始项目运行结果如图 14-6 所示。

My First AngularDart App

What do you need to do?

Nothing to do! Add items above.

图 14-6 运行结果

14.1.1 项目详情

项目的所有文件均采用下画线命名法。Angular 框架中的一个重要概念就是组件，组件通常是由 Dart 语言编写的组件类文件、HTML 语言编写的模板文件及 CSS 编写的样式文件构成。根据约定这些文件的名称必须以 component 作为后缀，例如：根组件关联的文件 app_component.dart、app_component.html 和 app_component.css。

项目中绝大多数开发工作将在 lib 目录下完成，web 目录则存放应用首页、全局样式文件和包含应用执行起点函数的文件，在 pubspec.yaml 文件中存放的是与项目配置相关的信息。项目的重要目录展开内容如图 14-7 所示。

pubspec.yaml 文件用于指示本项目的项目名、平台依赖、库依赖、开发依赖等信息，可用字段见章节库。依赖信息具体内容如下：

```
//chapter14/angulardart/pubspec.yaml
#项目名
name: angulardart
```

```
#描述信息
description: A web app that uses AngularDart Components
#version: 1.0.0
#homepage: https://www.example.com
#平台依赖信息
environment:
  sdk: '>=2.7.0 <3.0.0'
#库依赖信息
dependencies:
  angular: ^5.3.0
  angular_components: ^0.13.0
#开发依赖信息
dev_dependencies:
  angular_test: ^2.3.0
  build_runner: ^1.6.0
  build_test: ^0.10.8
  build_web_compilers: ^2.3.0
  pedantic: ^1.8.0
  test: ^1.6.0
```

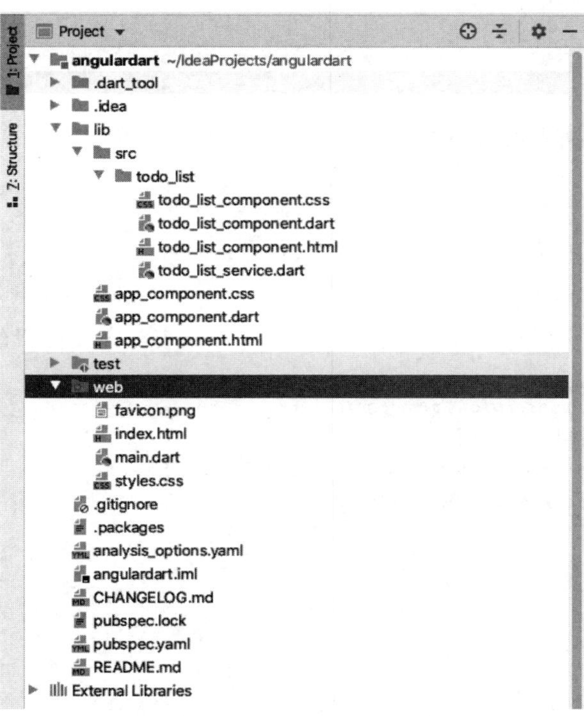

图 14-7　重要目录和文件

web 目录下的 index.html 文件是项目的首页，用于导入适用项目全局的样式文件、JavaScript 脚本文件等内容。代码如下：

```
//chapter14/angulardart/web/index.html
<!DOCTYPE html>
<html>
  <head>
    <title>angulardart</title>
    <meta charset="utf-8">
    <meta name="viewport" content="width=device-width, initial-scale=1">
    <link rel="stylesheet" href="styles.css">
    <link rel="icon" type="image/png" href="favicon.png">
    <script defer src="main.dart.js"></script>
  </head>
  <body>
    <my-app>Loading...</my-app>
  </body>
</html>
```

Angular 会将 Dart 程序编译成 JavaScript 以确保能在浏览器中运行，编译后的文件名为 main.dart.js。脚本引用代码如下：

```
<script defer src="main.dart.js"></script>
```

在页面主体部分插入项目根组件在注解中提供的选择器 my-app。

```
<body>
   <my-app>Loading...</my-app>
</body>
```

style.css 是项目全局样式文件，可以导入外部样式表或定义元素的全局样式。

```
//chapter14/angulardart/web/style.css
/* 导入外部样式表 */
@import URL(https://fonts.googleapis.com/css?family=Roboto);
@import URL(https://fonts.googleapis.com/css?family=Material+Icons);
/* 页面主体样式 */
body {
  max-width: 600px;
  margin: 0 auto;
  padding: 5vw;
}
/* 页面字体 */
* {
```

```
    font-family: Roboto, Helvetica, Arial, sans-serif;
}
```

main. dart 文件中的 main 函数是应用执行的起点，在函数体内部执行 angular 包定义的顶层函数 runApp。

runApp < T >(ComponentFactory < T > componentFactory,{InjectorFactory createInjector: _identityInjector})：以 componentFactory 为根启动一个新的 AngularDart 应用程序。

ComponentFactory 实例是使用@Component 注解的类的实现，例如：example. dart 文件代码如下：

```
@Component(
  selector: 'example',
  template: '...',
)
class Example{}
```

然后编译器在 example. template. dart 中生成 ExampleNgFactory，可以通过导入此生成的文件来对其进行访问。

InjectorFactory 实例是可选的，它是一个函数，用于向根目录提供服务，例如：路由服务。

main. dart 文件代码如下：

```
//chapter14/angulardart/web/main.dart
//导入 angular 库
import 'package:angular/angular.dart';
//导入项目根组件,该文件是由 AngularDart 编译器生成的
import 'package:angulardart/app_component.template.dart' as ng;
//main()是所有 dart 应用程序的执行起点函数
void main(){
  //runApp()是 Angular 应用程序的启动函数
  //将 AppComponent 作为根组件来启动应用程序
  //调用时采用类名加后缀 NgFactory
  runApp(ng.AppComponentNgFactory);
}
```

以 *. template. dart 结尾的文件都是由名为 angular_compiler 的编译器生成的文件，该编译器生成的文件存放在项目根目录. dart_tool/build/generated 目录下。

lib 目录是存放项目资源的文件夹，例如：组件、服务、模板、指令、样式文件等。在该目录下首先可以发现根组件的 3 个相关文件：

app_component. css 是根组件引用的样式文件，此时文件中没有任何实际作用的内容。

```
//chapter14/angulardart/lib/app_component.css
:host{
}
```

app_component.html 是根组件引用的模板文件，包含标题标签和子组件标签。

```
//chapter14/angulardart/lib/app_component.html
<h1>My First AngularDart App</h1>
<todo-list></todo-list>
```

app_component.dart 文件中在 AppComponent 类上使用组件注解@Component 以定义根组件。代码如下：

```
//chapter14/angulardart/lib/app_component.dart
import 'package:angular/angular.dart';
import 'src/todo_list/todo_list_component.dart';

@Component(
  selector: 'my-app',
  styleURLs: ['app_component.css'],
  templateURL: 'app_component.html',
  directives: [TodoListComponent],
)
class AppComponent{
}
```

在 lib 目录下还有一个 src 目录，所有的实现文件都应放在该目录下。src 目录下还有一个子组件目录 todo_list，该目录包含子组件 TodoListComponent 相关的文件。将与组件相关的文件放在单独的文件夹中是非常好的做法。

由于上述内容包含一些未学习内容，因此首先需要删除子组件目录 todo_list 及其包含的文件。

删除 app_component.html 文件中的子组件标签，修改结果如下：

```
<h1>My First AngularDart App</h1>
```

删除 app_component.dart 文件中与子组件相关的导入信息和注解信息，修改后代码如下：

```
//chapter14/angulardart/lib/app_component.dart
import 'package:angular/angular.dart';
//组件注解
@Component(
  selector: 'my-app',
```

```
)
//组件类
class AppComponent{
}
```

14.1.2 组件注解

AppComponent 是一个普通的类，因为有了注解@Component 而成为 angular 组件，按照约定组件类的名字以 Component 作为后缀。注解@Component 常用参数如表 14-1 所示。

表 14-1 组件注解参数

参　数	说　明
selector	在模板中指定一个 CSS 选择器，支持 element、[attribute]、. class、: not()和自定义 HTML 元素
template	组件视图的内联模板，HTML 字符串
templateURL	组件视图的外部模板 URL
styles	用于样式化组件视图的内联 CSS 样式
styleURLs	用于样式化组件视图的外部样式表 URL 列表
directives	组件模板中使用的指令列表
providers	此组件及其子级的依赖项注入提供程序的列表
pipes	组件模板中使用的管道列表
exports	导出在模板中可能使用的标识符列表。

14.1.3 组件模板

组件视图采用 HTML 语言，几乎支持所有合法的 HTML 元素，为了安全起见< script >元素不被支持。通过注解向组件提供组件视图时可以使用 template 或 templateURL 参数，template 参数接收的是字符串，而 templateURL 引用的是文件名。

使用单引号包裹单行 HTML 元素：

```
template: '< h1 > AngularDart 应用程序</h1 >',
```

使用三重单引号包裹多行 HTML 元素：

```
template: '''
  < h1 > AngularDart 应用程序</h1 >
  < p >根组件模板</p >''',
```

14.1.4 组件样式

组件样式采用标准 CSS 规则，Angular 会将组件样式和视图绑定。样式仅适用于当前组件的模板，而应用程序其他位置的元素均不受影响。可以使用 styles 或 styleURLs 参数为模板提供样式，styles 接收字符串列表，而 styleURLs 引用的是文件名。

styles 参数采用包含 CSS 样式的字符串列表：

```
styles: ['h1 {font-weight:normal;}','p {text-indent:2em;}']
```

重新运行应用程序查看页面信息。完整代码如下：

```
//chapter14/angulardart/lib/app_component.dart
import 'package:angular/angular.dart';
//组件注解
@Component(
  selector: 'my-app',
  template: '''
    <h1>AngularDart 应用程序</h1>
    <p>根组件模板</p>
  ''',
  styles: ['h1 {font-weight:normal;}','p {text-indent:2em;}']
)
//组件类
class AppComponent {
}
```

14.1.5 样式和模板文件

在实际开发中所需的模板结构复杂，CSS 样式也更加多样。为了结构清晰，通常会将模板和 CSS 样式分别放在不同的文件中。

将 CSS 样式放在 app_component.css 文件中。

```
//chapter14/angulardart/lib/app_component.css
h1 { font-weight:normal;}
p { text-indent:2em;}
```

将模板放在 app_component.html 文件中。

```
//chapter14/angulardart/lib/app_component.html
<h1>Angular Web</h1>
<p>模板</p>
```

在注解中使用styleURLs参数引用样式表文件，该参数可以接收多个样式表文件。使用templateURL参数引用模板文件，该参数只能接收单个模板文件。最终代码如下：

```
//chapter14/angulardart/lib/app_component.dart
import 'package:angular/angular.dart';
@Component(
  selector: 'my-app',
  styleURLs: ['app_component.css'],
  templateURL: 'app_component.html'
)
class AppComponent{
}
```

14.2 数据绑定

Angular支持数据绑定，它采用一种用于协调用户视图和应用程序数据值的机制。只要添加绑定标记到模板，Angular就会明确如何连接模板和组件。

数据绑定有4种形式，每种形式都有数据流动的方向：组件到DOM、DOM到组件、双向流动，如图14-8所示。绑定语法和类型如表14-2所示。

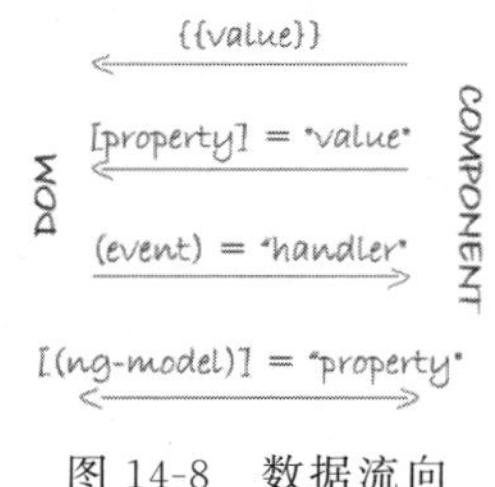

图14-8 数据流向

表14-2 数据绑定

数据方向	语法	类型
单向，从数据源到视图目标	{{expression}} [target]="expression" bind-target="expression"	插值 Property绑定 Attribute绑定 Class绑定 Style绑定
单向，从视图目标到数据源	(target)="statement" on-target="statement"	事件绑定
双向	[(target)]="expression"	双向数据绑定

数据绑定适用于DOM属性(property)和事件，而非属性(attribute)。本节仅介绍前3种数据绑定的使用，双向数据绑定将在后续合适的位置继续介绍。

创建新的Angular应用程序并命名为data_bind，lib目录下只留与根组件AppComponent相关文件即可，打开项目web目录下的index.html文件，选择合适的浏览器后打开即可运行本项目。

14.2.1 模板表达式和语句

在开始数据绑定前需要理解表中出现的两个重要单词：expression 和 statement。在 Angular 框架中它们分别被叫作模板表达式和模板语句。

模板表达式会产生一个值，Angular 执行表达式并将其值分配给绑定目标的属性，目标可能是 HTML 元素、组件或指令。

举个例子，在下述代码中的插值花括号({{}})包含的内容 1+1 就是模板表达式：

```
{{1 + 1}}
```

在属性绑定中，模板表达式会出现在等号右边的引号中：

```
[property] = "expression"
```

模板表达式与 Dart 语言中的表达式几乎相同，但也有部分内容是不被支持或有副作用的表达式：

(1) 赋值运算符(=、+=、-=、…)。

(2) new 或 const 关键字。

(3) 在表达式后使用分号(;)。

(4) 自增和自减符(++和--)。

(5) 位运算符|和 &。

(6) 字符串插值($ variableName 或 $ {expression})。

模板语句常见于事件绑定中，模板语句的作用是响应事件绑定目标引发的事件，通常用于修改组件属性或调用组件方法。下列代码将按钮的单击事件与语句 delete()绑定，该语句是组件中定义的方法：

```
<button (click) = "delete()">删除</button>
```

与模板表达式一样，模板语句与 Dart 语句类似。它支持赋值运算符(=)和多条语句，这意味着可以使用分号(;)。但也有一部分是被禁止的：

(1) new 或 const 关键字。

(2) 自增和自减符(++和--)。

(3) 分配运算符(+=和-=等)。

(4) 位运算符|和 &。

(5) 管道符(|)和安全导航运算符(?.)。

14.2.2 插值

插值是将组件属性的值传入模板的一种方式，它适用于对数据进行简要展示和处理的场景。语法如下：

```
{{expression}}
```

在模板中使用插值前需要在组件中定义属性。这里在组件 AppComponent 中定义标题(title)和名字(name)两个属性。示例代码如下：

```
//chapter14/data_bind/lib/app_component.dart
import 'package:angular/angular.dart';

@Component(
  selector: 'my-app',
  styleURLs: ['app_component.css'],
  templateURL: 'app_component.html',
)
class AppComponent {
  //用于插值
  String title = 'Angular Web';
  String name = 'Jobs';
}
```

Angular 使用双花括号({{}})作为插值符,插值使用模板表达式求值。将 title 属性通过插值放在 h1 元素之间,name 属性通过插值为输入框元素的 value 属性赋值。示例代码如下：

```
<h1>{{title}}</h1>
<input type="text" value="{{name}}">
```

插值符中的表达式也可以包含字符串,使用加号连接。在模板文件 app_component.html 中写入如下代码：

```
//chapter14/data_bind/lib/app_component.html
<h1>{{'标题' + title}}</h1>
<input type="text" value="{{name}}">
```

修改代码后会自动编译,刷新浏览器即可,运行效果如图 14-9 所示。

标题Angular Web

Jobs

图 14-9　插值

14.2.3 属性(property)绑定

编写属性绑定以设置视图元素的属性，绑定将属性设置为模板表达式的值。属性绑定使用中括号([])作为标记，语法如下：

```
<element [property] = "expression"></element>
```

最常见的属性绑定是将元素属性设置为组件属性值，在组件中定义 imgPath 和 isAble 属性。示例代码如下：

```
//chapter14/data_bind/lib/app_component.dart
import 'package:angular/angular.dart';

@Component(
  selector: 'my-app',
  styleURLs: ['app_component.css'],
  templateURL: 'app_component.html',
)
class AppComponent {
  //用于属性绑定
  String imgPath = '/favicon.png';
  bool isAble = true;
}
```

在模板文件中将组件的 imgPath 属性与 img 元素的 src 属性绑定，isAble 属性与按钮元素的 disabled 属性绑定。示例代码如下：

```
//chapter14/data_bind/lib/app_component.html
<img [src] = "imgPath">
<button [disabled] = "isAble">按钮处于禁用状态</button>
```

也可以使用 bind-前缀替代中括号：

```
<button bind-disabled = "!isAble">按钮处于可用状态</button>
```

运行效果如图 14-10 所示。

图 14-10 属性绑定

插值和属性绑定在很多情况下可以相互替代，将元素属性设置为非字符串值时，必须使用属性绑定。

14.2.4 属性(attribute)绑定

属性(attribute)绑定与属性(property)绑定语法类似,区别在于它使用 attr 前缀,后跟一个点(.)和该属性(attribute)的名字。然后,就可以使用具体值或模板表达式设置属性(attribute)值:[attr. attr-name]。

在模板文件中建立表格,并设置 colspan 属性。代码如下:

```
//chapter14/data_bind/lib/app_component.html
<table border = "1" style = "border - collapse:collapse;">
    <tr><td [attr.colspan] = "2">表格单元跨越两列</td><td>1,3</td><td>1,4</td></tr>
    <tr><td>2,1</td><td>2,2</td><td>2,3</td><td>2,4</td></tr>
</table>
```

运行结果如图 14-11 所示。

表格单元跨越两列		1,3	1,4
2,1	2,2	2,3	2,4

图 14-11 属性(attribute)绑定

属性(attribute)绑定主要使用场景是设置 aria 属性,对于无障碍阅读非常适用。例如:

```
<input type = "text" aria - label = "用户名"/>
```

14.2.5 类绑定

此处的类指元素属性 class,可以使用类绑定向元素的 class 属性添加或移除 CSS 类名。类绑定与属性绑定类似,区别在于类绑定使用 class 前缀,后跟点(.)和 class 名:[class. class-name]。

以下示例通过几种方式向元素添加和删除 uppercase 类。

直接将 CSS 名添加到属性 class 的字符串中:

```
<p class = "bold italic uppercase"> this is a test </p>
```

使用属性绑定[class],该绑定会使用组件属性 textStyle 替换元素 class 属性已指定的所有类名,组件属性 textStyle 是字符串形式,它可以包含多个类名。

```
<p class = "bold italic uppercase" [class] = "textStyle"> this is a test </p>
```

也可以通过类绑定[class. class-name]的形式绑定单个类名,表达式的值为 true 时添加该类名，值为 false 时移除该类名。

```
<!-- 使用组件属性添加或移除 class 名 uppercase -->
<p [class.uppercase] = "isUppercase"> this is a test </p>
<p [class.uppercase] = "!isUppercase"> this is a test </p>
```

在组件 AppComponent 中添加 textStyle 和 isUppercase 属性。代码如下：

```
//chapter14/data_bind/lib/app_component.dart
import 'package:angular/angular.dart';

@Component(
  selector: 'my-app',
  styleURLs: ['app_component.css'],
  templateURL: 'app_component.html',
)
class AppComponent {
  //用于 CSS 类绑定
  String textStyle = "bold italic";
  bool isUppercase = true;
}
```

在样式文件 app_component.css 中添加如下样式。代码如下：

```
//chapter14/data_bind/lib/app_component.css
.bold{
  /* 加粗字体 */
  font-weight:bold;
}
.italic{
  /* 使字体倾斜 */
  font-style:italic;
}
.uppercase{
  /* 对单词转大写 */
  text-transform: uppercase;
}
```

在模板文件 app_component.html 中添加以下内容。代码如下：

```
//chapter14/data_bind/lib/app_component.html
<p class = "bold italic uppercase"> this is a test </p>
<p class = "bold italic uppercase" [class] = "textStyle"> this is a test </p>
<p [class.uppercase] = "isUppercase"> this is a test </p>
<p [class.uppercase] = "!isUppercase"> this is a test </p>
```

运行结果如图 14-12 所示。

14.2.6 样式绑定

THIS IS A TEST

this is a test

THIS IS A TEST

this is a test

图 14-12　类绑定

可以使用样式(style)绑定设置内联样式，样式绑定语法类似属性绑定。以 style 为前缀，后跟一个点(.)和 CSS 样式属性的名字：[style. style-property]。

样式绑定在设置时通常会和条件运算符(expr1? expr2：expr3)配合使用。示例代码如下：

```
<p [style.color] = "isSpecial ? 'red': 'green'"> this is a test </p>
<p [style.background - color] = "canSave ? 'cyan': 'grey'" > this is a test </p>
```

某些样式具有单位，以下示例有条件地以 em 和%为单位设置字体大小。

```
<p [style.font - size.em] = "isSpecial ? 3 : 1" > this is a test </p>
<p [style.font - size. % ] = "!isSpecial ? 150 : 50" > this is a test </p>
```

在组件 AppComponent 中添加 isSpecial 和 canSave 属性。示例代码如下：

```
//chapter14/data_bind/lib/app_component.dart
import 'package:angular/angular.dart';

@Component(
  selector: 'my - app',
  styleURLs: ['app_component.css'],
  templateURL: 'app_component.html',
)
class AppComponent {
  //用于样式绑定
  bool isSpecial = false;
  bool canSave = true;
}
```

在模板文件 app_component. html 中添加如下内容：

```
//chapter14/data_bind/lib/app_component.html
<p [style.color] = "isSpecial ? 'red': 'green'"> this is a test </p>
<p [style.background - color] = "canSave ? 'cyan': 'grey'" > this is a test </p>
<p [style.font - size.em] = "isSpecial ? 3 : 1" > this is a test </p>
<p [style.font - size. % ] = "!isSpecial ? 150 : 50" > this is a test </p>
```

运行结果如图 14-13 所示。

样式属性名可以使用半字线(font-size)或小驼峰(fontSize)编写，它们都被 Angular 支持。但是只有半字线的形式能被 dart:html 库中的某些方法访问，因此首选半字线作为样式属性名。

this is a test

this is a test

this is a test

this is a test

图 14-13 样式绑定

14.2.7 事件绑定

为了响应用户的操作，时常需要监听某些事件，例如：按键、鼠标移动、单击和触摸等。为了让这些操作更加便利，可以使用 Angular 提供的事件绑定，在事件绑定中数据从元素传递到组件。

事件绑定语法构成：

```
<element (target) = "statement"></element>
```

括号里的名字 target 表示目标事件，以下示例中目标是按钮的单击事件。等号右边的双引号里包裹的是模板语句，以下示例中语句是组件的 onSave()方法。

```
<button (click) = "onSave()"> Save </button>
```

也可以以 on-前缀代替括号，这是规范形式：

```
<button on-click = "onSave()"> Save </button>
```

在事件绑定中，Angular 为目标事件设置事件处理程序。当事件触发时，处理程序执行模板语句。模板语句通常包含一个接收器，接收器响应该事件并执行操作。例如将 HTML 控件的值存储到组件属性。

绑定通过事件对象 $event 传递关于事件的所有信息，包括数据值。事件对象的结构由目标事件决定，如果目标事件是原生 DOM 元素事件，则 $even 是 DOM 事件对象，具有诸如 target 和 target.value 之类的属性。

在组件 AppComponent 中添加属性 value 用于存储值，添加 onSetValue()方法用于接收一个值，然后将值赋给组件属性 value。示例代码如下：

```
//chapter14/data_bind/lib/app_component.dart
import 'package:angular/angular.dart';

@Component(
```

```
    selector: 'my-app',
    styleURLs: ['app_component.css'],
    templateURL: 'app_component.html',
)
class AppComponent {
    //用于事件绑定
    String value;
    void onSetValue(var val){
      value = val;
    }
}
```

在模板文件 app_component.html 中添加的代码如下：

```
//chapter14/data_bind/lib/app_component.html
<input (input) = "onSetValue($event.target.value)">
<p>value:{{name}}</p>
```

此代码将 value 属性通过插值放在 p 元素中，用于显示 value 属性的值。然后将 input 事件绑定到输入框，当用户进行更改时，将触发 input 事件，并且绑定将执行模板语句 onSetValue($event.target.value)，其中 $event.target.value 用于获取更改后的值。

hello

value: hello

图 14-14 事件绑定

其执行结果是每次输入事件都会使 value 属性值被更新并在视图中显示，例如在输入框中输入 hello，运行效果如图 14-14 所示。

14.3 内置指令

指令是带有@Directive 注解的类，而组件是带有模板的指令。@Component 注解实际上是在@Directive 注解的基础上扩展了模板功能。从技术上来说组件就是指令，由于组件是 Angular 应用程序独特的组成部分，并且是 Angular 的核心，所以将组件单独拿出来做以区分。

Angular 模板是动态的，当 Angular 渲染它们时，将根据指令给出的指令转换 DOM。Angular 中指令分为属性指令和结构指令，当然也可以创建自定义指令。

14.3.1 属性指令

属性指令监听并修改其他 HTML 元素、属性(attribute)、属性(property)和组件的行为。它们通常应用于元素，就像它们就是 HTML 属性一样。

许多 Angular 包都有自己的属性指令，例如 Router 和 Forms 包。这里介绍常用的属性指令：

（1）NgClass：添加或移除一组 CSS 类。

（2）NgStyle：添加或移除一组 HTML 样式。

（3）NgModel：双向数据绑定到一个 HTML 表单元素，在小节表单中会详细说明。

使用指令需要在组件注解的参数 directives 中列出需要的指令，更新注解@Component。示例代码如下：

```
@Component(
  selector: 'my-app',
  styleURLs: ['app_component.css'],
  templateURL: 'app_component.html',
  directives: [NgClass,NgStyle],
)
```

1. NgClass

通过动态添加或删除 CSS 类可以控制元素的外观，而绑定到 NgClass 可以同时添加或删除多个 CSS 类。

在使用时，需要将 NgClass 绑定到 Map 类型的控件上。该 Map 对象的每个键都是一个 CSS 类名，如果应该添加该 CSS 类，则对应的值应设置为 true。如果应该移除该 CSS 类，则对应的值应设置为 false。为了动态控制是否添加或删除 CSS 类，值通常是一个结果为布尔值的表达式。

在组件 AppComponent 中添加 4 个布尔属性 isUppercase、isSpecial、isBold 和 isItalic，添加 Map 类型的控件 currentClasses，创建一个方法 setCurrentClasses()，该方法设置组件属性 currentClasses。从 Map 字面量中可以看出，键是 CSS 类名，值中构成表达式的属性是布尔值。示例代码如下：

```
//chapter14/built_in_directives/lib/app_component.dart
import 'package:angular/angular.dart';

@Component(
  selector: 'my-app',
  styleURLs: ['app_component.css'],
  templateURL: 'app_component.html',
  directives: [NgClass,NgStyle]
)
class AppComponent {
  //用于属性指令 ngClass
  bool isUppercase = true;
  bool isSpecial = false;
  bool isBold = true;
  bool isItalic = false;
```

```
  Map< String, bool > currentClasses = < String, bool >{};
  void setCurrentClasses(){
    currentClasses = < String, bool >{
      'bold': isBold,
      'italic': !isItalic,
      'uppercase': isUppercase
    };
  }
}
```

在样式文件中添加 CSS 类。代码如下：

```
//chapter14/built_in_directives/lib/app_component.css
.bold{
  /* 加粗字体 */
  font - weight:bold;
}
.italic{
  /* 使字体倾斜 */
  font - style:italic;
}
.uppercase{
  /* 对单词转大写 */
  text - transform: uppercase;
}
```

将 NgClass 指令绑定到 currentClasses 属性会相应地设置元素的 CSS 类。添加单击事件绑定，模板语句是 setCurrentClasses()。单击按钮后方法被执行，currentClasses 完成初始化。代码如下：

```
//chapter14/built_in_directives/lib/app_component.html
<div [ngClass] = "currentClasses"> this is a test </div>
<button (click) = "setCurrentClasses()">初始化组件属性 currentClasses </button>
```

可以在初始化组件或属性改变时调用 setCurrentClasses 方法，这里是通过按钮来初始化。单击按钮运行结果如图 14-15 所示。

2. NgStyle

可以根据组件的状态动态设置元素的内联样式，使用 NgStyle 指令可以同时设置多个内联样式。

THIS IS A TEST
初始化组件属性currentClasses

图 14-15 指令 NgClass

与 NgClass 类似，将 NgStyle 绑定到 Map 类型的控件，该 Map 对象的每个键都是一个 CSS 样式名，其对应的值是该样式的值。

在组件 AppComponent 中添加一个 setCurrentStyles()方法，该方法设置组件属性

currentStyles。在 Map 字面量的值部分使用了条件表达式，条件表达式根据布尔值返回一个字符串，该字符串是符合对应样式的值。示例代码如下：

```
//chapter14/built_in_directives/lib/app_component.dart
import 'package:angular/angular.dart';

@Component(
  selector: 'my - app',
  styleURLs: ['app_component.css'],
  templateURL: 'app_component.html',
  directives: [NgClass, NgStyle]
)
class AppComponent {
  //用于属性指令 ngStyle
  Map < String, String > currentStyles = < String, String >{};
  void setCurrentStyles() {
    currentStyles = < String, String >{
      'font - style': isItalic ? 'italic' : 'normal',
      'font - weight': isBold ? 'bold' : 'normal',
      'font - size': isSpecial ? '24px' : '36px'
    };
  }
}
```

将 NgStyle 属性绑定到 currentStyles 可以相应地设置元素的样式。添加单击事件绑定，模板语句是 setCurrentStyles()。单击按钮后方法被执行，currentStyles 完成初始化。代码如下：

```
//chapter14/built_in_directives/lib/app_component.html
<div [ngStyle] = "currentStyles"> this is a test </div>
<button (click) = "setCurrentStyles()">初始化组件属性 currentStyles </button>
```

可以在初始化组件或属性改变时调用 setCurrentStyles 方法，这里是通过按钮来初始化。单击按钮运行结果如图 14-16 所示。

14.3.2 结构指令

this is a test

初始化组件属性currentStyles

图 14-16 指令 NgStyle

结构指令用于 HTML 动态布局，通过添加、删除和控制它们所依附的宿主元素来构建或重构 DOM 结构。

常见的结构指令如下：

(1) NgIf：有条件地从 DOM 中添加或删除元素。

(2) NgFor：为列表中的每个项目重复一个模板。

(3) NgSwitch：仅显示多个可能的元素之一。

在组件注解的参数 directives 中列出需要的结构指令，更新注解@Component。代码如下：

```
@Component(
  selector: 'my-app',
  styleURLs: ['app_component.css'],
  templateURL: 'app_component.html',
  directives: [NgClass, NgStyle, NgIf, NgFor, NgSwitch, NgSwitchWhen, NgSwitchDefault],
)
```

1. NgIf

指令 NgIf 与布尔表达式绑定，根据表达式的值从 DOM 中添加或删除元素，使用 NgIf 指令的元素被称为宿主元素。

在组件 AppComponent 中添加属性 isNull 和切换其布尔值的 toggle 方法。示例代码如下：

```
//chapter14/built_in_directives/lib/app_component.dart
import 'package:angular/angular.dart';

@Component(
  selector: 'my-app',
  styleURLs: ['app_component.css'],
  templateURL: 'app_component.html',
  directives:[NgClass, NgStyle, NgIf, NgFor, NgSwitch, NgSwitchWhen, NgSwitchDefault]
)
class AppComponent {
  //用于结构指令 *ngIf
  bool isNull = true;
  //切换 isNull 的布尔值
  toggle(){
    isNull = isNull ? false : true;
  }
}
```

在模板文件 app_component.html 中添加的代码如下：

```
//chapter14/built_in_directives/lib/app_component.html
<button (click) = "toggle()">切换 isNull 的真假值</button>
<div *ngIf = "isNull"> isNull:{{isNull}}</div>
<div *ngIf = "!isNull">!isNull:{{isNull}}</div>
```

当 isNull 表达式返回 true 时，NgIf 将 div 元素添加到 DOM。如果表达式为 false，则 NgIf 从 DOM 中删除 div 元素。

需要说明的是 NgIf 指令工作模式与显示或隐藏不是一回事，显示或隐藏通常会使用类或样式绑定来控制元素的可见性。在模板文件中添加的代码如下：

```
//chapter14/built_in_directives/lib/app_component.html
<!-- isNull 初始值为 true -->
<div [class.hidden] = "!isNull">使用类绑定显示</div>
<div [class.hidden] = "isNull">使用类绑定隐藏</div>
<div [style.display] = "isNull? 'block' : 'none'">使用样式显示</div>
<div [style.display] = "isNull? 'none' : 'block'">使用样式隐藏</div>
```

样式文件中添加如下 CSS 类，代码如下：

```
//chapter14/built_in_directives/lib/app_component.css
.hidden{
  /* 隐藏元素 */
  display:none;
}
```

使用类或样式绑定隐藏元素时，该元素及其所有后代仍保留在 DOM 中。这些元素的所有组件都保留在内存中，Angular 可能会继续检查更改。应用可能会占用大量计算资源，从而降低性能。

使用 NgIf 指令时，如果表达式为 false，则 Angular 将从 DOM 中删除该元素及其后代。它销毁了它们的组件，有可能释放大量资源，从而带来了响应速度更快的用户体验。

显示或隐藏适用于子元素少的元素，而 NgIf 适用于大型组件树。

NgIf 指令常用于防止 null 值，如果表达式尝试访问 null 值，Angular 将抛出错误。因此可以通过判断属性是否为 null 来避免错误。

```
<div *ngIf = "name != null"> Hello, {{name}}</div>
```

2. NgFor

NgFor 是一种重复指令，一种以相同元素结构显示当前列表中每个条目的方法。通常定义一个 HTML 元素块，Angular 使用该 HTML 元素块作为模板来呈现列表中的每个条目。在元素块中也可以包含子组件声明的选择器，在后续会介绍。

首先在 app_component.dart 文件中添加 JieQi 类，在组件中定义 jieqiList 列表，并使用字面量为其初始化值。示例代码如下：

```
//chapter14/built_in_directives/lib/app_component.dart
import 'package:angular/angular.dart';

@Component(
  selector: 'my-app',
```

```
    styleURLs: ['app_component.css'],
    templateURL: 'app_component.html',
    directives: [NgClass,NgStyle,NgIf,NgFor,NgSwitch,NgSwitchWhen,NgSwitchDefault]
)
class AppComponent {
    //用于结构指令 * ngFor
    //使用字面量初始化泛型为 JieQi 的列表
    List < JieQi > jieqiList = [
      JieQi(1,'雨水'),
      JieQi(2,'谷雨'),
      JieQi(3,'白露'),
      JieQi(4,'寒露'),
      JieQi(5,'霜降'),
      JieQi(6,'小雪'),
      JieQi(7,'大雪')];
}
//JieQi 类
class JieQi{
  int id;
  String label;
  JieQi(this.id,this.label);
}
```

将 NgFor 应用于< div >的示例代码如下：

```
//chapter14/built_in_directives/lib/app_component.html
< div * ngFor = "let jieqi of jieqiList">
    {{jieqi.label}}
</div >
```

分配给 * ngFor 的字符串不是模板表达式，这是一种 Angular 可以解析的微语法。字符串 let jieqi of jieqiList 的意思是：将 jieqiList 列表中的单个条目存储在本地循环变量 jieqi 中，并可以在每次迭代的 HTML 模板中使用循环变量 jieqi。

1）模板输入变量

jieqi 之前的 let 关键字创建了一个名为 jieqi 的模板输入变量。NgFor 指令会迭代组件的 jieqiList 属性返回的 jieqiList 列表，并在每次迭代期间将 jieqi 设置为列表中的当前项目。

要访问 jieqi 的属性，需在 NgFor 宿主元素（或其子元素）中引用 jieqi 输入变量。这里在插值中引用 jieqi。

```
< div * ngFor = "let jieqi of jieqiList">{{jieqi.label}}</div >
```

2）索引

NgFor 指令上下文的 index 属性在每次迭代中返回该条目的从 0 开始的索引。可以在模板输入变量中捕获索引并在模板中使用它。

示例在名为i的变量中捕获索引，并显示该索引和jieqi名。示例代码如下：

```
//chapter14/built_in_directives/lib/app_component.html
<div *ngFor="let jieqi of jieqiList;let i=index;trackBy:trackById">
    ({{i}}) - {{jieqi.label}}
</div>
```

3）追踪TrackBy

NgFor指令的性能可能会很差，尤其是对于大型列表而言。对一项进行很小的更改，删除一项或添加一项，都会触发一系列DOM操作。

例如，重新查询服务器可能会重置列表jieqiList中的所有JieQi对象，且其中大多数是重复的，因为每个JieQi对象的id是固定的。此时Angular只能检测到新对象引用的新列表，因此只能拆除旧的DOM元素并插入所有新的DOM元素。

Angular可以通过使用TrackBy避免这种混乱，使用时为其设置一个函数，该函数必须符合追踪函数的类型定义。示例代码如下：

```
Object TrackByFn(int index,dynamic item)
```

它是一个为索引处的条目返回唯一键的函数。默认情况下条目本身被用作键实例化新的模板，如果数据发生变化，Angular将会视为不同的对象来重新渲染数据。如果可以确定迭代对象中条目的唯一性，则可以提供遵循TrackByFn类型定义的函数优化性能。

向该组件添加一个方法，该方法返回NgFor应该跟踪的值id。示例代码如下：

```
//chapter14/built_in_directives/lib/app_component.dart
Object trackById(int index, dynamic o) {
  return o is JieQi ? o.id : o;
}
```

在微语法表达式中，将TrackBy设置为此方法。示例代码如下：

```
//chapter14/built_in_directives/lib/app_component.html
<div *ngFor="let jieqi of jieqiList;let i=index">
    {{jieqi.id}},{{jieqi.label}}
</div>
```

3. NgSwitch

NgSwitch就像Dart中的switch语句。它可以根据switch条件显示几个可能元素中的一个元素，Angular仅将所选元素放入DOM。

NgSwitch实际上需要3个指令协作：NgSwitch、NgSwitchCase和NgSwitchDefault。

（1）NgSwitch是控制器，将其绑定到返回switch值的表达式。NgSwitch是属性指令，而不是结构指令。它的作用是更改其伴随指令的行为，而不会直接操作 DOM。

(2) NgSwitchCase 和 NgSwitchDefault 是结构指令，它们在 DOM 中添加或删除了元素。当 NgSwitchCase 的绑定值等于 switch 值时，会将其元素添加到 DOM。如果所有 NgSwitchCase 中都没有匹配值，则 NgSwitchDefault 将其元素添加到 DOM 中，NgSwitchDefault 可以省略。

在组件中添加属性 color，该属性的作用是提供 switch 值。示例代码如下：

```
//chapter14/built_in_directives/lib/app_component.dart
String color = 'Green';
```

在模板中添加代码如下：

```
//chapter14/built_in_directives/lib/app_component.html
<div [ngSwitch] = "color">
    <span * ngSwitchCase = "'Red'">红色</span>
    <span * ngSwitchCase = "'Green'">绿色</span>
    <span * ngSwitchCase = "'Black'">黑色</span>
    <span * ngSwitchDefault >蓝色</span>
</div>
```

在这里 color 值是一个字符串，但 switch 值可以是任何类型。

14.4 模板引用变量

8min

模板引用变量通常是对模板中一个 DOM 元素的引用，它还可以引用 Angular 组件或指令等。

在组件中添加属性 number，并添加方法 callPhone()，方法 callPhone()的作用是将传入的值赋给 number。示例代码如下：

```
//chapter14/template_reference_variable/lib/app_component.dart
import 'package:angular/angular.dart';
@Component(
  selector: 'my-app',
  styleURLs: ['app_component.css'],
  templateURL: 'app_component.html',
)
class AppComponent {
  String number;
  callPhone(var tel) {
    number = tel;
  }
}
```

在元素上使用＃加变量名声明一个引用变量，而后在模板中的任何位置都可以使用模板引用变量。此示例中在元素<input>中声明模板引用变量phone，在元素<button>单击事件的模板语句callPhone(phone.value)中引用。示例代码如下：

```
//chapter14/template_reference_variable/lib/app_component.html
<!-- 声明引用变量 phone -->
<input #phone placeholder="输入电话号码" type="tel">
<!-- 使用引用变量 phone -->
<button (click)="callPhone(phone.value)">拨打</button>
<p>{{number}}</p>
```

14.4.1 赋值

大多数情况下，Angular会将模板引用变量的值设置为声明该变量的元素。在上一个示例中，模板引用变量phone指的是输入框。

单击按钮处理程序会将输入值传递给组件的callPhone方法，但是有一些指令可以更改该行为并将其值设置为其他值。例如：NgForm指令可以将其自身赋值给模板引用变量。

指令NgForm是由包angular_forms提供的，这里跟着步骤做就可以了，包angular_forms的更多内容将在后续做详细说明。首先添加依赖项并执行pub get命令：

```
dependencies:
  angular: ^5.3.0
  angular_components: ^0.13.0
  angular_forms: ^2.1.2
```

在组件的指令列表中添加包含angular_forms包所有指令的常量值formDirectives。属性name用于与输入控件进行双向数据绑定，属性submitMessage用于存储表单的值，方法onSubmit接收一个NgForm参数，并设置属性submitMessage的值。示例代码如下：

```
//chapter14/template_reference_variable/lib/app_component.dart
import 'package:angular/angular.dart';
import 'package:angular_forms/angular_forms.dart';

@Component(
  selector: 'my-app',
  styleURLs: ['app_component.css'],
  templateURL: 'app_component.html',
  directives: [formDirectives],
)
class AppComponent {
  String name;
```

```
    String submitMessage = '';
    void onSubmit(NgForm form) {
      submitMessage = '提交表单的值是 ${form.value}';
    }
}
```

在模板中建立表单，在元素 form 上定义模板引用变量 newForm，并将 NgForm 指令的导出对象 ngForm 赋值给 newForm。示例代码如下：

```
//chapter14/template_reference_variable/lib/app_component.html
<form (ngSubmit) = "onSubmit(newForm)" #newForm = "ngForm">
    <input ngControl = "name"
            required
            [(ngModel)] = "name">
    <button type = "submit"
            [disabled] = "!newForm.form.valid">提交</button>
</form>
<div [hidden] = "!newForm.form.valid">
    {{submitMessage}}
</div>
```

在此示例中，模板引用变量 newForm 出现 3 次，并被大量 HTML 分隔。newForm 的值是对 Angular 表单指令 NgForm 的引用，该指令能够跟踪表单中每个控件的值和有效性。原生< form >元素没有 form 属性，但是 NgForm 指令有。如果 newForm.form.valid 无效，将禁用提交按钮，并在表单有效时将整个表单控制树传递给组件的 onSubmit 方法。

14.4.2 说明

模板引用变量(#phone)与在 *ngFor 中看到的模板输入变量(let phone)不同，模板引用变量的范围是整个模板，除非在结构指令控制的嵌入式视图中声明了引用变量。在嵌入式视图中声明的模板引用变量仅对被结构指令嵌入的模板部分可见。

注意：在外部声明的模板引用变量可以在嵌入式视图中引用，反之则不能。请勿在同一模板中多次定义相同的变量名，运行时它们的值将不可预测。

14.5 服务

24min

服务最常用的场景是提供数据，数据通常来自服务器，因此数据服务总是异步的。只要组件需要，就可以通过依赖注入将数据提供给组件。为了方便演示，本节提供的数据将由本地提供，而不是通过服务器获取。

14.5.1 定义实体类

在目录 lib/src 下创建 Employee 类，包含 id、name 和 salary 属性。示例代码如下：

```
//chapter14/service/lib/src/employee.dart
class Employee{
  final int id;
  String name;
  num salary;
  Employee(this.id,this.name,this.salary);
}
```

14.5.2 创建服务

在目录 lib/src 下创建 employee_service.dart 文件，导入 employee.dart 文件并定义 EmployeeService 类。在类上使用@Injectable()注解，表示可被注入。示例代码如下：

```
//chapter14/service/lib/src/employee_service.dart
import 'package:angular/angular.dart';
import 'employee.dart';

@Injectable()
class EmployeeService{
}
```

在类中定义 getAll()方法，并用字面量定义一个 Employee 列表。因为在实际使用中服务通常是异步的，所以可以将返回值使用 Future 封装，并在 getAll()方法上使用 async 标记。示例代码如下：

```
//chapter14/service/lib/src/employee_service.dart
import 'package:angular/angular.dart';
import 'employee.dart';

@Injectable()
class EmployeeService{
  Future<List<Employee>> getAll() async{
    var emps = [
      Employee(1, 'Mr. Nice',3000),
      Employee(2, 'Narco',3105),
      Employee(3, 'Bombasto',3988),
      Employee(4, 'Celeritas',3401),
      Employee(5, 'Magneta',9971),
      Employee(6, 'RubberMan',4533),
```

```
        Employee(7, 'Dynama',6720),
        Employee(8, 'Dr IQ',4907),
        Employee(9, 'Magma',5278),
        Employee(10, 'Tornado',7800)
      ];
      return emps;
    }
}
```

14.5.3 使用服务

在 AppComponent 组件中通过文件导入 EmployeeService 和 Employee 类。

```
import 'src/employee.dart';
import 'src/employee_service.dart';
```

在组件注解@Component 的 providers 参数中标识服务提供者，该参数接收一个列表。通常使用 ClassProvider 方法来加载服务。示例代码如下：

```
@Component(
  selector: 'my-app',
  styleURLs: ['app_component.css'],
  templateURL: 'app_component.html',
  directives: [NgFor],
  providers: [ClassProvider(EmployeeService)],
)
```

添加列表属性 emps、EmployeeService 类型的私有属性_employeeService，并在组件的构造函数中初始化该属性。示例代码如下：

```
final EmployeeService _employeeService;
List<Employee> emps;
AppComponent(this._employeeService);
```

添加 getEmployees()方法，方法用于将服务中获取的数据传递给 emps 属性。由于数据获取是异步的，因此方法需要用 async 标记，并且在使用服务获取数据的语句前面使用 await 关键字。示例代码如下：

```
void getEmployees() async{
  emps = await _employeeService.getAll();
}
```

让组件实现生命周期函数 ngOnInit()并调用 getEmployees()方法，该函数用于初始化组件数据。示例代码如下：

```
class AppComponent implements OnInit{
  @override
  void ngOnInit(){
   getEmployees();
  }
}
```

此时组件 AppComponent 的完整示例代码如下：

```
//chapter14/service/lib/app_component.dart
import 'package:angular/angular.dart';
import 'src/employee.dart';
import 'src/employee_service.dart';
@Component(
  selector: 'my-app',
  styleURLs: ['app_component.css'],
  templateURL: 'app_component.html',
  directives: [NgFor],
  providers: [ClassProvider(EmployeeService)],
)
class AppComponent implements OnInit{
  //定义一个 EmployeeService 类型的私有变量
  final EmployeeService _employeeService;
  List<Employee> emps;
  //注入服务
  AppComponent(this._employeeService);
  //覆写 ngOnInit 方法
  @override
  void ngOnInit() {
    //在组件初始化时调用 getEmployees 方法
    getEmployees();
  }
  //使用服务对象_employeeService 获取雇员列表
  void getEmployees() async{
    emps = await _employeeService.getAll();
  }
}
```

在模板文件中添加元素块，用于迭代 emps 属性并显示数据。模板代码如下：

```
//chapter14/service/lib/app_component.html
<ul class="emps">
    <li *ngFor="let emp of emps">
```

```
        < span class = "badge">{{emp.id}}</span >
        {{emp.name}}
        < span class = "badge salary">{{emp.salary}}</span >
    </li >
</ul >
```

定义元素块样式，在 app_component.css 文件中添加的样式代码如下：

```
//chapter14/service/lib/app_component.css
.emps {
  margin: 0 0 2em 0;
  list - style - type: none;
  padding: 0;
  width: 15em;
}
.emps li {
  cursor: pointer;
  position: relative;
  left: 0;
  background - color: #EEE;
  margin: .5em;
  padding: .3em 0;
  height: 1.6em;
  border - radius: 4px;
}
.emps li:hover {
  color: #607D8B;
  background - color: #EEE;
  left: .1em;
}
.emps .text {
  position: relative;
  top: - 3px;
}
.emps .badge {
  display: inline - block;
  font - size: small;
  color: white;
  padding: 0.8em 0.7em 0 0.7em;
  background - color: #607D8B;
  line - height: 1em;
  position: relative;
  left: - 1px;
  top: - 4px;
  height: 1.8em;
```

```
  margin-right: .8em;
  border-radius: 4px 0 0 4px;
}
.emps .salary {
  float:right;
  margin-right:0;
  margin-left: .8em;
  border-radius: 0 4px 4px 0;
}
```

运行结果如图 14-17 所示。

1 Mr. Nice 3000
2 Narco 3105
3 Bombasto 3988
4 Celeritas 3401
5 Magneta 9971
6 RubberMan 4533
7 Dynama 6720
8 Dr IQ 4907
9 Magma 5278
10 Tornado 7800

图 14-17　雇员列表

55min

14.6　子组件

前面所有的内容都是在模板生成的根组件 AppComponent 中完成的，在本节将介绍如何创建组件，并将其作为根组件 AppComponent 的子组件。

14.6.1　创建组件

通常会将组件及其模板和样式文件放在以其名字命名的文件夹中，在 src 目录下建立 employee 目录。根据约定与组件相关的 HTML、CSS、dart 文件需要以_component 作为文件名后缀。在 employee 目录下建立 employee_component.dart 文件，指定其选择器为 employee。示例代码如下：

```
//chapter14/child_component/lib/src/employee/employee_component.dart
import 'package:angular/angular.dart';
@Component(
  selector:'employee',
)
class EmployeeComponent{
}
```

这里就不单独建立模板文件了，使用内联模板，向@Component 注解提供 template 参数和值。示例代码如下：

```
//chapter14/child_component/lib/src/employee/employee_component.dart
import 'package:angular/angular.dart';
@Component(
  selector:'employee',
  template: '''
      <div>雇员 ID:5<div>
      <div>雇员名:Magneta<div>
      <div>雇员薪资:9971<div>
      ''',
)
class EmployeeComponent{
}
```

这就是一个组件的基本组成部分，在实际使用中可能会复杂一些。

14.6.2 添加到父组件

首先需要在 AppComponent 组件中引入子组件。

```
import 'src/employee/employee_component.dart';
```

因为组件实际上是带有模板的指令，所以应当将子组件 EmployeeComponent 添加到指令列表 directives 中。一个组件可以包含多个子组件。更新 AppComponent 组件的@Component 注解。示例代码如下：

```
//chapter14/child_component/lib/app_component.dart
import 'package:angular/angular.dart';
import 'src/employee/employee_component.dart';

@Component(
  selector: 'my-app',
  styleURLs: ['app_component.css'],
  templateURL: 'app_component.html',
```

```
  directives: [coreDirectives,EmployeeComponent],
)
class AppComponent {
}
```

前面在使用内置指令时，都是将每个指令单独列出，实际上可以使用 coreDirectives 代替所有的内置指令，例如：NgClass、NgIf、NgFor 等。

然后在模板文件中添加子组件定义的选择器 employee。示例代码如下：

```
//chapter14/child_component/lib/app_component.html
<employee></employee>
```

运行结果如图 14-18 所示。

14.6.3 输入输出属性

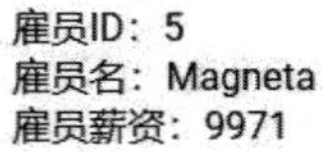

雇员ID: 5
雇员名: Magneta
雇员薪资: 9971

图 14-18 子组件视图

在子组件中模板内容是自定义的，与父组件没有产生联系。在实际开发中子组件的某些数据可能是由父组件提供的，这就需要提供输入属性做支持。

1. 输入属性

前面的内容主要集中于绑定声明右侧模板表达式和语句中的组件成员，该位置的成员是数据绑定源。本节重点介绍绑定的目标，这些目标是绑定声明左侧的指令属性。这些指令属性必须声明为输入或输出。

创建 Employee 类，代码如下：

```
//chapter14/child_component/lib/src/employee.dart
class Employee{
  final int id;
  String name;
  num salary;
  Employee(this.id,this.name,this.salary);
}
```

创建服务 EmployeeService，代码如下：

```
//chapter14/child_component/lib/src/employee_service.dart
import 'package:angular/angular.dart';
import 'employee.dart';

@Injectable()
class EmployeeService{
  Future<List<Employee>> getAll() async{
```

```
    var emps = [
      Employee(1, 'Mr. Nice',3000),
      Employee(2, 'Narco',3105),
      Employee(3, 'Bombasto',3988),
      Employee(4, 'Celeritas',3401),
      Employee(5, 'Magneta',9971),
      Employee(6, 'RubberMan',4533),
      Employee(7, 'Dynama',6720),
      Employee(8, 'Dr IQ',4907),
      Employee(9, 'Magma',5278),
      Employee(10, 'Tornado',7800)
    ];
    return emps;
  }
}
```

在子组件 EmployeeComponent 中导入 employee.dart 文件。代码如下：

```
import '../employee.dart';
```

在子组件 EmployeeComponent 中定义一个 Employee 类型的属性 emp，并使用@Input 注解将其标记为输入属性。代码如下：

```
@Input()
Employee emp;
```

添加核心指令常量 coreDirectives 到指令列表。使用 NgIf 指令判断属性 emp 是否为空，如果不为空，则在模板中使用插值访问属性 emp 的数据。示例代码如下：

```
//chapter14/child_component/lib/src/employee/employee_component.dart
import 'package:angular/angular.dart';
import '../employee.dart';

@Component(
  selector:'employee',
  template: '''
    <div *ngIf="emp != null">
       <div>雇员 ID:{{emp.id}}</div>
       <div>雇员名:{{emp.name}}</div>
       <div>雇员薪资:{{emp.salary}}</div>
    </div>
     ''',
  directives: [coreDirectives],
)
```

```
class EmployeeComponent{
  @Input()
  Employee emp;
}
```

绑定的目标是绑定标记([]、()或[()])内的属性或事件,数据源位于引号("")或插值符号({{}})内。绑定目标有了,现在需要绑定数据源。已知在父组件中有一个列表属性emps,因此可以将列表中被选中的条目作为输入数据源。在父组件中声明接收选中条目的Employee类型的属性selected,声明通过单击事件为属性selected赋值的onSelect()方法。代码如下:

```
Employee selected;
void onSelect(Employee emp){
  selected = emp;
}
```

根组件AppComponent的完整代码如下:

```
//chapter14/child_component/lib/app_component.dart
import 'package:angular/angular.dart';

import 'src/employee.dart';
import 'src/employee_service.dart';
import 'src/employee/employee_component.dart';

@Component(
  selector: 'my-app',
  styleURLs: ['app_component.css'],
  templateURL: 'app_component.html',
  directives: [coreDirectives,EmployeeComponent],
  providers: [ClassProvider(EmployeeService)],
)
class AppComponent implements OnInit{
  //定义一个 EmployeeService 类型的私有变量
  final EmployeeService _employeeService;
  List<Employee> emps;
  //注入服务
  AppComponent(this._employeeService);
  //覆写 ngOnInit 方法
  @override
  void ngOnInit(){
    //在组件初始化时调用 getEmployees 方法
    getEmployees();
  }
```

```
  //使用服务对象_employeeService 获取雇员列表
  void getEmployees() async{
    emps = await _employeeService.getAll();
  }

  //被选中的雇员
  Employee selected;
  //更新选中的雇员
  void onSelect(Employee emp){
    selected = emp;
  }
}
```

更新模板为每个条目绑定单击事件，并为子组件 employee 添加属性绑定。代码如下：

```
//chapter14/child_component/lib/app_component.html
<ul class = "emps">
    <li *ngFor = "let emp of emps"
        (click) = "onSelect(emp)" >
        <span class = "badge">{{emp.id}}</span>
        {{emp.name}}
        <span class = "badge salary">{{emp.salary}}</span>
    </li>
</ul>

<employee [emp] = "selected"></employee>
```

更新样式文件以获得更好的交互体验。样式代码如下：

```
//chapter14/child_component/lib/app_component.css
.emps {
  margin: 0 0 2em 0;
  list-style-type: none;
  padding: 0;
  width: 15em;
}
.emps li {
  cursor: pointer;
  position: relative;
  left: 0;
  background-color: #EEE;
  margin: .5em;
  padding: .3em 0;
  height: 1.6em;
  border-radius: 4px;
```

```
}
.emps li:hover {
  color: #607D8B;
  background-color: #EEE;
  left: .1em;
}
.emps .text {
  position: relative;
  top: -3px;
}
.emps .badge {
  display: inline-block;
  font-size: small;
  color: white;
  padding: 0.8em 0.7em 0 0.7em;
  background-color: #607D8B;
  line-height: 1em;
  position: relative;
  left: -1px;
  top: -4px;
  height: 1.8em;
  margin-right: .8em;
  border-radius: 4px 0 0 4px;
}
.emps .salary {
  float:right;
  margin-right:0;
  margin-left: .8em;
  border-radius: 0 4px 4px 0;
}
```

运行项目，单击列表中的任何条目，子组件视图都会更新为该条目的内容，这里单击最后一个条目，运行结果如图 14-19 所示。

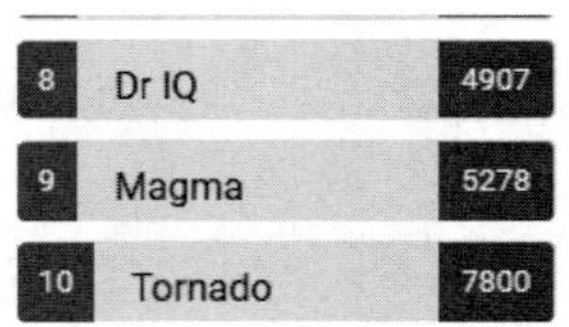

图 14-19 输入属性

2. 输出属性

输出属性是事件绑定的目标，意味着它会触发某种事件。指令通常使用 StreamController 对象引发自定义事件，创建 StreamController 对象并将其 stream 作为属性公开。通过调用 StreamController.add(payload)触发一个事件，并传递携带有效载荷的消息，它可以是任何类型和结构，父指令通过事件绑定到该属性来监听事件，并通过 $event 对象访问事件传递的信息。

现在通过示例演示输出属性的定义与使用，思路是在子组件中通过手动提供数据并触发事件。父组件通过与子组件的输出属性绑定的语句接收消息，并传给相关属性以供模板显示。

在子组件中使用注解@Output()声明输出属性,对外部来说是事件。示例代码如下:

```
//chapter14/child_component/lib/src/employee/employee_component.dart
import 'dart:async';

import 'package:angular/angular.dart';
import '../employee.dart';

@Component(
  selector:'employee',
  templateURL: 'employee_component.html',
  directives: [coreDirectives],
)
class EmployeeComponent{
  //声明输入属性
  @Input()
  Employee emp;
  //声明 StreamController 对象,该对象可以用于触发事件
  final _str = StreamController<String>();
  //声明输出属性 add,对外部来说就是事件
  @Output()
  get add => _str.stream;
  //用于处理单击事件的函数
  void addStr(var str){
    //发送数据到事件
    _str.add(str);
  }
}
```

该组件定义了一个 StreamController 属性,并通过 add 属性的 getter 方法公开了该控制器的 stream 对象。

将组件中原有的 HTML 元素放置在模板文件中并添加新的 HTML 元素,输入框用于手动输入数据,按钮的单击事件用于将输入框的值传递给 addStr 方法,方法 addStr 调用 StreamController 对象的 add 方法以触发一个事件。模板代码如下:

```
//chapter14/child_component/lib/src/employee/employee_component.html
<div *ngIf="emp != null">
    <div>雇员 ID:{{emp.id}}</div>
    <div>雇员名:{{emp.name}}</div>
    <div>雇员薪资:{{emp.salary}}</div>
</div>

<input #input>
<button (click)="addStr(input.value)">添加</button>
```

当用户单击添加按钮时，将调用组件的 addStr()方法以触发事件并传递数据。

在父组件中提供接收数据的属性 receiver，并提供响应 employee 子组件 add 事件的函数 onAdd。代码如下：

```
//chapter14/child_component/lib/app_component.dart
//用于接收数据
String receiver;
//响应 add 事件的处理程序
onAdd(var event){
  receiver = event;
}
```

在父组件模板中添加元素代码如下：

```
//chapter14/child_component/lib/app_component.html
<p *ngIf = "receiver != null">{{'子组件提供的数据:' + receiver}}</p>
<employee [emp] = "selected" (add) = "onAdd( $event)"></employee>
```

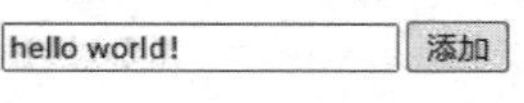

图 14-20 输出属性

在输入框中随意输入数据，单击“添加”按钮，add 事件将会触发，Angular 调用父组件的 onAdd 方法，并传递 $event 对象到其内部赋值给 receiver 属性，因为控制器 StreamController 的泛型是 String，所以对象 $event 的数据是 String 类型的。运行效果如图 14-20 所示。

3. 起别名

通过向注解@Input 和@Output 传递一个 String 类型的参数，可以为输入和输出属性起别名。示例代码如下：

```
@Input('employee')
Employee emp;

final _str = StreamController<String>();
@Output('addString')
get add => _str.stream;
void addStr(var str){
  _str.add(str);
}
```

在属性绑定和事件绑定中使用别名。代码如下：

```
<employee [employee] = "selected" (addString) = "onAdd( $event)"></employee>
```

14.6.4 双向数据绑定

有时希望既要显示组件属性值，又要在用户进行更改时更新该属性。对于元素来说既要设置元素属性又要监听元素更改事件。输入和输出属性为这样的场景提供了技术支持。

双向数据绑定结合了属性绑定和事件绑定，其绑定语法[(x)]也结合了属性绑定语法[]和事件绑定语法()。

绑定示例代码如下：

```
<element [(x)] = "property" ></element>
```

准确来说，双向数据绑定是属性绑定和事件绑定的语法糖。与之等价的代码如下：

```
<element [x] = "property" (xChange) = "property = $ event"></element>
```

在命名时属性名可以是符合规范的标识符，事件名是属性名加后缀 Change。语法[()]很容易被验证，为元素提供一个可设置的属性 x，并为其提供名为 xChange 的相应事件。

通过示例演示其工作原理，在 lib/src 目录下新建 sizer_component.dart 文件，并创建组件 SizerComponent。代码如下：

```
import 'dart:async';
import 'dart:math';
import 'package:angular/angular.dart';
const minSize = 8;
const maxSize = minSize * 5;
@Component(
  selector: 'my-sizer',
  template: '''
    <div>
      <button (click) = "dec()" [disabled] = "size <= minSize"> - </button>
      <button (click) = "inc()" [disabled] = "size >= maxSize"> + </button>
      <label [style.font-size.px] = "size"> FontSize: {{size}}px </label>
    </div>''',
  exports: [minSize, maxSize],
)
class SizerComponent {
  int _size = minSize * 2;
  int get size => _size;
  @Input()
  void set size(/* String|int */ val) {
    int z = val is int ? val : int.tryParse(val);
    if (z != null) _size = min(maxSize, max(minSize, z));
  }
  final _sizeChange = StreamController<int>();
  @Output()
```

```
  Stream<int> get sizeChange => _sizeChange.stream;
  void dec() => resize(-1);
  void inc() => resize(1);
  void resize(int delta) {
    size = size + delta;
    _sizeChange.add(size);
  }
}
```

在 AppComponent 中导入 sizer_component.dart 文件，添加 SizerComponent 到指令列表，并提供绑定参数 fontSizePx。代码如下：

```
//chapter14/child_component/lib/app_component.dart
import 'package:angular/angular.dart';

import 'src/employee.dart';
import 'src/employee_service.dart';
import 'src/employee/employee_component.dart';
import 'src/sizer_component.dart';

@Component(
  selector: 'my-app',
  styleURLs: ['app_component.css'],
  templateURL: 'app_component.html',
  directives: [coreDirectives,EmployeeComponent,SizerComponent],
  providers: [ClassProvider(EmployeeService)],
)
class AppComponent implements OnInit{
  //...
  //用于双向数据绑定
  int fontSizePx = 10;
}
```

在根组件模板中添加新元素：

```
//chapter14/child_component/lib/app_component.html
<my-sizer [(size)]="fontSizePx" #mySizer></my-sizer>
<div [style.font-size.px]="mySizer.size">Resizable Text</div>
```

单击加号或减号按钮可以控制字体大小，运行结果如图 14-21 所示。

- + FontSize: 19px
Resizable Text

图 14-21 双向数据绑定

14.7 表单

37min

表单是业务类应用程序必备的内容，可以使用表单完成注册、登录、预定航班及其他数据的输入任务。

在编写表单时，对输入数据做校验以有效指导用户和良好的输入体验是非常重要的。本节将介绍通过 angular_forms 包对表单元素进行双向数据绑定、更改跟踪、验证和错误处理。

Angular 表单由 angular_forms 包提供指令和样式，它向 HTML 基础表单提供高级功能，将包添加到依赖项并执行 pub get 命令：

```
//chapter14/forms/pubspec.yaml
dependencies:
  angular: ^5.3.0
  angular_components: ^0.13.0
  angular_forms: ^2.1.2
```

14.7.1 建立数据模型

当用户输入表单数据时，需要捕获更改信息以更新模型的实例。模型通常是表单中各个输入项的集合，这里模型 Employee 包含了 id、age、name 和 department 4 个字段。示例代码如下：

```
//chapter14/forms/lib/src/employee.dart
class Employee{
  int id,age;
  String name,department;
  Employee(this.id,this.name,this.age,this.department);
  //覆写 toString 方法
  String toString(){
    return '$id: $name ($age) 部门:$department';
  }
}
```

14.7.2 建立表单

表单分为两个部分：基于 HTML 的模板和处理数据及用户交互的组件类。

在组件中添加代码如下：

```
//chapter14/forms/lib/app_component.dart
import 'dart:convert';
```

```
import 'package:angular/angular.dart';
import 'package:angular_forms/angular_forms.dart';
import 'src/employee.dart';

@Component(
  selector: 'my-app',
  styleURLs: ['app_component.css'],
  templateURL: 'app_component.html',
  directives: [coreDirectives,formDirectives],
)
class AppComponent {
  //表单中下拉列表需要的选项列表
  static const List<String> deps = ['设计部','技术部','财务部','行政部'];
  //初始化表单模型数据
  Employee model = Employee(1,'姚环',23,deps[0]);
}
```

组件内容主要导入了模型 Employee，并且为 model 和 deps 提供了模拟数据。submitted 用于记录表单提交状态，onSubmit 方法用于处理表单的提交请求，这里并没有通过 onSubmit 方法将表单数据提交到服务器，只是将数据编码为 json 并存储在属性 data 中。

在模板文件中编写表单代码如下：

```
//chapter14/forms/lib/app_component.html
<div class="container">
    <h1>雇员表单</h1>
    <form>
        <div class="form-group">
            <label for="name">名字   * </label>
            <input type="text" class="form-control" id="name" required>
        </div>
        <div class="form-group">
            <label for="age">年龄</label>
            <input type="number" class="form-control" id="age">
        </div>
        <div class="form-group">
            <label for="department">所属部门   * </label>
            <select class="form-control" id="department" required>
                <option *ngFor="let dep of deps" [value]="dep">{{dep}}</option>
            </select>
        </div>
        <div class="row">
            <div class="col-auto">
                <button type="submit" class="btn btn-primary">提交</button>
            </div>
```

```
            <small class="col text-right">* 必需</small>
        </div>
    </form>
</div>
```

表单列出了 3 个表单组件，它们分别展示模型 Employee 的 name、age 和 department 字段。用户可以在输入框中提供内容，或在下拉列表中选择条目。

控件 input 具有 required 属性，表示必填。在表单中还会发现用于控制结构的元素上设置有 form-group、form-control 样式类，以及为提交按钮设置的一些样式类。样式类 form-group 是将其包裹的元素作为一组，form-control 样式类应用于基础表单控件，label 元素中的 for 属性的值是与其关联的表单控件的 id 名。这些样式类是在 Bootstrap 框架中定义的，在项目的 index.html 文件中的<head>元素中引用如下样式文件。引用代码如下：

```
<link rel="stylesheet"
href="https://maxcdn.bootstrapcdn.com/bootstrap/4.0.0-beta/css/bootstrap.min.css"
integrity="sha384-/Y6pD6FV/Vv2HJnA6t+vslU6fwYXjCFtcEpHbNJ0lyAFsXTsjBbfaDjzALeQsN6M"
crossorigin="anonymous">
```

运行项目结果如图 14-22 所示。

雇员表单

名字*

年龄

所属部门*

设计部

Submit　　*必需

图 14-22　表单

14.7.3　表单指令

此时并没有将表单控件与 Employee 模型数据产生联系，这种联系指模型数据能够在对应表单控件中显示，还要能通过输入控件更新模型数据。实现这样的功能首先想到的是双向数据绑定，而 angular_forms 包提供了指令 NgModel，该指令使得表单控件与模型实现双向数据绑定，其语法如下：

```
<element [(ngModel)]="expression"></element>
```

NgModel 指令会创建 NgControl 实例,并将其绑定到表单控件元素。表单 NgControl 实例跟踪控件的值、用户交互和验证状态,并使视图与模型保持同步。NgModel 指令旨在用作独立值,如果希望将其作为 Angular 表单系统的一部分,则必须使用 NgControl 指令为其指定一个名字,该名字可以是整个表单域任何唯一的值。NgControl 指令的作用是将控件注册到 Angular 表单系统,在内部 Angular 创建 NgFormControl 实例,每个 NgFormControl 都以分配给 ngControl 指令的名称注册。

向表单控件添加 NgModel 和 NgControl 指令,修改后的部分模板代码如下:

```
//chapter14/forms/lib/app_component.html
<div class = "form - group">
            <label for = "name">名字   * </label>
            <input type = "text" class = "form - control" id = "name" required
                    [(ngModel)] = "model.name" ngControl = "name">
        </div>
        <div class = "form - group">
            <label for = "age">年龄</label>
            <input type = "number" class = "form - control" id = "age"
                    [(ngModel)] = "model.age" ngControl = "age">
        </div>
        <div class = "form - group">
            <label for = "department">所属部门   * </label>
            <select class = "form - control" id = "department" required
                    [(ngModel)] = "model.department" ngControl = "department">
                <option * ngFor = "let dep of deps" [value] = "dep">{{dep}}</option>
            </select>
</div>
```

现在运行项目,结果如图 14-23 所示。

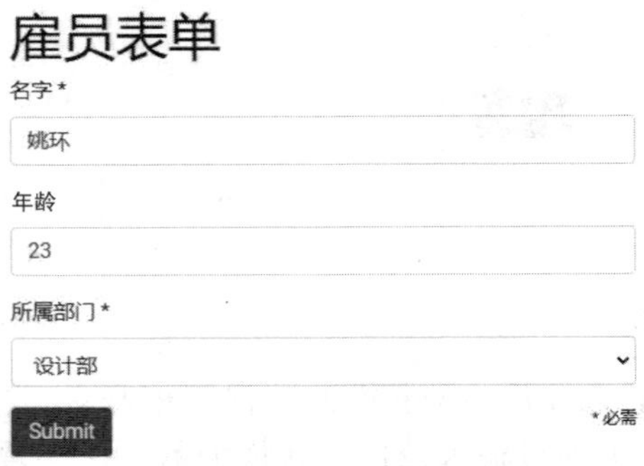

图 14-23 表单指令

NgControl 指令使得 Angular 表单的每个控件都能够跟踪自己的状态,并通过以下属性使得状态可供检查。

(1) dirty 和 pristine：指示控件的值是否已更改。

(2) touched 和 untouched：表示控件是否已被访问。

(3) valid：反映控件值的有效性。

其中控件的 valid 属性最常用，为了提供良好的视觉反馈，使用 Bootstrap 表单样式类 is-valid 和 is-invalid 标识控件值是否有效。

在名字的 input 元素上添加模板引用变量，变量名叫 name。模板引用变量 name 通过语法 #name="ngForm"绑定到与输入元素关联的 NgModel。指令的 exportAs 属性是可以在模板中使用的名称，用于将该指令分配给变量。因为 ngModel 指令的 exportAs 属性值为 ngForm，所以将模板引用变量 name 的值设置为 ngForm。

使用引用变量 name 和 CSS 类绑定有条件地分配适当的表单有效性 CSS 类。模板代码如下：

```
<div class = "form - group">
    <label for = "name">名字   * </label>
    <input type = "text" class = "form - control" id = "name" required
        [(ngModel)] = "model.name" ngControl = "name"
        #name = "ngForm"
        [class.is - valid] = "name.valid"
        [class.is - invalid] = "!name.valid">
</div>
```

14.7.4 提交表单

此时填写完符合要求的数据后提交表单，在表单底部的提交按钮不会执行任何操作，但由于其类型为 submit，因此会触发表单提交事件。

为了使表单可以正常使用，将组件的 onSubmit 方法与表单的 ngSubmit 事件绑定。代码如下：

```
<form (ngSubmit) = "onSubmit()" #empForm = "ngForm">
```

Angular 会自动创建一个 NgForm 指令并附加在 form 元素上，NgForm 指令的 exportAs 属性值为 ngForm，因此可以将其赋值给模板引用变量。NgForm 指令通过附加功能补充了 form 元素，它包含使用 ngModel 和 ngControl 指令为表单元素创建的控件，并监视它们的属性，包括有效性。

通过引用变量 empForm 将表单的整体有效性绑定到提交按钮的 disabled 属性。代码如下：

```
<button [disabled] = "!empForm.form.valid" type = "submit" class = "btn btn - primary">提
交</button>
```

Angular 表单是某些层次结构中控件的集合，可以通过 NgForm 指令的属性 value 访问整个表单的数据，属性 value 是一个反映表单数据结构的 JSON 对象。添加 onSubmit 方法并提供一个接收表单值的属性 data。示例代码如下：

```
//chapter14/forms/lib/app_component.dart
import 'dart:convert';

import 'package:angular/angular.dart';
import 'package:angular_forms/angular_forms.dart';
import 'src/employee.dart';

@Component(
  selector: 'my-app',
  styleURLs: ['app_component.css'],
  templateURL: 'app_component.html',
  directives: [coreDirectives,formDirectives],
)
class AppComponent {
  //表单中下拉列表需要的选项列表
  static const List<String> deps = ['设计部','技术部','财务部','行政部'];
  //初始化表单模型数据
  Employee model = Employee(1,'姚环',23,deps[0]);
  //用于存储表单数据
  String data;
  //用于更新表单数据
  void onSubmit(val){
    //接收表单中的数据并编码为 json
    data = json.encode(val);
  }
}
```

修改表单上 ngSubmit 事件绑定的信息，并在表单外通过插值显示 data 属性的值。模板代码如下：

```
//chapter14/forms/lib/app_component.html
<div class="container">
    <h1>雇员表单</h1>
    <form (ngSubmit)="onSubmit(empForm.value)" #empForm="ngForm">
        <div class="form-group">
            <label for="name">名字   *</label>
            <input type="text" class="form-control" id="name" required
                   [(ngModel)]="model.name" ngControl="name"
                   #name="ngForm"
                   [class.is-valid]="name.valid"
                   [class.is-invalid]="!name.valid">
```

```
        </div>
        <div class = "form - group">
            <label for = "age">年龄</label>
            <input type = "number" class = "form - control" id = "age"
                    [(ngModel)] = "model.age" ngControl = "age">
        </div>
        <div class = "form - group">
            <label for = "department">所属部门   * </label>
            <select class = "form - control" id = "department" required
                    [(ngModel)] = "model.department" ngControl = "department">
                <option *ngFor = "let dep of deps" [value] = "dep">{{dep}}</option>
            </select>
        </div>
        <div class = "row">
            <div class = "col - auto">
                <button [disabled] = "!empForm.form.valid" type = "submit" class = "btn btn -
primary">提交</button>
            </div>
            <small class = "col text - right"> *  必需</small>
        </div>
    </form>
    <p>{{data}}</p>
</div>
```

运行项目，单击 Submit 按钮，运行效果如图 14-24 所示。

图 14-24 表单数据处理

14.8 Angular 架构回顾

5min

至此已经学习了 Angular 框架的基础知识，现在来回顾它的各个组成部分：模板、指令、注解、组件、服务、数据绑定、依赖注入和提供者。

(1) 模板：通常由 HTML 元素构成，它告诉 Angular 如何呈现组件。模板与普通 HTML 又有一些区别，例如：可在元素上使用属性绑定、事件绑定；可以将元素与 NgIf 或 NgFor 等指令配合使用；甚至可以包含自定义元素。其中自定义元素是在 Angular 组件的注解中指定的自定义选择器名。

(2) 指令：模板是动态的，当 Angular 渲染它们时，将根据给定指令转换 DOM。指令包含组件、属性指令和结构指令。

(3) 注解：注解告诉 Angular 如何处理一个类或属性。注解@Component 将一个类标识为组件，注解@Directive 将一个类标识为指令，注解@Injectable 将一个类标识为服务，注解@Pipe 将一个类标识为管道，注解@Input 将一个属性标识为输入属性，注解@Output 将一个属性标识为输出属性。

(4) 组件：组件用于控制其模板呈现的视图，在组件类中定义应用程序逻辑，该类通过属性和方法的 API 与视图进行交互。

(5) 服务：可以涵盖的内容很多，主要包括提供应用程序需要的功能和数据。

(6) 数据绑定：一种协调模板和组件的机制，将绑定标记添加到 HTML 模板，以告诉 Angular 如何连接模板和组件。数据绑定包含插值、属性绑定、事件绑定和双向数据绑定。

(7) 依赖注入：依赖注入是一种为类的新实例提供所需的完整依赖关系的方法，大多数依赖项是服务。最常见的场景是使用依赖注入为组件提供所需的服务，Angular 通过查看构造函数参数的类型来判断组件需要哪些服务。注入器维护其已创建的服务实例的容器，在 Angular 创建组件时，会先向注入器询问组件所需的服务，如果请求的服务实例不在容器中，则注入程序将创建一个服务并将其添加到容器中，然后再将服务返回到 Angular。解析并返回所有请求的服务后，Angular 可以使用这些服务作为参数来调用组件的构造函数。

(8) 提供者：在使用依赖注入前，必须使用注入器注册一个服务的提供者，提供者可以创建或返回服务，并且通常是服务类本身。可以向组件注册提供者，也可以在启动应用程序时通过根注入器注册。注册提供者的常见方法是在@Component 注解的 providers 参数值中指定。当组件需要某个服务时，如果注入器中没有对应服务，注入器将使用注册的提供者创建该服务。

第15章

Angular 高级

15.1 属性指令

28min

一个属性指令常用于更改 DOM 元素的外观或行为。

Angular 中有 3 种指令：

(1) 组件：带有模板的指令。

(2) 结构指令：通过添加和删除 DOM 元素更改 DOM 布局。

(3) 属性指令：更改元素、组件或其他指令的外观或行为。

属性指令有 2 种类型：

(1) 基于类：使用类实现功能齐全的属性指令。

(2) 函数式：使用顶层函数实现的无状态属性指令。

15.1.1 基于类的属性指令

创建基于类的属性指令需要先编写带有@Directive 注解的控制类，并且需向注解提供属性选择器。控制器类的构造函数接收一个 HTML 元素作为参数，并通过操作该元素实现所需的指令行为。开始前先创建新项目，只留下根组件 AppComponent。

1. 编写指令代码

在项目 lib/src 目录下创建 fontsize_directive.dart 文件，并编写 FontsizeDirective 指令。代码如下：

```
//chapter15/attribute_directive/lib/src/fontsize_directive.dart
import 'dart:html';
import 'package:angular/angular.dart';
//指令注解
//selector 为属性选择器
@Directive(selector: '[myFontsize]')
class FontsizeDirective {
  //指令的构造函数
  //注入宿主元素对象
```

```
  FontsizeDirective(Element el) {
    //修改元素的样式：字体大小
    el.style.fontSize = '24px';
  }
}
```

@Directive注解中的CSS选择器用于标识在模板中与指令关联的HTML元素。CSS选择器使用方括号包裹标识符来表示属性选择器，这里的属性选择器是[myFontsize]。Angular会将指令行为应用于模板中所有具有myFontsize属性的元素。

尽管属性命名为fontsize比myFontsize更简洁，但对于自定义属性采用前缀实际上是最好的做法，这里的前缀是my。这样可以避免与标准HTML属性冲突，也可以减少与第三方命名冲突的概率。前缀不能使用ng，该前缀仅为Angular内部使用，使用该前缀可能导致错误。

@Directive注解后边是指令的控制器类，名为FontsizeDirective。Angular为每个匹配的元素创建指令的控制器类的新实例，并将HTML元素(Element)注入控制器类的构造函数。在函数体中通过操作元素对象控制元素的外观或行为。

2. 应用属性指令

要使用指令FontsizeDirective，需在模板中为元素提供myFontsize属性。此模板将指令作为属性应用于段落(<p>)元素，在Angular术语中p元素是属性myFontsize的宿主。

在根组件模板中添加以下内容：

```
//chapter15/attribute_directive/lib/app_component.html
<p myFontsize>Hello World!</p>
```

现在，在根组件AppComponent中导入指令文件，并将FontsizeDirective指令添加到指令列表中。这样当Angular在根组件模板中遇到myFontsize属性时就会识别指令。代码如下：

```
//chapter15/attribute_directive/lib/app_component.dart
import 'package:angular/angular.dart';
import 'src/fontsize_directive.dart';

@Component(
  selector: 'my-app',
  styleURLs: ['app_component.css'],
  templateURL: 'app_component.html',
  directives: [FontsizeDirective],
)
class AppComponent {
}
```

运行应用程序，在浏览器中 myFontsize 属性将段落文本的字体大小置为 24px，运行结果如图 15-1 所示。

AngularDart App 2

Hello World!

图 15-1 基于类的属性指令

Angular 在 p 元素上找到了 myFontsize 属性，它创建了指令 FontsizeDirective 的新实例，并对 p 元素的引用注入指令的构造函数中，该构造函数将 p 元素的字体大小修改为 24px。

3. 响应用户事件

现在 myFontsize 仅使用固定值设置元素的字体大小，为让其更加动态，可以监听用户行为。例如：将鼠标移入或移除元素，并通过修改字体大小来做出响应。

声明_el 私有属性并在构造函数中使用宿主元素对象初始化。除此之外还需要记录元素初始的字体大小，使用_size 属性完成此工作。代码如下：

```
Element _el;
String _size;

FontsizeDirective(Element el){
  _el = el;
  _size = _el.style.fontSize;
}
```

添加一个辅助方法，该方法设置宿主元素的字体大小。代码如下：

```
_fontsize([String size]){
  _el.style.fontSize = size ?? _size;
}
```

添加两个事件处理程序，使它们在鼠标进入或离开时做出响应，每个事件处理程序都由@HostListener 注解修饰。在处理程序体中使用辅助方法控制宿主元素的字体大小。示例代码如下：

```
@HostListener('mouseenter')
void onMouseEnter() {
  _fontsize('36px');
}

@HostListener('mouseleave')
void onMouseLeave() {
  _fontsize();
}
_fontsize([String size]){
  _el.style.fontSize = size ?? _size;
}
```

@HostListener 注解用于监听指令或组件宿主元素的事件，它接收一个 DOM 事件名作为参数。这里的宿主元素是 p，事件名分别是 mouseenter 和 mouseleave。

完整的指令代码如下：

```
//chapter15/attribute_directive/lib/src/fontsize_directive.dart
import 'dart:html';
import 'package:angular/angular.dart';
//指令注解
//selector 为属性选择器
@Directive(selector: '[myFontsize]')
class FontsizeDirective {
  //缓存宿主元素对象
  Element _el;
  //缓存宿主元素初始字体大小
  String _size;

  //指令的构造函数
  //注入宿主元素对象
  FontsizeDirective(Element el) {
    _el = el;
    _size = _el.style.fontSize;
  }
  //辅助方法
  _fontsize([String size]){
    //若提供参数 size 则采用该值,否则使用初始字体大小
    _el.style.fontSize = size ?? _size;
  }
  //监听宿主元素的鼠标移入事件
  @HostListener('mouseenter')
  void onMouseEnter() {
    //这里使用固定字体大小
    _fontsize('24px');
  }

  //监听宿主元素的鼠标移出事件
  @HostListener('mouseleave')
  void onMouseLeave() {
    _fontsize();
  }
}
```

刷新浏览器，运行程序，此时元素中的字体大小为其默认值，当鼠标移到元素上时，元素的字体大小变为 24px。

4. 为指令赋值

目前字体大小都是在指令中硬编码的，在本节将使用指令动态设置字体大小。

首先向指令类添加输入属性 fontsize。示例代码如下：

```
//输入属性
@Input()
String fontsize;
```

使用注解@Input 标记 fontsize 属性，使 fontsize 属性可用于属性绑定。若属性没有注解@Input 标记，则 Angular 会拒绝绑定。

修改鼠标移入事件的处理程序的逻辑代码。示例代码如下：

```
//监听宿主元素的鼠标移入事件
@HostListener('mouseenter')
void onMouseEnter() {
  //若 fontsize 不为空则使用该值，否则使用 24px
  _fontsize(fontsize ?? '24px');
}
```

在元素上应用 myFontsize 属性指令时，可以使用输入属性 fontsize 自定义字体大小。最新模板代码如下：

```
//chapter15/attribute_directive/lib/app_component.html
<p myFontsize fontsize="13px">Hello World!</p>
<p myFontsize [fontsize]="'17px'">Hello World!</p>
```

刷新浏览器，将鼠标移入 p 元素，p 元素内的字体大小将根据提供的值变化。

5. 绑定别名

现在已经可以动态地设置字体大小，但是此时需要提供两个属性。幸运的是，可以根据需要为指令属性加上别名，此时可以使用属性指令的名字作为输入属性的别名。示例代码如下：

```
@Input('myFontsize')
String fontsize;
```

在指令内部，该属性是 fontsize，在指令外部它是 myFontsize。此时就将预想的属性名称和绑定语法结合在了一起，最新模板代码如下：

```
//chapter15/attribute_directive/lib/app_component.html
<p [myFontsize]="'19px'">Hello World!</p>
```

指令 FontsizeDirective 的最新版本代码如下：

```
//chapter15/attribute_directive/lib/src/fontsize_directive.dart
import 'dart:html';
import 'package:angular/angular.dart';
//指令注解
//selector 为属性选择器
@Directive(selector: '[myFontsize]')
class FontsizeDirective {
  //缓存宿主元素对象
  Element _el;
  //缓存宿主元素初始字体大小
  String _size;

  //输入属性
  @Input('myFontsize')
  String fontsize;

  //指令的构造函数
  //传入元素对象
  FontsizeDirective(Element el) {
    _el = el;
    _size = _el.style.fontSize;
  }
  //辅助方法
  _fontsize([String size]){
    //若提供参数 size 则采用该值,否则使用初始字体大小
    _el.style.fontSize = size ?? _size;
  }
  //监听宿主元素的鼠标移入事件
  @HostListener('mouseenter')
  void onMouseEnter() {
    //若 fontsize 不为空则使用该值,否则使用 24px
    _fontsize(fontsize ?? '24px');
  }

  //监听宿主元素的鼠标移出事件
  @HostListener('mouseleave')
  void onMouseLeave() {
    _fontsize();
  }
}
```

15.1.2 函数式指令

函数式指令是呈现一次的无状态指令,可以使用注解@Directive 修饰一个顶层函数来创建函数式指令。

创建函数式属性指令，代码如下：

```
//chapter15/attribute_directive/lib/src/auto_id_directive.dart
import 'dart:html';
import 'package:angular/angular.dart';
//计数变量
int _id = 0;
//在顶层函数使用属性指令注解
//注入宿主元素，并注入宿主属性 auto-id 的值
@Directive(selector:'[auto-id]')
void autoIdDirective(Element el,@Attribute('auto-id') String prefix){
  //以 auto-id 属性的值为前缀加计数变量_id 的值
  //为宿主元素的 id 属性赋值
  el.id = '$prefix${_id++}';
}
```

指令的选择器是[auto-id]，表示指令将应用于带有属性 auto-id 的元素。与基于类的构造函数一样第一个参数是宿主元素对象。@Attribute 注解用于注入宿主元素中指定属性的属性值，这里注入宿主元素 auto-id 属性的值。在函数内部将属性 auto-id 的值作为前缀，附加计数变量_id 的值作为宿主元素 id 属性的值。

编写功能指令时，需遵循以下规则：

（1）使函数返回类型为 void。

（2）在@Directive 注解中，仅使用 selector 参数。

在组件中导入函数式指令文件，并将指令添加到指令列表中。示例代码如下：

```
//chapter15/attribute_directive/lib/app_component.dart
import 'package:angular/angular.dart';
import 'src/fontsize_directive.dart';
import 'src/auto_id_directive.dart';

@Component(
  selector: 'my-app',
  styleURLs: ['app_component.css'],
  templateURL: 'app_component.html',
  directives: [FontsizeDirective,autoIdDirective],
)
class AppComponent {
}
```

在根组件模板中添加代码如下：

```
//chapter15/attribute_directive/lib/app_component.html
<div #d1 auto-id="div-">Auto-ID:{{d1.id}}</div>
<div #d2 auto-id="div-">Auto-ID:{{d2.id}}</div>
```

```
<p #p1 auto-id="p-">Auto-ID:{{p1.id}}</p>
<div #d3 auto-id="div-">Auto-ID:{{d3.id}}</div>
<div #d4 auto-id="div-">Auto-ID:{{d4.id}}</div>
<p #p2 auto-id="p-">Auto-ID:{{p2.id}}</p>
```

尽管功能指令是无状态的，但它们可能受全局状态影响，运行效果如图 15-2 所示。

```
Auto-ID:div-0
Auto-ID:div-1

Auto-ID:p-2

Auto-ID:div-3
Auto-ID:div-4

Auto-ID:p-5
```

图 15-2 函数式指令

38min

15.2 组件样式

Angular 应用程序采用标准 CSS 设置模板元素的样式，即可以在应用程序中使用 CSS 样式表、选择器、规则和媒体查询。

Angular 将组件样式与组件绑定在一起，即组件样式只对组件模板内的元素有效，这样可以避免多个组件的样式相互干扰，为样式的模块化设计提供便利。

前面已经介绍了在组件中使用 styles 参数提供内联样式，使用 styleURLs 参数提供样式表文件。在这里主要介绍一些特殊选择器，它们可以用于设置父组件、子组件甚至全局样式。

15.2.1 :host

使用:host 伪类选择器，可以在组件的宿主元素中应用目标样式，而不是在组件模板内定位元素。宿主元素是指在定义组件时提供的选择器，:host 选择器是访问宿主元素的唯一途径，因为它不是组件模板的一部分，该宿主元素位于父组件的模板中。

组件定义代码如下：

```
//chapter15/component_styles/lib/src/child/child_component.dart
import 'package:angular/angular.dart';

@Component(
  selector: 'child',
  styleURLs: ['child_component.css'],
  templateURL: 'child_component.html',
  directives: [],
```

```
)
class ChildComponent{
}
```

在模板中添加元素代码如下：

```
//chapter15/component_styles/lib/src/child/child_component.html
<p>child 组件</p>
```

定义宿主元素样式代码如下：

```
//chapter15/component_styles/lib/src/child/child_component.css
//匹配宿主元素
:host{
  //宿主元素以块元素显示
  display: block;
   margin:6px;
  //向宿主元素添加 1 像素实线黑色边框
  border: 1px solid black;
  //将宿主元素设置为绿色
  background-color:green;
}
```

在根组件 AppComponent 中引入子组件，并将其添加到指令列表。代码如下：

```
//chapter15/component_styles/lib/app_component.dart
import 'package:angular/angular.dart';
import 'src/child/child_component.dart';

@Component(
  selector: 'my-app',
  styleURLs: ['app_component.css'],
  templateURL: 'app_component.html',
  directives: [ChildComponent],
)
class AppComponent {
}
```

在根组件模板中添加元素代码如下：

```
//chapter15/component_styles/lib/app_component.html
<div>AppComponent 中</div>
<child></child>
```

运行结果如图 15-3 所示。

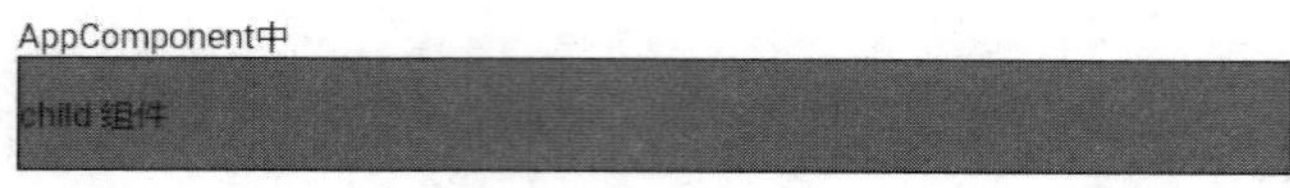

图 15-3 宿主元素样式

15.2.2 :host()

使用形式:host()是在:host 的基础上提供另一个选择器,使用该选择器筛选宿主元素,满足条件的宿主元素将被匹配。

以下选择器仍然将宿主元素作为目标,筛选条件是 CSS 类中带有 class 名 active 的宿主元素。代码如下:

```
//chapter15/component_styles/lib/src/child/child_component.css
//匹配带有 class 名 active 的宿主元素
:host(.active) {
  //将宿主元素边框宽度设置为 3 像素
  border - width: 3px;
}
```

在根组件模板中添加一个 CSS 类中带有 class 名 active 的 child 元素。模板代码如下:

```
//chapter15/component_styles/lib/app_component.html
< div > AppComponent 中</div >
< child ></child >
< child class = "active"></child >
```

运行结果如图 15-4 所示。

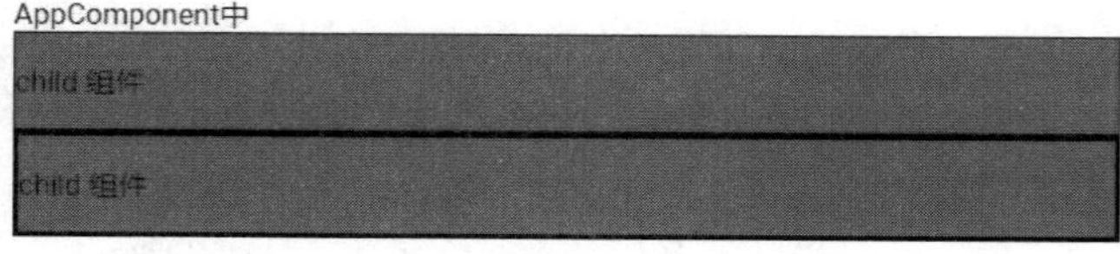

图 15-4 样式应用于部分宿主元素

15.2.3 :host-context()

有时需要根据组件视图外部的某些条件更改组件模板内元素的外观。伪类选择器:host-context()工作方式与:host()的形式相同,:host-context()接收一个选择器作为参数,在组件宿主元素的所有祖先元素中查找该选择器,直到文档根,若宿主元素的任意祖先元素满足该选择器的查找条件,则该宿主元素被匹配。:host-context()选择器需与另一个选择器结合使用,当找到满足条件的祖先元素后，再在组件模板中匹配第二个选择器所指定的元素。

以下示例仅在宿主元素的任意祖先元素具有 CSS 类 theme 的情况下，将背景颜色样式应用于组件模板内的所有 h2 元素。代码如下：

```
//chapter15/component_styles/lib/src/child/child_component.css
/* 匹配祖先元素带有 CSS 类 theme 的宿主元素，再匹配满足条件的宿主元素下的 h2 元素 */
:host-context(.theme) h2{
  /* 设置 h2 元素的背景颜色 */
  background-color: #eef;
}
```

更新组件元素代码如下：

```
//chapter15/component_styles/lib/src/child/child_component.html
<p>child 组件</p>
<h2>child h2 </h2>
```

在根组件模板中添加元素代码如下：

```
<div class="theme">
    <h2>AppComponent .theme h2 </h2>
    <child></child>
</div>
```

为了便于理解，为祖先元素拥有类 theme 的 child 元素添加同级元素 h2，这样就可以比较样式应用的范围。

运行结果如图 15-5 所示。

图 15-5 样式应用于部分宿主元素的子元素

15.2.4 ::ng-deep

组件样式通常仅适用于组件自己模板中的 HTML，使用::ng-deep 选择器可将样式向下强制通过子组件树应用于所有子组件模板中满足条件的元素。

::ng-deep 选择器适用于嵌套组件的任何深度，并且适用于该组件的视图子级和内容子级。在根组件样式文件中添加样式代码如下：

```
//chapter15/component_styles/lib/app_component.css
:host ::ng-deep h2{
  font-style: italic;
}
```

该样式将应用于根组件 AppComponent 模板中的所有 h2 元素，以及所有子组件模板中的所有 h2 元素，运行结果如图 15-6 所示。

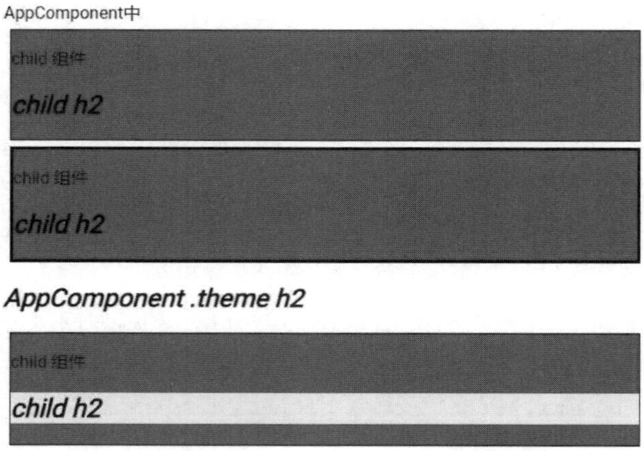

AppComponent中

child 组件

child h2

child 组件

child h2

AppComponent .theme h2

child 组件

child h2

图 15-6 样式应用于子组件中的元素

15.2.5 样式导入

可以在一个 CSS 文件中导入另一个 CSS 文件，规则适用于标准 CSS @import 规则。创建外部 CSS 文件 external.css，并添加样式代码如下：

```
//chapter15/component_styles/lib/src/child/external.css
p{
  /* 字体颜色:白色 */
  color: white;
  /* 背景颜色:木色 */
  background-color: bURLywood;
}
```

在组件 ChildComponent 样式文件中采用相对路径导入该文件的 URL：

```
//chapter15/component_styles/lib/src/child/child_component.css
@import 'external.css';
```

刷新浏览器，运行结果如图 15-7 所示。

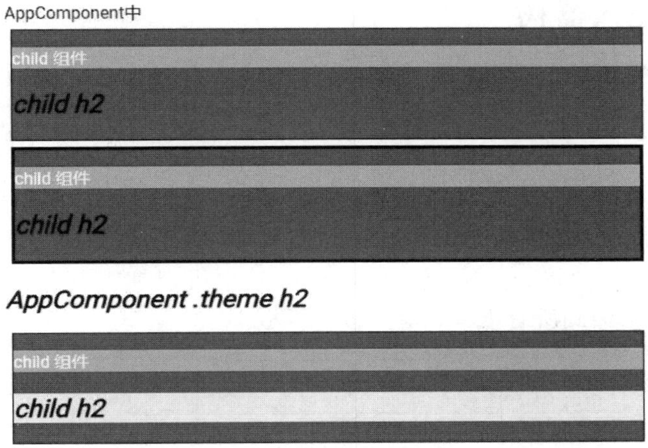

图 15-7　导入外部样式

15.2.6　视图封装

默认情况下组件样式被封装，仅影响组件模板中的 HTML 元素或自定义元素。可以使用特殊的选择器来影响组件视图之外的元素，也可以完全禁用组件的视图封装。

禁用视图封装会使组件的样式变为全局样式，为此需将组件的元数据 encapsulation 参数设置为 ViewEncapsulation. None。代码如下：

```
@Component(
  //...
  encapsulation: ViewEncapsulation.None,
)
```

ViewEncapsulation 枚举可以具有两个值：

(1) Emulated：默认值，Angular 通过预处理 CSS 来模拟 shadow DOM 的行为，以便有效地将 CSS 范围限定在组件的视图中。

(2) None：Angular 不进行视图封装，相反它使组件的样式变为全局样式。前面讨论的范围规则、隔离和保护不适用，这本质上与将组件的样式粘贴到项目 web 目录下的 styles. css 文件中相同。

26min

15.3 依赖注入

Angular 具有分层的依赖注入系统：实际上有一个与应用程序的组件树平行的注入器树，可以在该组件树的任何级别重新配置注入器。

15.3.1 注入器树

Angular 应用程序是一棵组件树，每个组件实例都有其自己的注入器。注入器树与组件树平行。

一个组件的注入器可能是组件树中更高级别的祖先注入器的代理，这可以提高效率并节省资源。几乎察觉不到它们之间的差异，需要关注的是每个组件都有自己的注射器。

在前面的例子中，根部是 AppComponent 组件，它包含一些子组件。其中之一是 EmployeeListComponent，EmployeeListComponent 也可以拥有其他组件。总之，Angular 从根组件开始逐渐延伸到所有组件，如图 15-8 所示。

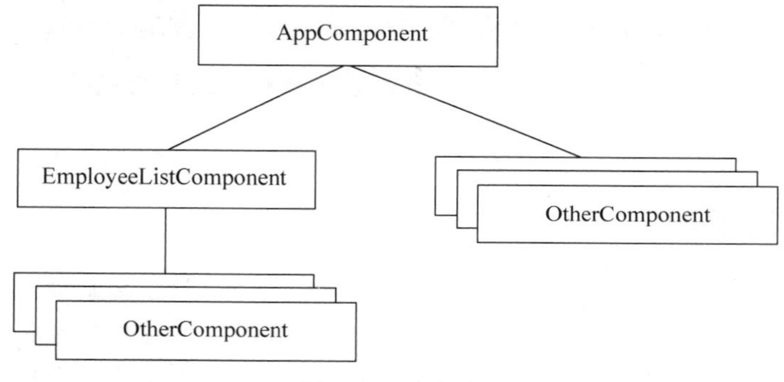

图 15-8 组件树

当组件请求依赖时，Angular 会尝试通过在该组件自己的注入器中注册的提供程序来满足该依赖。如果组件的注入器缺少提供者，它将把请求传递到其父组件的注入器，如果该注入器仍然无法满足请求，则继续向上传递。请求一直冒泡，直到 Angular 找到可以处理该请求的注入器或耗尽注入器为止。如果注入器树中没有符合请求的依赖，则 Angular 会抛出错误。

也可以限制冒泡，使用 viewProviders 参数代替 Providers，依赖项注入提供者列表将仅限于此组件及其子组件的模板。

15.3.2 服务隔离

虽然可以在根组件的注入器中提供所有服务，以便在所有组件中都可以通过注入来使用服务。出于架构原因，更好的做法是对服务的访问限制为它所属的应用程序域。

在项目 lib/src 目录下创建 employee 目录，并创建数据模型 Employee，代码如下：

```
//chapter15/dependency-injection/lib/src/employee/employee.dart
class Employee{
  final int id;
  String name;
  num salary;
  Employee(this.id,this.name,this.salary);
}
```

创建服务 EmployeeService，它的 getAll 方法能够返回使用字面量构建的 Employee 类型的列表 emps，代码如下：

```
//chapter15/dependency-injection/lib/src/employee/employee_service.dart
import 'package:angular/angular.dart';
import 'employee.dart';

@Injectable()
class EmployeeService{
  var emps = [
    Employee(5, 'Magneta',9971),
    Employee(6, 'RubberMan',4533),
    Employee(7, 'Dynama',6720),
    Employee(8, 'Dr IQ',4907),
    Employee(9, 'Magma',5278),
    Employee(10, 'Tornado',7800)
  ];
  List<Employee> getAll(){
    return emps;
  }
}
```

创建组件 EmployeeListComponent，它从服务 EmployeeService 获取 emps，并在模板中展示 emps 列表。

这里没有在 AppComponent 组件的提供者列表中添加 EmployeeService，而是在 EmployeeListComponent 的提供者列表中添加 EmployeeService，代码如下：

```
//chapter15/dependency-injection/lib/src/employee/employee_list_component.dart
import 'package:angular/angular.dart';

import 'employee.dart';
import 'employee_service.dart';

@Component(
  selector:'employee-list',
  templateURL: 'employee_list_component.html',
```

```
  directives: [coreDirectives],
  providers: [ClassProvider(EmployeeService)],
)
class EmployeeListComponent{
  final EmployeeService _employeeService;
  List<Employee> emps;
  //注入服务 EmployeeService
  EmployeeListComponent(this._employeeService){
    emps = _employeeService.getAll();
  }
}
```

向模板添加元素代码如下：

```
//chapter15/dependency-injection/lib/src/employee/employee_list_component.html
<div>
    <h3>雇员列表</h3>
    <table *ngIf="emps != null">
        <tr>
            <th>雇员 ID</th>
            <th>雇员名</th>
            <th>雇员薪资</th>
        </tr>
        <!-- 根据 emps 列表迭代雇员 -->
        <tr *ngFor="let emp of emps">
            <td>{{emp.id}}</td>
            <td>{{emp.name}}</td>
            <td>{{emp.salary}}</td>
        </tr>
    </table>
</div>
```

通过在 EmployeeListComponent 元数据中提供 EmployeeService，并且仅在 EmployeeListComponent 及其子组件树中提供该服务。这意味着服务 EmployeeService 只在能使用到它的地方被提供，而根组件 AppComponent 下的其他组件无法访问它。

15.3.3 多个编辑会话

许多应用程序允许用户同时处理多个打开的任务。例如在薪资分配应用程序中，会计人员可能同时打开多个雇员的薪资分配表。

要打开雇员的薪资分配表，会计人员需单击雇员名称，以便打开一个用于编辑该薪资分配表的组件。每个选定的雇员薪资分配表都会在其自己的组件实例中被打开，并且可以同时打开多个薪资分配表。

每个薪资分配表组件都具有以下特征：

（1）薪资分配表组件是其自身的薪资分配表的编辑会话。

（2）可以更改薪资分配表而不会影响其他组件的数据。

（3）可以保存更改后的薪资或取消。

为模型类添加工厂命名构造函数 Employee.copy，用于复制原对象。示例代码如下：

```
//chapter15/dependency-injection/lib/src/employee/employee.dart
class Employee{
  final int id;
  String name;
  num salary;
  Employee(this.id,this.name,this.salary);
  //返回原对象的副本
  factory Employee.copy(Employee e) =>
  Employee(e.id,e.name,e.salary);
}
```

向服务 EmployeeService 添加 saveSalary 方法，用于更新列表 emps 中匹配雇员的薪资。代码如下：

```
//chapter15/dependency-injection/lib/src/employee/employee_service.dart
import 'package:angular/angular.dart';
import 'employee.dart';
@Injectable()
class EmployeeService{
  var emps = [
    Employee(5, 'Magneta',9971),
    Employee(6, 'RubberMan',4533),
    Employee(7, 'Dynama',6720),
    Employee(8, 'Dr IQ',4907),
    Employee(9, 'Magma',5278),
    Employee(10, 'Tornado',7800)
  ];
  List<Employee> getAll(){
    return emps;
  }
  //保存匹配雇员的薪资信息
  void saveSalary(Employee emp){
    //判断 emps 列表是否包含传入的对象
    //如果包含就返回该对象,否则返回空
    var employee = emps.firstWhere((e){
      return e.id == emp.id;
    },orElse: () => null);

    if(employee != null ){
```

```
      //修改对应雇员的薪资
      employee.salary = emp.salary;
    }
  }
}
```

添加服务 EmployeeSalaryService，它缓存单个 Employee，跟踪该对象的更改，并可以保存或恢复它。它还委派给应用程序范围内的单例 EmployeeService，并通过注入获得。代码如下：

```
//chapter15/dependency-injection/lib/src/employee/employee_salary_service.dart
import 'employee.dart';
import 'employee_service.dart';

class EmployeeSalaryService{
  final EmployeeService _employeeService;
  Employee _currentEMP, _originalEMP;

  //注入应用程序范围内的单例 EmployeeService
  EmployeeSalaryService(this._employeeService);

  //employee 的 setter 方法
  void set employee(Employee emp) {
    //使用_originalEMP 缓存原对象
    _originalEMP = emp;
    //复制原对象到_currentEMP
    _currentEMP = Employee.copy(emp);
  }

  //employee 的 getter 方法 返回最新雇员信息
  Employee get employee => _currentEMP;

  //恢复薪资信息
  void restoreSalary(){
    //恢复原对象
    employee = _originalEMP;
  }

  //保存薪资信息
  void saveSalary(){
    //修改单例 EmployeeService 中的数据
    _employeeService.saveSalary(_currentEMP);
  }
}
```

添加组件 EmployeeSalaryComponent，它负责展示薪资分配表，并提供与服务 EmployeeSalaryService 的交互。代码如下：

```
//chapter15/dependency-injection/lib/src/employee/employee_salary_component.dart
import 'dart:async';

import 'package:angular/angular.dart';
import 'package:angular_forms/angular_forms.dart';

import 'employee.dart';
import 'employee_salary_service.dart';

@Component(
  selector:'employee-salary',
  templateURL: 'employee_salary_component.html',
  styleURLs: ['employee_salary_component.css'],
  directives: [coreDirectives,formDirectives],
  providers: [ClassProvider(EmployeeSalaryService)],
)
class EmployeeSalaryComponent{
  final EmployeeSalaryService _employeeSalaryService;
  String message = '';
  //注入服务 EmployeeSalaryService
  EmployeeSalaryComponent(this._employeeSalaryService);
  //声明输入对象
  @Input()
  void set employee(Employee emp){
    _employeeSalaryService.employee = emp;
  }
  Employee get employee => _employeeSalaryService.employee;

  //保存数据
  Future<void> onSaved() async {
    await _employeeSalaryService.saveSalary();
    await flashMessage('已保存');
  }

  //取消更改,还原对象信息
  Future<void> onCanceled() async {
    _employeeSalaryService.restoreSalary();
    await flashMessage('已取消');
  }

  final _close = StreamController<Null>();
  //声明输出对象
```

```
  @Output()
  Stream<Null> get close => _close.stream;
  //关闭薪资分配表
  //因为只需触发事件而不需要传递数据,因此向 add 传递 null 对象
  void onClose() => _close.add(null);

  //刷新消息
  void flashMessage(String msg) async {
    message = msg;
    await Future.delayed(Duration(milliseconds: 500));
    message = '';
  }
}
```

添加薪资分配表对应的模板代码如下:

```
//chapter15/dependency-injection/lib/src/employee/employee_salary_component.html
<div class="salary">
    <div class="msg" [class.canceled]="message === 'Canceled'">{{message}}</div>
    <fieldset>
        <span id="name">{{employee.name}}</span>
        <label id="id"> ID: {{employee.id}}</label>
    </fieldset>
    <fieldset>
        <label>
            薪资: <input type="number" [(ngModel)]="employee.salary" class="num">
        </label>
    </fieldset>
    <fieldset>
        <button (click)="onSaved()">保存</button>
        <button (click)="onCanceled()">取消</button>
        <button (click)="onClose()">关闭</button>
    </fieldset>
</div>
```

将组件 EmployeeSalaryComponent 添加到组件 EmployeeListComponent 的指令列表,添加存储处于编辑状态的雇员列表 selectedEmployees,并提供向该列表添加和移除雇员信息的方法 showSalary 与 closeSalary。代码如下:

```
//chapter15/dependency-injection/lib/src/employee/employee_list_component.dart
import 'package:angular/angular.dart';

import 'employee.dart';
import 'employee_service.dart';
import 'employee_salary_component.dart';
```

```
@Component(
  selector:'employee-list',
  templateURL: 'employee_list_component.html',
  directives: [coreDirectives,EmployeeSalaryComponent],
  providers: [ClassProvider(EmployeeService)],
)
class EmployeeListComponent{
  final EmployeeService _employeeService;
  List<Employee> emps;
  //注入服务 EmployeeService
  EmployeeListComponent(this._employeeService){
    emps = _employeeService.getAll();
  }
  //存储当前处于编辑状态的雇员列表
  final List<Employee> selectedEmployees = [];

  //通过将单个雇员添加到 selectedEmployees 列表以显示相关薪资分配表
  void showSalary(Employee emp){
    //判断 selectedEmployees 列表是否已存在该雇员
    if(!selectedEmployees.any((e) => e.id == emp.id)){
      selectedEmployees.add(emp);
    }
  }
  //通过移除 selectedEmployees 列表中的相应元素关闭薪资分配表
  void closeSalary(int index){
    selectedEmployees.removeAt(index);
  }
}
```

更新组件 EmployeeListComponent 的模板，向雇员列表的每个条目添加单击事件并绑定到 showSalary 方法，使用列表 selectedEmployees 迭代多个薪资分配表，向每个 EmployeeSalaryComponent 组件实例注入当前迭代雇员信息，并将实例的 close 事件绑定到 showSalary 方法。代码如下：

```
//chapter15/dependency-injection/lib/src/employee/employee_list_component.html
<div>
    <h3>雇员列表</h3>
    <table *ngIf="emps != null">
        <tr>
            <th>雇员 ID</th>
            <th>雇员名</th>
            <th>雇员薪资</th>
        </tr>
```

```
            <!-- 根据 emps 列表迭代雇员 -->
            <!-- 添加单击事件绑定到 showSalary 方法 -->
            <tr *ngFor="let emp of emps" (click)="showSalary(emp)">
                <td>{{emp.id}}</td>
                <td>{{emp.name}}</td>
                <td>{{emp.salary}}</td>
            </tr>
        </table>
    </div>
    <!-- 根据 selectedEmployees 列表迭代多个子组件实例 -->
    <employee-salary *ngFor="let selected of selectedEmployees;let i = index"
                    [employee]="selected" (close)="closeSalary(i)">
    </employee-salary>
```

任意单击雇员列表中的两个条目，运行效果如图 15-9 所示。

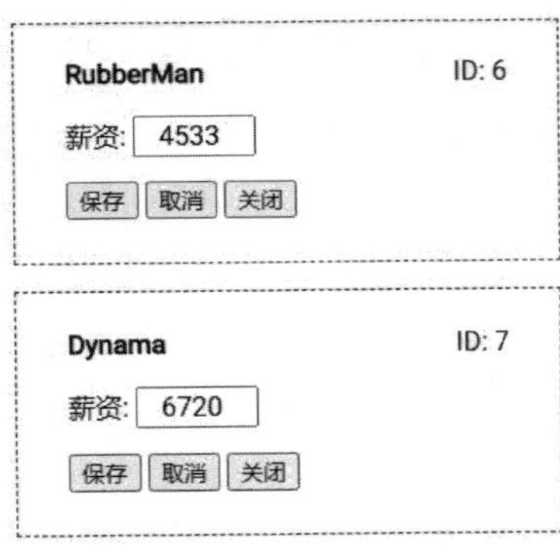

图 15-9　多编辑会话

45min

15.4　生命周期挂钩

Angular 会创建和渲染组件及其子级，在数据绑定属性更改时对组件进行检查，并在组件从 DOM 中删除之前将其销毁。

Angular 提供了一组生命周期挂钩函数，它们为组件或指令提供了关键时刻的可视性及发生时的操作能力。

指令具有相同的生命周期挂钩集，减去特定于组件内容和视图的挂钩，如图 15-10 所示。

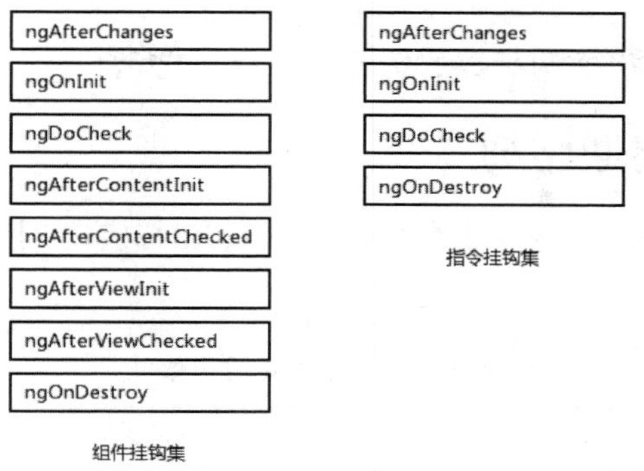

图 15-10 组件和指令挂钩集

15.4.1 组件生命周期挂钩

指令和组件实例都具有生命周期，因为由 Angular 负责创建、更新和销毁它们。通过实现 Angular 核心库中一个或多个生命周期挂钩接口，开发人员可以利用生命周期中的关键时刻完成一些操作。

每个接口都有一个挂钩函数，以 ng 为前缀，其名称为接口名。例如，OnInit 接口具有一个名为 ngOnInit 的挂钩函数，Angular 在创建组件后不久会调用该方法。示例代码如下：

```
//chapter15/lifecycle-hooks/lib/app_component.dart
import 'package:angular/angular.dart';
import 'package:angular_forms/angular_forms.dart';

@Component(
  selector: 'my-app',
  template:'''<p *ngFor="let msg of msgs">{{msg}}</p>''',
  directives: [coreDirectives,formDirectives],
)
class AppComponent implements OnInit{
  List<String> msgs = [];
  AppComponent(){
    msgs.add('AppComponent 构造函数');
  }
  @override
  void ngOnInit(){
    msgs.add('AppComponent ngOnInit');
  }
}
```

AppComponent 构造函数

AppComponent ngOnInit

图 15-11　初始化周期函数

运行结果如图 15-11 所示。

没有指令或组件将实现所有生命周期挂钩函数，并且某些挂钩仅对组件有意义。Angular 仅在实现了指令或组件挂钩函数的情况下才调用它。

15.4.2　生命周期序列

通过调用构造函数创建组件或指令后，Angular 在特定时刻按以下顺序调用生命周期挂钩函数，如表 15-1 所示。

表 15-1　生命周期函数

挂钩函数	目的和时间
ngAfterChanges	当 Angular 设置或更新数据绑定的输入属性时响应。 在 ngOnInit 之前及当一个或多个数据绑定输入属性更改时调用
ngOnInit	在 Angular 首次显示数据绑定属性并设置指令或组件的输入属性后，初始化指令或组件 在 ngOnChanges 首次调用之后调用一次
ngDoCheck	检测 Angular 无法或无法自行检测到的变化并采取措施。 在每次更改检测运行期间调用，紧接在 ngOnChanges 和 ngOnInit 之后
ngAfterContentInit	在 Angular 将外部内容投射到组件的视图中后做出响应。 在首次 ngDoCheck 调用之后调用一次。 仅组件挂钩
ngAfterContentChecked	Angular 检查投影到组件中的内容后响应。 在 ngAfterContentInit 和每个后续的 ngDoCheck 调用之后调用。 仅组件挂钩
ngAfterViewInit	在 Angular 初始化组件的视图和子视图之后响应。 在第一次 ngAfterContentChecked 调用之后调用一次。 仅组件挂钩
ngAfterViewChecked	在 Angular 检查组件的视图和子视图之后响应。 在 ngAfterViewInit 和随后的每个 ngAfterContentChecked 之后调用。 仅组件挂钩
ngOnDestroy	在 Angular 销毁指令/组件之前进行清理。取消订阅可观察对象并分离事件处理程序，以避免内存泄漏。 在 Angular 销毁指令/组件之前调用

15.4.3　其他生命周期挂钩

除了这些组件挂钩之外，其他 Angular 子系统可能还有自己的生命周期挂钩。例如，路由器还具有自己的路由器生命周期挂钩，使得开发人员可以利用路由导航中的特定时刻。路由的生命周期挂钩 routerOnActivate 等同于组件的生命周期挂钩 ngOnInit，两者都有前缀以避免冲突，并且都在初始化组件时正确运行。

第三方库也可以实现自己的挂钩，以使开发人员可以更好地控制这些库的使用方式。

15.4.4 生命周期练习

该示例通过在根目录 AppComponent 的控制下呈现演示了组件的各个生命周期挂钩。它遵循一种通用模式：父组件充当子组件的测试平台，子组件实现多个生命周期挂钩方法。

首先定义一个日志服务，其作用是打印各个生命周期传递的消息。示例代码如下：

```
//chapter15/lifecycle-hooks/lib/src/log_service.dart
import 'package:angular/angular.dart';
@Injectable()
class LogService{
  //缓存日志
  List<String> logs = [];
  //添加日志
  void log(String msg){
    logs.add(msg);
  }
  //安排视图刷新以确保显示及时
  tick() => Future(() {});
}
```

然后定义组件 HooksComponent，注入日志服务 LogService，这里不要使用 providers 参数提供服务，而要使用父组件 AppComponent 注入器中的服务。使其实现所有生命周期挂钩函数，并在实现方法中向日志服务提供消息。示例代码如下：

```
//chapter15/lifecycle-hooks/lib/src/hooks_component.dart
import 'package:angular/angular.dart';
import 'package:lifecycle_hooks/src/log_service.dart';
import 'package:angular_forms/angular_forms.dart';

@Component(
  selector: 'hooks',
  template:'''
  <p>通过修改属性,触发 ngDoCheck、AfterContentChecked 和 AfterViewChecked</p>
  <input [(ngModel)] = "name">''',
  directives: [coreDirectives,formDirectives],
)
class HooksComponent implements
    AfterChanges,
    OnInit,
    DoCheck,
```

```
    AfterContentInit,
    AfterContentChecked,
    AfterViewInit,
    AfterViewChecked,
    OnDestroy{
  final LogService _log;
  //输入属性
  @Input()
  String name;
  //记录与视图相关的挂钩函数的执行次数
  int _afterContentCheckedCounter = 1;
  int _afterViewCheckedCounter = 1;
  int _afterChangesCounter = 1;
  int _doCheckCounter = 1;
  //构造函数,注入服务 LogService
  HooksComponent(this._log){
    _log.log('HooksComponent 构造函数');
  }

  @override
  void ngAfterChanges() {
    _log.log('HooksComponent ngAfterChanges ( ${_afterChangesCounter++})');
  }

  @override
  void ngAfterContentChecked() {
    _log.log('HooksComponent ngAfterContentChecked( ${_afterContentCheckedCounter++})');
  }

  @override
  void ngAfterContentInit() {
    _log.log('HooksComponent ngAfterContentInit');
  }

  @override
  void ngAfterViewChecked() {
    _log.log('HooksComponent ngAfterViewChecked( ${_afterViewCheckedCounter++})');
  }

  @override
  void ngAfterViewInit() {
    _log.log('HooksComponent ngAfterViewInit');
  }

  @override
```

```
  void ngDoCheck() {
    _log.log('HooksComponent ngDoCheck(${_doCheckCounter++})');
  }

  @override
  void ngOnDestroy() {
    _log.log('HooksComponent ngOnDestroy');
  }

  @override
  void ngOnInit() {
    _log.log('HooksComponent ngOnInit');
  }
}
```

在 AppComponent 中注入服务 LogService，并将 HooksComponent 添加到指令列表。添加 toggleChild 函数用于创建或销毁 HooksComponent 组件实例。示例代码如下：

```
//chapter15/lifecycle-hooks/lib/app_component.dart
import 'package:angular/angular.dart';
import 'package:angular_forms/angular_forms.dart';

import 'src/hooks_component.dart';
import 'src/log_service.dart';

@Component(
  selector: 'my-app',
  styleURLs: ['app_component.css'],
  templateURL: 'app_component.html',
  directives: [coreDirectives,formDirectives,HooksComponent],
  providers: [ClassProvider(LogService)],
)
class AppComponent{
  final LogService _log;
  String name = 'lei';
  AppComponent(this._log);
  //logs 用于指向服务 LogService 中的 logs
  List<String> get logs => _log.logs;
  //组件创建和销毁控制变量
  bool isShow = false;
  //组件创建或销毁控制方法
  toggleChild(){
    isShow = !isShow;
    //用于触发数据更新以便更新视图
    _log.tick();
```

```
    }
}
```

向模板添加元素代码如下：

```
//chapter15/lifecycle-hooks/lib/app_component.html
<button (click) = "toggleChild()">{{isShow ? '销毁':'创建'}}hooks 实例</button>
<div *ngIf = "isShow">
    <p>通过修改属性 name,更改传入 HooksComponent 实例的输入属性 name</p>
    <p>触发组件 HooksComponent 的钩子函数 ngAfterChanges</p>
    <input [(ngModel)] = "name">
</div>

<hr>

<hooks *ngIf = "isShow" [name] = "name"></hooks>

<h4>-- 生命周期挂钩函数执行日志 --</h4>
<div *ngFor = "let msg of logs;let i = index">{{msg}}</div>
```

运行结果如图 15-12 所示。

销毁hooks实例

通过修改属性name，更改传入HooksComponent实例的输入属性name

触发组件HooksComponent的钩子函数ngAfterChanges

lei2

通过修改属性，触发ngDoCheck、AfterContentChecked和AfterViewChecked

lei2

-生命周期挂钩函数执行日志-

HooksComponent 构造函数
HooksComponent ngAfterChanges (1)
HooksComponent ngOnInit
HooksComponent ngDoCheck(1)
HooksComponent ngAfterContentInit
HooksComponent ngAfterContentChecked(1)
HooksComponent ngAfterViewInit
HooksComponent ngAfterViewChecked(1)
HooksComponent ngDoCheck(2)
HooksComponent ngAfterContentChecked(2)
HooksComponent ngAfterViewChecked(2)
HooksComponent ngAfterChanges (2)
HooksComponent ngDoCheck(3)
HooksComponent ngAfterContentChecked(3)
HooksComponent ngAfterViewChecked(3)

图 15-12　生命周期函数实例

应当充分使用提供的操作，控制属性更改和组件状态变化，仔细分析以便更准确了解各个生命周期挂钩的使用与调用时机。

15.5 管道

19min

每个应用程序都会获取数据，通常将其原始 toString 值直接传递到视图，但在某些时候会导致不好的用户体验。例如，在大多数使用情况下，用户喜欢以简单的格式（如 1988 年 4 月 15 日）查看日期，而不是原始的字符串格式（Fri 1988 年 4 月 15 日 00:00:00 GMT-0700，太平洋夏令时间）。

有些值可以通过一些小改动而变得易于阅读。应用程序中可能对某些值进行许多相同的转换，几乎可以将它们视为样式。实际上，可以像对待样式一样在 HTML 模板中应用它们。

本节介绍管道，这是一种可以在 HTML 中对显示值进行转换的方法。

15.5.1 使用管道

使用管道前需要提供 pipes 参数，常量 commonPipes 包含了所有内置管道。

```
pipes:[commonPipes],
```

在模板中使用格式示例代码如下：

```
{{expr | pipeName}}
```

管道将数据作为输入，并将其转换为所需的输出。在插值表达式中，将表达式 expr 的值通过管道运算符(|)传递到右侧的管道函数，所有管道都以这种方式工作。

Angular 内置大量管道：

(1) DatePipe：日期管道，用于格式化日期。

(2) UpperCasePipe：将字符串转换为大写。

(3) LowerCasePipe：将字符串转换为小写。

(4) CurrencyPipe：将数字转换为本地货币的表示形式。

(5) PercentPipe：将数字转换为百分比的表示形式。

在组件中定义日期和字符串属性，代码如下：

```
//chapter15/pipes/lib/app_component.dart
import 'package:angular/angular.dart';
@Component(
  selector: 'my-app',
```

```
  styleURLs: ['app_component.css'],
  templateURL: 'app_component.html',
  pipes: [commonPipes],
)
class AppComponent {
  DateTime birthday = DateTime(2020,6,25);
  String str = 'Lower Upper';
  int number = 99;
}
```

在模板中添加元素，代码如下：

```
//chapter15/pipes/lib/app_component.html
<h6>原日期格式:{{birthday}}</h6>
<p>date:{{birthday | date}}</p>
<h6>原字符串:{{str}}</h6>
<p>uppercase:{{str | uppercase}}</p>
<p>lowercase:{{str | lowercase}}</p>
<h6>原数字:{{number}}</h6>
<p>currency:{{number | currency}}</p>
<p>percent:{{number | percent}}</p>
```

运行结果如图 15-13 所示。

原日期格式：2020-06-25 00:00:00.000

date:Jun 25, 2020

原字符串：Lower Upper

uppercase:LOWER UPPER

lowercase:lower upper

原数字：99

currency:USD99

percent:9,900%

图 15-13　内置管道

15.5.2　参数化管道

管道可以接收任意数量的可选参数来微调其输出。要将参数添加到管道，需在管道名称后加上冒号(:)，然后输入参数值，例如：currency:'EUR'。如果管道接收多个参数，则需用冒号分隔值，例如：slice:1:5。

日期管道的参数构成元素的使用格式如表 15-2 所示。

表 15-2 日期参数格式

构成元素	数字表示	两位数表示
年	y(2020)	yy(20)
月	M(6)	MM(06)
日	d(9)	dd(09)
小时(12)	h(1 PM)	hh(01 PM)
小时(24)	H(13)	HH(13)
分钟	m(3)	mm(03)
秒	s(9)	ss(09)

参数值可以是任何有效的模板表达式,例如:字符串或组件属性。向组件添加日期格式切换方法 toggleFormat,它控制组件的 format 属性在格式 yMMdd 和格式 MM/dd/yy 之间切换。示例代码如下:

```
//chapter15/pipes/lib/app_component.dart
//切换控制变量
bool toggle = true;
//格式化字符串
get format => toggle ? 'yMMdd' : 'MM/dd/yy';
//切换控制方法
void toggleFormat() {
  toggle = !toggle;
}
```

可以直接将字符串 MM/dd/yy 用作日期管道的参数,也可以将管道的 format 参数绑定到组件的 format 属性。向模板添加一个按钮,并将其单击事件绑定到组件的 toggleFormat()方法。向模板添加代码如下:

```
//chapter15/pipes/lib/app_component.html
<p>birthday:{{ birthday | date:"MM/dd/yy" }}</p>
<p>birthday:{{ birthday | date:format }}</p>
<button (click) = "toggleFormat()">切换格式</button>
```

刷新浏览器,单击"切换格式"按钮,日期在"20200625"和"06/25/20"之间交替显示。

15.5.3 管道链

可以将管道以潜在有用的组合方式连接在一起,即可以同时使用多个管道。可以将 birthday 连接到 DatePipe 并连接到 UpperCasePipe。示例代码如下:

```
{{birthday | date | uppercase}}
```

此示例连接了与上述相同的管道，并向 date 管道传递了一个参数。示例代码如下：

```
{{birthday | date:'fullDate' | uppercase}}
```

向模板添加元素代码如下：

```
//chapter15/pipes/lib/app_component.html
<p>birthday:{{ birthday | date}}</p>
<p>birthday:{{ birthday | date | uppercase}}</p>

<p>birthday:{{ birthday | date:'fullDate'}}</p>
<p>birthday:{{ birthday | date:'fullDate' | uppercase}}</p>
```

15.5.4 自定义管道

可以编写自定义管道，管道是使用@Pipe 注解修饰的类，@Pipe 注解的参数是管道名。管道类必须实现 PipeTransform 接口的 transform 方法，该方法可以接收多个参数，第一个参数是通过管道符传入的值，其他参数是在使用管道时提供的参数。方法内部执行转换操作，并返回转换后的值。参数和返回值可以是任何类型。

这是一个名为 RoundAreaPipe 的自定义管道，它的作用是根据圆的半径计算圆的面积。示例代码如下：

```
//chapter15/pipes/lib/src/round_area_pipe.dart
import 'package:angular/angular.dart';

@Pipe('roundArea')
class RoundAreaPipe extends PipeTransform{
  //半径作为参数并返回圆的面积
  num transform(num r){
    var pi = 3.14;
    return pi * r * r;
  }
}
```

使用管道时，首先在组件中导入该管道文件，然后在 pipes 列表中列出该管道。示例代码如下：

```
//chapter15/pipes/lib/app_component.dart
import 'package:angular/angular.dart';
import 'src/round_area_pipe.dart';

@Component(
```

```
    selector: 'my-app',
    styleURLs: ['app_component.css'],
    templateURL: 'app_component.html',
    directives: [coreDirectives,formDirectives],
    pipes: [commonPipes,RoundAreaPipe],
)
class AppComponent {
}
```

随后就可以在模板中添加元素，代码如下：

```
//chapter15/pipes/lib/app_component.html
<p>半径为 9 厘米的圆的面积为{{9 | roundArea}}平方厘米</p>
```

15.5.5 管道和变更检测

Angular 通过在每个 DOM 事件(每次击键、鼠标移动、计时器滴答和服务器响应)之后运行的更改检测进程来寻找对数据绑定值的更改，这可能占用大量资源，Angular 尽可能地降低影响。使用管道时，Angular 选择一种更简单、更快速的变更检测算法。

在组件中添加泛型为字符串的 list 列表，在 reset 方法中为该列表初始化值。在构造函数中调用 reset 方法，添加方法 addStr 用于向列表 list 添加字符串。示例代码如下：

```
//chapter15/pipes/lib/app_component.dart
List<String> list;
AppComponent(){
  reset();
}
void addStr(String str){
  list.add(str);
}
void reset(){
  list = ['Bombasto','RubberMan','Magneta','Magma'];
}
```

使用默认的主动更改检测策略来监视和更新 list 列表中每个字符串的显示。在模板中添加元素代码如下：

```
//chapter15/pipes/lib/app_component.html
<input type="text" #box
       (keyup.enter)="addStr(box.value); box.value=''"
       placeholder="请输入字符串">
<button (click)="reset()">重置列表</button>
<h6>未使用管道</h6>
```

```
<div *ngFor="let str of list">
    {{str}}
</div>
```

然后定义管道 PrefixPipe，它的作用是根据提供的前缀 prefix，返回 list 列表中所有满足条件的字符串组成的新列表。示例代码如下：

```
//chapter15/pipes/lib/src/prefix_pipe.dart
import 'package:angular/angular.dart';

@Pipe('prefix')
class PrefixPipe extends PipeTransform{
  //第一个参数是字符串类型的列表，第二个参数表示字符串前缀
  List<String> transform(List<String> list,String prefix){
    //返回符合前缀 prefix 的字符串的新列表
   return list.where((str) => str.startsWith(prefix)).toList();
  }
}
```

然后将管道 PrefixPipe 添加到组件的 pipes 列表中，并在模板中添加元素，代码如下：

```
//chapter15/pipes/lib/app_component.html
<h6>使用管道 prefix</h6>
<div *ngFor="let str of (list | prefix:'M')">
    {{str}}
</div>
```

刷新浏览器，输入字符串并回车，没有使用管道的模板循环会在添加字符串后立即更新显示，使用管道的模板循环则不会更新显示。

15.5.6 纯与不纯

管道有两类：纯管道和不纯的管道。默认情况下，管道是纯管道。到目前为止，所使用的每个管道都是纯的。通过将管道的 pure 标志设置为 false，可以使其变为不纯的管道。

可以像这样使 PrefixPipe 变为不纯的管道，示例代码如下：

```
@Pipe('prefix', pure:false)
```

先来了解纯与不纯之间的区别。

1. 纯的管道

Angular 仅在检测到输入值的纯更改时才执行纯管道。在 Angular 中，纯更改仅由对象引用的更改引起。

Angular 忽略复合对象内的更改。例如更改 List 或 Map 对象中的数据，则不会调用纯

管道。这似乎很严格，但速度很快。对象引用检查很快，比深入检查符合对象内部的差异快得多，因此 Angular 可以快速确定是否可以跳过管道执行和视图更新。

因此，当可以使用默认变更检测策略时，最好使用纯管道。如果不能，则可以使用不纯的管道。

2. 不纯的管道

Angular 在每个组件更改检测周期内执行不纯管道。每次击键或移动鼠标时，都会频繁调用不纯的管道。考虑到这一点，应格外小心地使用不纯的管道。它会消耗大量计算资源，长时间运行的管道可能会破坏用户体验。

定义一个不纯的管道 PrefixImpurePipe，在这里直接使用 PrefixImpurePipe 继承 PrefixPipe 就可以了，完整示例代码如下：

```
//chapter15/pipes/lib/src/prefix_pipe.dart
@Pipe('prefixImpure',pure: false)
class PrefixImpurePipe extends PrefixPipe{}
```

从继承 PrefixPipe 来证明内部没有做任何更改，唯一的区别是 pipe 元数据中的 pure 标志。然后将 PrefixImpurePipe 添加到组件的 pipes 列表，在模板中添加元素，代码如下：

```
//chapter15/pipes/lib/app_component.html
<h6>使用不纯的管道 prefixImpure</h6>
<div *ngFor="let str of (list | prefixImpure:'M')">
    {{str}}
</div>
```

此时向 list 列表中添加字符串，不纯的管道 prefixImpure 就会检查更改，带有 M 开头的字符串将会被显示在视图中。

3. 异步管道

AsyncPipe 是不纯的管道，可以使用 Future 或 Stream 作为输入，并且会自动订阅输入。AsyncPipe 也是有状态的。管道维护对输入 Stream 的订阅，并不断地从流中传递值到管道。

在组件中添加流控制器 st，添加一个接收 st 的 Stram 对象的变量 message，在方法 addMes 中由流控制器 st 发送数据到 message。示例代码如下：

```
//chapter15/pipes/lib/app_component.dart
//Stream 控制器
StreamController<String> st = StreamController<String>();
//返回 st 的流对象
Stream<String> get message => st.stream;
//触发事件并传递数据
addMes(String str){
  st.add(str);
}
```

在模板中通过输入控件的 enter 事件调用 addMes 方法,并在插值中使用异步管道。在模板中添加元素代码如下:

```
//chapter15/pipes/lib/app_component.html
<h6>Async 管道</h6>
<input type="text" #mes
       (keyup.enter)="addMes(mes.value); mes.value=''"
       placeholder="请输入消息并回车">
<p>message:{{message | async}}</p>
```

每当在输入控件中输入数据并回车,message 信息就会更新。

47min

15.6 路由

当用户执行应用程序任务时,Angular 路由器可以从一个视图导航到下一个视图。本指南涵盖了路由器的主要功能,并通过可实时运行的小应用程序的演变来说明它们。

15.6.1 路由基础

1. 添加依赖项

路由器功能位于 angular_router 包中,将包添加到 pubspec 依赖项,并执行 pub get 命令:

```
//chapter15/router/pubspec.yaml
dependencies:
  angular: ^5.3.0
  angular_components: ^0.13.0
  angular_router: ^2.0.0-alpha+22
```

在任何需要使用路由器功能的 Dart 文件中,导入路由器库,示例代码如下:

```
import 'package:angular_router/angular_router.dart';
```

2. 添加全局路由提供者

在应用程序的引导函数中指定 routerProvidersHash,使 Angular 知道应用程序使用了路由功能。示例代码如下:

```
//chapter15/router/web/main.dart
import 'package:angular/angular.dart';
import 'package:angular_router/angular_router.dart';
import 'package:router/app_component.template.dart' as ng;

import 'main.template.dart' as self;
```

```
const useHashLS = false;
@GenerateInjector(
  //在生产环境使用 routerProviders
  routerProvidersHash,
)
final InjectorFactory injector = self.injector$Injector;

void main() {
  runApp(ng.AppComponentNgFactory, createInjector: injector);
}
```

默认的路由定位策略 LocationStrategy 采用路径定位策略 PathLocationStrategy，所以在生产环境中使用 routerProviders。在开发环境中使用 routerProvidersHash，因为 webdev serve 不支持深层次的链接，即无法通过链接直接访问首页以外的页面。

3. 设置 base href

添加<base href>元素到应用程序的 index.html 文件。当引用 CSS 文件、脚本和图像时，浏览器使用 href 的值作为相对 URL 的前缀。在页面间导航时，路由也会使用 href 的值作为相对 URL 的前缀。

在开发环境中动态设置<base>元素，以便可以在开发过程中使用任何官方推荐的工具来运行和测试应用程序。示例代码如下：

```
//chapter15/router/web/index.html
<!DOCTYPE html>
<html>
  <head>
    <title>router</title>
    <meta charset="utf-8">
    <meta name="viewport" content="width=device-width, initial-scale=1">
    <link rel="stylesheet" href="styles.css">
    <link rel="icon" type="image/png" href="favicon.png">
    <script>
      //警告：不要在生产环境中动态设置<base href>
      (function () {
        var m = document.location.pathname.match(/^(\/[-\w]+)+\/web($|\/)/);
        document.write('<base href="' + (m ? m[0] : '/') + '" />');
      }());
    </script>
    <script defer src="main.dart.js"></script>
  </head>
  <body>
    <my-app>Loading...</my-app>
  </body>
</html>
```

在生产环境中，将 script 脚本替换为 base 元素，其中 href 设置为应用程序的根路径。如果路径为空，则使用"/"。示例代码如下：

```
<head>
  <base href="/">
</head>
```

创建任务和雇员列表组件，TaskListComponent 组件代码如下：

```
//chapter15/router/lib/src/task/task_list_component.dart
import 'package:angular/angular.dart';

@Component(
  selector: 'task-list',
  template: '''
    <h2>任务列表</h2>
    <p>具体任务</p>
  ''',
)
class TaskListComponent{}
```

EmployeeListComponent 组件代码如下：

```
//chapter15/router/lib/src/employee/employee_list_component.dart
import 'package:angular/angular.dart';

@Component(
  selector: 'employee-list',
  template: '''
    <h2>雇员列表</h2>
    <p>雇员信息</p>
  ''',
)
class EmployeeListComponent{}
```

上述两个组件模板中没有实质性内容。

4．路由

路由告诉路由器用户单击链接或直接粘贴 URL 到浏览器时显示哪些视图。

1）路由路径

RoutePath 类的构造函数声明代码如下：

```
RoutePath({
  String path,
  this.parent,
```

```
    this.useAsDefault = false,
    this.additionalData,
  }) : this.path = URL.trimSlashes(path);
```

参数 path 表示路径。parent 是 RoutePath 类型，表示父路由路径。useAsDefault 表示是否用作默认路由。additionalData 表示可以是任意类型的附加数据。

通常会将所有的路由路径定义并封装在一个文件中。为每个应用视图定义一个路由路径，并将这些路由路径作为类变量封装在类 RoutePaths 中。示例代码如下：

```
//chapter15/router/lib/src/route_paths.dart
import 'package:angular_router/angular_router.dart';

class RoutePaths{
  static final tasks = RoutePath(path:'tasks');
  static final employees = RoutePath(path:'employees');
}
```

通过在单独的文件中定义路由路径，可以避免导航结构复杂的应用程序中路由定义之间的循环依赖性。

2）路由定义

路由器根据路由定义列表协调应用导航，路由定义将路由路径与组件相关联，组件负责处理到路径的导航及关联视图的渲染。

路由定义由 RouteDefinition 类负责，其工厂构造函数等价于如下代码：

```
factory RouteDefinition({
  String path,
ComponentFactory < Object > component,
  bool useAsDefault,
  dynamic additionalData,
  RoutePath routePath})
  : this.path = Url.trimSlashes(path ?? routePath?.path),
  this.useAsDefault = useAsDefault ?? routePath?.useAsDefault ?? false,
  this.additionalData = additionalData ?? routePath?.additionalData;
```

参数 path 表示路径。component 表示与路径关联的组件实现，ComponentFactory 是指用@Component 注解的类 T 背后的支持实现。useAsDefault 表示是否用作默认路由，additionalData 表示可以是任意类型的附加数据。routePath 表示定义的路径路由。若提供了 routePath 参数，且 path、useAsDefault 或 additionalData 未指定，它们的值将被 routePath 中的数据覆盖。

定义一组路由，并将它们作为类变量封装在类 Routes 中。示例代码如下：

```
//chapter15/router/lib/src/routes.dart
import 'package:angular_router/angular_router.dart';

import 'employee/employee_list_component.template.dart' as employee_list_template;
import 'task/task_list_component.template.dart' as task_list_template;
import 'route_paths.dart';
export 'route_paths.dart';

class Routes{
  //导航到任务列表组件的路由
  static final tasks = RouteDefinition(
    routePath: RoutePaths.tasks,
    component: task_list_template.TaskListComponentNgFactory,
  );
  //导航到雇员列表组件的路由
  static final employees = RouteDefinition(
    routePath: RoutePaths.employees,
    component: employee_list_template.EmployeeListComponentNgFactory,
  );
  //返回已定义路由列表
  static final all = <RouteDefinition>[
    tasks,
    employees,
  ];
}
```

将 AppComponent 作为路由组件，导入路由定义文件，在指令列表中添加路由器指令常量 routerDirectives。出现了新的@Component 注解参数 exports，它的值是一个列表，在该列表中定义的标识符可以在模板中引用，其使用方式和组件属性一样。这里将静态类 RoutePaths 和 Routes 导出到模板，使得模板中可以引用它们。示例代码如下：

```
//chapter15/router/lib/app_component.dart
import 'package:angular/angular.dart';
import 'package:angular_router/angular_router.dart';

import 'src/routes.dart';

@Component(
  selector: 'my-app',
  styleURLs: ['app_component.css'],
  templateURL: 'app_component.html',
  directives: [routerDirectives],
  exports: [RoutePaths,Routes],
)
class AppComponent{}
```

5. 导航

路由包中定义了 3 个指令用于与导航相关的操作：

RouterLink 指令需要在锚标签 a 上使用。将路由路径通过属性绑定到 routerLink，该指令会为锚标签添加 href 属性并设值。可以直接为 routerLink 属性指定链接，也可以绑定到路由路径。示例代码如下：

```
<!-- 指定明确的链接 -->
<a routerLink = "/employees">雇员列表</a>
<!-- 指定路由路径 -->
<a [routerLink] = "RoutePaths.employees.toURL()">雇员列表</a>
```

RouterLinkActive 与 RouterLink 指令配合使用，当连接的路由处于活动状态时，将指定 CSS 类添加到绑定的元素。可以将其和 RouterLink 指令绑定在同一个元素，也可以绑定到 RouterLink 指令绑定的元素的祖先元素上。示例代码如下：

```
<!-- 与 RouterLink 指令绑定在同一元素上 -->
<a routerLink = "/employees" routerLinkActive = "active - route">雇员列表</a>
<!-- 绑定在 RouterLink 指令绑定元素的祖先元素上 -->
<div routerLinkActive = "active - route">
   <a routerLink = "/employees">雇员列表</a>
</div>
```

RouterOutlet 指令作用对象是 router-outlet 元素。在 DOM 中，路由器通过在 router-outlet 元素之后插入视图元素作为同级来显示视图。使用时需要将需要展示的路由绑定到输入属性 routes 上。

```
<router - outlet [routes] = "Routes.all"></router - outlet>
```

更新 AppComponent 模板，使其具有两个路由的导航栏和控制视图的 router-outlet 元素。示例代码如下：

```
//chapter15/router/lib/app_component.html
<h1>Angular 路由</h1>
<nav>
    <a [routerLink] = "RoutePaths.tasks.toURL()"
       [routerLinkActive] = "'active - route'">任务列表</a>
    <a [routerLink] = "RoutePaths.employees.toURL()"
       [routerLinkActive] = "'active - route'">雇员列表</a>
</nav>
<router - outlet [routes] = "Routes.all"></router - outlet>
```

在样式文件中添加 CSS 类，代码如下：

```
//chapter15/router/lib/app_component.css
/* 导航链接样式 */
nav a {
  padding: 5px 10px;
  text-decoration: none;
  margin-right: 10px;
  margin-top: 10px;
  display: inline-block;
  background-color: #eee;
  border-radius: 4px;
}
nav a:visited, a:link {
  color: #607D8B;
}
nav a:hover {
  color: #039be5;
  background-color: #CFD8DC;
}
nav a.active {
  color: #039be5;
}
/* 路由处于活跃状态的链接样式 */
.active-route {color: #039be5}
```

路由器将显示视图作为< router-outlet >元素的同级元素插入 DOM 中，其中显示视图是当前活跃路由相关联的组件的模板，且该路由必须存在于绑定到< router-outlet >元素输入属性 routes 的路由集合中。这里 Routes.all 包含所有已定义的路由。

15.6.2 常用配置

1. 设置默认路由

启动应用程序后，浏览器的初始 URL 与任何已配置的路由都不匹配，这意味着应用启动后不会显示任何组件，只会显示导航栏。用户必须单击导航栏中的一个链接，才能触发对应组件视图的显示。

如果希望应用启动后就显示某个组件，可以在路由定义中将参数 useAsDefault 的值配置为 true。示例代码如下：

```
static final employees = RouteDefinition(
  routePath: RoutePaths.employees,
  component: employee_list_template.EmployeeListComponentNgFactory,
  useAsDefault: true
);
```

刷新浏览器，应用程序显示雇员列表，URL 路径为“/”。

2. 重定向路由

可以通过匹配路径重定向到另一个路由，重定向路由的构造函数代码如下：

```
factory RouteDefinition.redirect({
  String path,
  String redirectTo,
  bool useAsDefault,
  additionalData,
  RoutePath routePath,
})
```

redirectTo 参数的值为 URL，表示将要跳转的地址。删除路由定义中的配置参数 useAsDefault，添加重定向路由，该路由会匹配到 URL 路径"/"，然后跳转到"/#/employees"。示例代码如下：

```
static final all = <RouteDefinition>[
  tasks,
  employees,
  RouteDefinition.redirect(
    path: '',
    redirectTo: RoutePaths.employees.toURL(),
  )
];
```

3. 通配符路由

目前已经创建了到"/#/tasks"和"/#/employees"的路由，如果向路由器提供了其他的路径，则可以通过通配符路由匹配未找到的路由。

首先创建组件 NotFoundComponent，用于拦截未定义的路由。示例代码如下：

```
//chapter15/router/lib/src/not_found_component.dart
import 'package:angular/angular.dart';

@Component(
  selector: 'not-found',
  template: '<h2>页面未找到</h2>',
)
class NotFoundComponent {}
```

完整的路由定义代码如下：

```
//chapter15/router/lib/src/routes.dart
import 'package:angular_router/angular_router.dart';

import 'employee/employee_list_component.template.dart' as employee_list_template;
import 'task/task_list_component.template.dart' as task_list_template;
```

```
import 'not_found_component.template.dart' as not_found_template;
import 'route_paths.dart';
export 'route_paths.dart';

class Routes{
  static final tasks = RouteDefinition(
    routePath: RoutePaths.tasks,
    component: task_list_template.TaskListComponentNgFactory,
  );

  static final employees = RouteDefinition(
    routePath: RoutePaths.employees,
    component: employee_list_template.EmployeeListComponentNgFactory,
  );

  static final all = <RouteDefinition>[
    tasks,
    employees,
    RouteDefinition.redirect(
      path: '',
      redirectTo: RoutePaths.employees.toURL(),
    ),
    RouteDefinition(
      path: '.+',
      component: not_found_template.NotFoundComponentNgFactory,
    )
  ];
}
```

正则表达式“.+”匹配所有非空路径。因为路由器会根据路由定义列表 all 中的顺序依次匹配路由，因此应当将通配符路由放在路由定义列表 all 中的最后一个。

15.6.3 函数导航

本节实现的功能是通过单击雇员列表中的条目跳转到所对应雇员信息页，跳转是通过与条目单击事件绑定的函数完成的。

首先创建 Employee 类，代码如下：

```
//chapter15/router/lib/src/employee/employee.dart
class Employee{
  final int id;
  String name;
  num salary;
  Employee(this.id,this.name,this.salary);
}
```

创建提供数据服务的类 EmployeeService，代码如下：

```
//chapter15/router/lib/src/employee/employee_service.dart
import 'employee.dart';
class EmployeeService{
  //定义雇员列表
  var emps = [
    Employee(5, 'Magneta',9971),
    Employee(6, 'RubberMan',4533),
    Employee(7, 'Dynama',6720),
    Employee(8, 'Dr IQ',4907),
    Employee(9, 'Magma',5278),
    Employee(10, 'Tornado',7800)
  ];
  //获取所有雇员信息
  List<Employee> getAll(){
    return emps;
  }
  //根据 id 获取单个雇员信息
  Employee getById(int id){
    return emps.firstWhere((emp) => emp.id == id);
  }
}
```

将服务 EmployeeService 添加到组件 EmployeeListComponent 的提供者列表。

```
providers:[ClassProvider(EmployeeService)],
```

添加路由路径 employee，代码如下：

```
//chapter15/router/lib/src/route_paths.dart
import 'package:angular_router/angular_router.dart';
//定义常量 idParam，它的值为 id
const idParam = 'id';
class RoutePaths{
  //定义任务列表页的路由路径对象 tasks
  static final tasks = RoutePath(path:'tasks');
  //定义雇员列表页的路由路径对象 employees
  static final employees = RoutePath(path:'employees');
  //定义雇员信息页的路由路径对象 employee
  //并需在路径中携带参数 id
  static final employee = RoutePath(path: '${employees.path}/:$idParam');
}
//解析参数列表中参数 id 的值
int getId(Map<String,String> parameters){
```

```
  final id = parameters[idParam];
  return id == null ? null : int.parse(id);
}
```

在路由路径对象 employee 的 path 参数值中，“${employees.path}”代表引用路由路径对象 employees 的路径；“/”代表分隔符；“:$idParam”表示一个参数，参数以冒号(:)开始，从代码可以看出 $idParam 的值为 id，此处 id 是一个占位符。例如：查看 id 为 5 的雇员信息，实际 URL 中其路径是“/employees/5”。

如果需要传递多个参数，则每个参数都应以冒号开始加上占位符，在参数间必须使用分割符，且参数的实际值不能包含与分割符一样的字符串序列。常用的分隔符是斜杠(/)或逗号(,)。例如：示例代码中包含 id 和 other 两个参数，它们之间使用逗号作为分隔符。

```
RoutePath(path: '${employees.path}/:$idParam,:other');
```

方法 getId 是一个帮助函数，用于解析参数列表中参数 id 的值。

添加组件 EmployeeComponent，用于展示雇员信息。示例代码如下：

```
//chapter15/router/lib/src/employee/employee_component.dart
import 'package:angular/angular.dart';
import 'package:angular_router/angular_router.dart';

import '../route_paths.dart';
import 'employee.dart';
import 'employee_service.dart';

@Component(
  selector: 'employee',
  templateURL: 'employee_component.html',
  styleURLs: ['employee_component.css'],
  directives: [coreDirectives],
  providers: [ClassProvider(EmployeeService)],
)
//实现 OnActivate 接口
class EmployeeComponent implements OnActivate {
  final EmployeeService _employeeService;
  final Router _router;
  Employee emp;
  //注入服务 EmployeeService 和路由器 Router
  EmployeeComponent(this._employeeService, this._router);

  //实现路由器的生命周期挂钩函数 onActivate
  @override
  void onActivate(_, RouterState current){
```

```
    //解析当前路由中参数 id 的值
    var id = getId(current.parameters);
    //如果 id 值存在,则获取对应 id 的雇员信息
    if (id != null) emp = _employeeService.getById(id);
  }

  //导航到雇员列表
  Future<NavigationResult> goBack(){
    //使用路由器对象导航到指定路由
    _router.navigate(RoutePaths.employees.toURL());
  }
}
```

在提供者列表中添加并注入服务 EmployeeService。注入了路由器对象 Router,该对象的 navigate 方法用于导航到新的路由,它接收一个必选参数 path 和一个可选参数 navigationParams。

```
Future<NavigationResult> navigate(
  String path, [
  NavigationParams navigationParams,
]);
```

方法 navigate 会导航到与参数 path 匹配的路由,对象 navigationParams 用于向路由指定可选参数列表。可以向 NavigationParams 的构造函数提供参数 queryParameters 以构建新示例。示例代码如下:

```
Future<NavigationResult> goBack() => _router.navigate(
  RoutePaths.heroes.toURL(),
   NavigationParams(queryParameters: {idParam: '${emp.id}'}));
```

查询参数 queryParameters 是一个 Map 集合,可以提供多个参数。以 id 为 5 的雇员为例,其最终跳转路径等价于"/employees? id=5"。查询参数是以问号? 引导的,参数间使用符号 & 分隔。

将 OnActivate 接口添加到组件的实现接口列表中,实现路由器的生命周期函数 onActivate,当组件通过路由激活时会触发该生命周期函数。它会接收两个参数,示例代码如下:

```
void onActivate(RouterState previous, RouterState current);
```

两个参数都是表示路由器状态的 RouterState 对象,该对象包含路由的路径 path、路由定义列表、参数集合 parameters、查询参数集合 queryParameters。第一个 RouterState 代表上一个路由的状态,第二个 RouterState 代表当前路由的状态,因为用不到上一个路由的状态,因此在代码中使用占位符"_"代替。

更新组件 EmployeeComponent 的模板，示例代码如下：

```
//chapter15/router/lib/src/employee/employee_component.html
<div *ngIf="emp!= null">
    <h2>{{emp.name}}</h2>
    <div>
        <label>id: </label>{{emp.id}}</div>
    <div>
        <label>薪资: </label>{{emp.salary}}
    </div>
    <button (click)="goBack()">返回</button>
</div>
```

更新组件 EmployeeComponent 的样式，代码如下：

```
//chapter15/router/lib/src/employee/employee_component.css
label {
  display: inline-block;
  width: 3em;
  margin: .5em 0;
  color: #607D8B;
  font-weight: bold;
}

button {
  margin-top: 20px;
  font-family: Arial;
  background-color: #eee;
  border: none;
  padding: 5px 10px;
  border-radius: 4px;
  cursor: pointer; cursor: hand;
}
button:hover {
  background-color: #cfd8dc;
}
```

导入组件 EmployeeComponent 的实现类，并添加路由定义 employee。代码如下：

```
//chapter15/router/lib/src/routes.dart
import 'employee/employee_component.template.dart' as employee_template;
//...
class Routes {
  //...
  static final employee = RouteDefinition(
```

```
    routePath: RoutePaths.employee,
    component: employee_template.EmployeeComponentNgFactory,
  );

  static final all = <RouteDefinition>[
    //...
    employee,
    //...
  ];
}
```

更新组件 EmployeeListComponent 的模板代码如下：

```
//chapter15/router/lib/src/employee/employee_list_component.html
<ul *ngIf="emps != null" class="list">
    <li *ngFor="let emp of emps"
        (click)="onSelect(emp)">
        <span class="badge">{{emp.id}}</span>
        <span class="text">{{emp.name}}</span>
    </li>
</ul>
```

更新组件 EmployeeListComponent 的代码如下：

```
//chapter15/router/lib/src/employee/employee_list_component.dart
import 'package:angular/angular.dart';
import 'package:angular_router/angular_router.dart';

import '../route_paths.dart';
import 'employee_service.dart';
import 'employee.dart';

@Component(
  selector: 'employee-list',
  templateURL: 'employee_list_component.html',
  styleURLs: ['employee_list_component.css'],
  directives: [coreDirectives],
  providers: [ClassProvider(EmployeeService)],
)
class EmployeeListComponent{
  final EmployeeService _employeeService;
  final Router _router;
  List<Employee> emps;
  EmployeeListComponent(this._employeeService,this._router){
    //初始化雇员列表 emps
```

```
    emps = _employeeService.getAll();
  }

  //根据指定 id 构建导航到雇员信息页的路由
  String _empURL(int id) =>
      RoutePaths.employee.toURL(parameters: {idParam: '$id'});

  //导航到指定路由
  Future<NavigationResult> _gotoDetail(int id) =>
    _router.navigate(_empURL(id));

  //响应雇员列表条目的单击事件
  void onSelect(Employee emp) {
    _gotoDetail(emp.id);
  }
}
```

组件注入了服务 EmployeeService 和路由器 Router，并在构造函数中通过服务的 getAll 方法对雇员列表 emps 进行了初始化。

当用户单击雇员列表中的条目时会执行 onSelect 方法，该方法又会调用 gotoDetail 方法，该方法的功能是通过路由器的 navigate 方法导航到雇员信息页，其参数通过 empURL 方法获得，empURL 方法内部通过 RoutePath 对象的 toURL 方法构建 URL，并指定参数值。

更新组件 EmployeeListComponent 的样式，代码如下：

```
//chapter15/router/lib/src/employee/employee_list_component.css
.list {
  margin: 0 0 2em 0;
  list-style-type: none;
  padding: 0;
  width: 15em;
}
.list li {
  cursor: pointer;
  position: relative;
  left: 0;
  background-color: #EEE;
  margin: .5em;
  padding: .3em 0;
  height: 1.6em;
  border-radius: 4px;
}
.list li:hover {
  color: #607D8B;
```

```
  background-color: #EEE;
  left: .1em;
}
.list .text {
  position: relative;
  top: -3px;
}
.list .badge {
  display: inline-block;
  font-size: small;
  color: white;
  padding: 0.8em 0.7em 0 0.7em;
  background-color: #607D8B;
  line-height: 1em;
  position: relative;
  left: -1px;
  top: -4px;
  height: 1.8em;
  margin-right: .8em;
  border-radius: 4px 0 0 4px;
}
```

保存更改并刷新浏览器，单击雇员列表中的条目，观察视图更改和路由变化。

15.6.4 子路由

前面已经介绍了与路由相关的所有重要内容，在本节中将创建一个与应用程序根组件无关的路由组件，它将管理与任务相关的路由。

将 TaskListComponent 作为路由组件，它拥有自己的路由集 routes 和 router-outlet，就像 AppComponent 一样。通过该路由组件导航到 TaskHomeComponent 和 TaskComponent。TaskComponent 组件接收一个参数并展示相关任务的信息。TaskHomeComponent 组件则单纯作为子路由的默认路由。

创建服务 TaskService，用于提供与任务相关的数据和功能。代码如下：

```
//chapter15/router/lib/src/task/task_service.dart
class Task{
  int id;
  String task;
  Task(this.id,this.task);
}

class TaskService{
  var tasks = [
    Task(5, '业务分析'),
```

```
    Task(6, '产品分析'),
    Task(7, '结构设计'),
    Task(8, '视觉设计')
  ];
  List<Task> getAll(){
    return tasks;
  }
  Task getById(int id){
    return tasks.firstWhere((task) => task.id == id);
  }
}
```

创建组件 TaskHomeComponent，它将用作子路由的默认路由。代码如下：

```
//chapter15/router/lib/src/task/task_home_component.dart
import 'package:angular/angular.dart';

@Component(
  selector: 'task-home',
  template: '''
  <div>
    任务中心
  </div>
  ''',
  directives: [coreDirectives],
)
class TaskHomeComponent{}
```

创建组件 TaskComponent，它将用于展示单个任务信息。代码如下：

```
//chapter15/router/lib/src/task/task_component.dart
import 'package:angular/angular.dart';
import 'package:angular_router/angular_router.dart';

import 'task_service.dart';
import 'route_paths.dart';

@Component(
  selector: 'task',
  template: '''
    <div *ngIf="task!= null">
        <div>
            <label>id: </label>{{task.id}}</div>
        <div>
            <label>任务: </label>{{task.task}}
```

```
        </div>
        <button (click) = "goBack()">返回</button>
      </div>
    ''',
    styleURLs: ['task_component.css'],
    directives: [coreDirectives],
)
class TaskComponent implements OnActivate{
  final TaskService _taskService;
  final Router _router;
  Task task;
  //注入服务 TaskService 和 Router
  TaskComponent(this._taskService,this._router);
  //实现路由器生命周期函数 onActivate
  @override
  void onActivate(RouterState previous, RouterState current) {
    var id = getId(current.parameters);
    task = _taskService.getById(id);
  }
  //跳转到子路由的默认路由
  Future<NavigationResult> goBack(){
    _router.navigate(RoutePaths.home.toURL());
  }
}
```

样式代码如下：

```
//chapter15/router/lib/src/task/task_component.css
label {
  display: inline-block;
  width: 3em;
  margin: .5em 0;
  color: #607D8B;
  font-weight: bold;
}

button {
  margin-top: 20px;
  font-family: Arial;
  background-color: #eee;
  border: none;
  padding: 5px 10px;
  border-radius: 4px;
  cursor: pointer; cursor: hand;
}
```

```
button:hover {
  background-color: #cfd8dc;
}
```

定义子路由的路由路径。代码如下：

```
//chapter15/router/lib/src/task/route_paths.dart
import 'package:angular_router/angular_router.dart';
import '../route_paths.dart' as _parent;
export '../route_paths.dart' show idParam,getId;
class RoutePaths{
  //定义子路由的默认路由
  static final home = RoutePath(
    path:'',
    parent: _parent.RoutePaths.tasks,
  );
  //定义子路由下导航到任务信息页的路由
  static final task = RoutePath(
      path:':${_parent.idParam}',
      parent: _parent.RoutePaths.tasks,
  );
}
```

该文件导入了父路由定义的路由路径文件，其别名为_parent。将常量 idParam 和方法 getId 导出，使得引入当前路由路径文件的库可以使用。

在路由路径定义中出现了新的参数 parent，它表示当前路径的父路径，该参数的值引用当前路由组件 TaskListComponent 在父路由中定义的路由路径 tasks。

添加子路由的路由定义。代码如下：

```
//chapter15/router/lib/src/task/routes.dart
import 'package:angular_router/angular_router.dart';

import 'task_component.template.dart' as task_template;
import 'task_home_component.template.dart' as task_home_template;
import 'route_paths.dart';
export 'route_paths.dart';

class Routes{
  //导航到子路由下 TaskHomeComponent 组件的路由定义
  //并设置为子路由下的默认路由
  static final home = RouteDefinition(
    routePath: RoutePaths.home,
    component: task_home_template.TaskHomeComponentNgFactory,
    useAsDefault: true,
```

```
  );
  //导航到子路由下 TaskComponent 组件的路由定义
  static final task = RouteDefinition(
    routePath: RoutePaths.task,
    component: task_template.TaskComponentNgFactory,
  );
  //引用所有已定义路由
  static final all = <RouteDefinition>[
    home,
    task,
  ];
}
```

更新 TaskListComponent 组件。代码如下：

```
//chapter15/router/lib/src/task/task_list_component.dart
import 'package:angular/angular.dart';
import 'package:angular_router/angular_router.dart';

import 'task_service.dart';
import 'routes.dart';
@Component(
  selector: 'task-list',
  template: '''
  <ul *ngIf="tasks != null" class="list">
      <li *ngFor="let task of tasks"
          (click)="onSelect(task)">
          <span class="badge">{{task.id}}</span>
          <span class="text">{{task.task}}</span>
      </li>
  </ul>
  <router-outlet [routes]="Routes.all"></router-outlet>
  ''',
  styleURLs: ['task_list_component.css'],
  directives: [coreDirectives, routerDirectives],
  providers: [ClassProvider(TaskService)],
  exports: [RoutePaths, Routes],
)
class TaskListComponent {
  final TaskService _taskService;
  final Router _router;
  List<Task> tasks;
  TaskListComponent(this._taskService, this._router){
    //初始化任务列表 tasks
    tasks = _taskService.getAll();
```

```
  }
  //响应任务列表条目的单击事件
  void onSelect(Task task) {
    _gotoDetail(task.id);
  }
  //根据指定 id 构建导航到任务信息页的路由
  String _empURL(int id) =>
      RoutePaths.task.toURL(parameters: {idParam: '$id'});
  //导航到指定路由
  Future<NavigationResult> _gotoDetail(int id) =>
      _router.navigate(_empURL(id));
}
```

更新组件的样式，代码如下：

```
//chapter15/router/lib/src/task/task_list_component.css
.list {
  margin: 0 0 2em 0;
  list-style-type: none;
  padding: 0;
  width: 15em;
}
.list li {
  cursor: pointer;
  position: relative;
  left: 0;
  background-color: #EEE;
  margin: .5em;
  padding: .3em 0;
  height: 1.6em;
  border-radius: 4px;
}
.list li:hover {
  color: #607D8B;
  background-color: #EEE;
  left: .1em;
}
.list .text {
  position: relative;
  top: -3px;
}
.list .badge {
  display: inline-block;
  font-size: small;
  color: white;
```

```
    padding: 0.8em 0.7em 0 0.7em;
    background-color: #607D8B;
    line-height: 1em;
    position: relative;
    left: -1px;
    top: -4px;
    height: 1.8em;
    margin-right: .8em;
    border-radius: 4px 0 0 4px;
  }
```

保存文件并刷新浏览器，进入任务列表，单击其中的任意条目，观察子路由变化情况。本节的重点应该关注在路由路径定义文件上，其他部分与15.6.3节的内容相似。通常不会在开发中使用子路由，它的适用场景是重用路由组件，即组合多个现有路由器。

15.6.5 生命周期函数

默认情况下，任何用户都可以在应用程序中任意进行导航，对于完善的应用程序这样做并不总是正确的，例如：

（1）用户无权导航到目标组件。

（2）用户必须先经过权限验证。

（3）在显示目标组件前初始化一些数据。

（4）用户可能在离开组件前有表单数据未保存。

路由提供了一系列生命周期函数，以帮助开发者处理这些情况，如表15-3所示。

表15-3 路由生命周期函数

挂钩函数	目的和时间
onActivate	当路由激活组件时调用，常用于数据初始化。例如：从网络请求中获取数据
onDeactivate	当路由停用组件时调用
canNavigate	控制当前路由是否可跳转
canReuse	设置当前组件实例是否可复用。注意：父组件可复用，子组件实例才可复用
canDeactivate	允许有条件地停用路由
canActivate	允许有条件地启用路由

所有的生命周期函数都包含在同名接口中，对应的接口名分别为OnActivate、OnDeactivate、CanNavigate、CanReuse、CanDeactivate、CanActivate。

首先在组件TaskListComponent上使用mixin类CanReuse，其作用是保证TaskHomeComponent组件实例能被重用。代码如下：

```
//chapter15/router/lib/src/task/task_list_component.dart
class TaskListComponent with CanReuse{
//...
}
```

在组件中 TaskHomeComponent 实现所有路由周期函数对应的接口，并实现周期函数。示例代码如下：

```
//chapter15/router/lib/src/task/task_home_component.dart
import 'package:angular/angular.dart';
import 'package:angular_router/angular_router.dart';

@Component(
  selector: 'task-home',
  template: '''
  <div>
    任务中心
  </div>
  ''',
  directives: [coreDirectives],
)
class TaskHomeComponent with CanReuse implements CanActivate, CanDeactivate, CanNavigate,
OnActivate,OnDeactivate {

  //在路由激活组件时调用
  @override
  void onActivate(RouterState previous, RouterState current) {
    //常用于初始化数据
    print('TaskHomeComponent: onActivate: ${previous?.toURL()} -> ${current?.toURL()}');
  }

  //在路由停用组件之前调用
  @override
  void onDeactivate(RouterState current, RouterState next) {
    //常用于停用组件实例前处理必须完成的任务
    print('TaskHomeComponent: onDeactivate: ${current?.toURL()} -> ${next?.toURL()}');
  }

  //允许有条件地激活新路由
  @override
  Future<bool> canActivate(RouterState current, RouterState next) async{
    print('TaskHomeComponent: canActivate: ${current?.toURL()} -> ${next?.toURL()}');
    //在实际使用中需根据条件控制是否允许激活当前路由
    //这里没有添加控制条件,直接返回 true,表示允许
    return true;
  }

  //允许有条件地停用路由
  @override
  Future<bool> canDeactivate(RouterState current, RouterState next) async{
```

```
    print('TaskHomeComponent: canDeactivate: ${current?.toURL()} -> ${next?.toURL()}');
    //在实际使用中需根据条件控制是否允许停用当前路由
    //这里没有添加控制条件,直接返回 true,表示允许
    return true;
  }

  //允许有条件地阻止导航
  @override
  Future<bool> canNavigate() async{
    print('TaskHomeComponent: canNavigate');
    //在实际使用中需根据条件控制是否允许导航
    //这里没有添加控制条件,直接返回 true,表示允许
    return true;
  }

  //允许重用现有的组件实例
  @override
  Future<bool> canReuse(RouterState current, RouterState next) async{
    print('TaskHomeComponent: canReuse: ${current?.toURL()} -> ${next?.toURL()}');
    //在实际使用中需根据条件控制是否允许重用此组件实例
    //这里没有添加控制条件,直接返回 true,表示允许
    return true;
  }
}
```

刷新浏览器,打开控制台。单击导航中的任务列表,再单击列表中任意一个条目,再单击"返回"按钮。运行结果如图 15-14 所示。

```
TaskHomeComponent: canActivate: /employees -> /tasks
TaskHomeComponent: onActivate: /employees -> /tasks
TaskHomeComponent: canNavigate
TaskHomeComponent: canDeactivate: /tasks -> /tasks/7
TaskHomeComponent: onDeactivate: /tasks -> /tasks/7
TaskHomeComponent: canReuse: /tasks -> /tasks/7
TaskHomeComponent: canActivate: /tasks/7 -> /tasks
TaskHomeComponent: onActivate: /tasks/7 -> /tasks
>
```

图 15-14 路由生命周期函数

15.7 结构指令

前面已经介绍了结构指令的使用,这里将介绍结构指令的更多细节。

15.7.1 星号前缀

在使用结构指令时会注意到指令前面会带上星号(*),本节将介绍它的具体用途。

1. NgIf

使用 * ngIf 根据模板表达式判断是否显示元素及内容。代码如下：

```
<div *ngIf = "name != null">名字:{{name}}</div>
```

星号实际上是语法糖,在内部处理时,Angular 会将其分解为<template>元素,并将宿主元素<div>包裹在<template>元素中。代码如下：

```
<template [ngIf] = "name != null">
    <div>名字:{{name}}</div>
</template>
```

此时 * ngIf 指令转移到<template>元素上,并且成为其属性指令[ngIf]。其余部分将按原结构包裹在 template 元素中。

2. NgFor

* ngFor 完整功能的使用示例代码如下：

```
<div *ngFor = "let emp of emps;let i = index;
let odd = odd;trackBy:trackById" [class.odd] = "odd">
    Id: {{emp.id}} Name:{{emp.name}} Salary:{{emp.salary}}
</div>
```

Angular 会将其分解为<template>元素,代码如下：

```
<template ngFor let-emp [ngForOf] = "emps" let-i = "index" let-odd = "odd"
          [ngForTrackBy] = "trackById">
  <div [class.odd] = "odd"> Id: {{emp.id}} Name:{{emp.name}} Salary:{{emp.salary}}</div>
</template>
```

这显然比 NgIf 使用起来更加复杂,需要说明的是在 * ngFor 后边字符串之外的部分仍然会停留在宿主元素中,此示例[ngClass]="odd"就停留在<div>中。

Angular 微语法可以在一个字符串中简洁有效地配置指令,而微语法解析器将字符串转换为元素<template>上的属性：

(1) let 关键字声明在模板中引用的模板输入变量。此示例中输入变量是 emp、i 和 odd。解析器将 let emp、let i 和 let odd 转换为名为 let-emp、let-i 和 let-odd 的模板输入变量。

(2) 微语法解析器将 of 和 trackby 更名为 Of 和 TrackBy,并且使用 ngFor 作为前缀,最终名称为 ngForOf 和 ngForTrackBy。它们是 NgFor 两个输入属性的名称，这样，指令就可以知道列表是 emps，而跟踪函数是 trackById。

（3）当 NgFor 指令遍历列表时，它会设置和重置自身上下文对象的属性。这些属性包括 index 和 odd，以及一个特殊属性 $implicit。

（4）将 let-i 和 let-odd 变量定义为 let i=index 和 let odd=odd。Angular 将它们设置为上下文的 index 和 odd 属性的当前值。

（5）未向 let-hero 指定上下文属性，因为它的来源是隐式的。Angular 将 let-hero 设置为上下文的 $implicit 属性的值，$implicit 属性的值由 NgFor 使用当前迭代的 emp 为其进行初始化。

使用 let 关键字声明模板输入变量，变量的作用范围仅限于重复模板的单个实例。可以在模板包含的其他结构指令中使用相同的变量名。

宿主元素可以使用多个属性指令，但是只能使用一个结构指令。因为结构指令会对宿主元素及其后代进行复杂处理，当多个指令对同一宿主元素进行处理时，哪一个应该优先成为一个难题。因此常规的做法是将多个结构指令应用于不同的宿主元素上。

3. NgSwitch

从前面的知识已经了解到 NgSwitch 实际上是一组协作指令：NgSwitch、NgSwitchCase、NgSwitchDefault。

先来看一个示例，代码如下：

```
<div [ngSwitch] = "color">
    <span *ngSwitchCase = "'Red'">红色</span>
    <span *ngSwitchCase = "'Green'">绿色</span>
    <span *ngSwitchCase = "'Black'">黑色</span>
    <span *ngSwitchDefault >蓝色</span>
</div>
```

NgSwitch 是一个属性指令，用于控制其他两个协作指令的行为。因此使用[ngSwitch]而不是 *ngSwitch。

NgSwitchCase 和 NgSwitchDefault 是结构性指令，因此使用形式为 *ngSwitchCase 和 *ngSwitchDefault。与其他结构指令一样，它们也可简化为<template>元素的形式。代码如下：

```
<div [ngSwitch] = "color">
    <template [ngSwitchCase] == "'Red'">
        <span>红色</span>
    </template>
    <template [ngSwitchCase] == "'Green'">
        <span>绿色</span>
    </template>
    <template [ngSwitchCase] == "'Black'">
```

```
        <span>黑色</span>
    </template>
    <template ngSwitchDefault>
        <span>蓝色</span>
    </template>
</div>
```

在应用结构指令时可以手动指定为<template>元素的形式,但推荐使用星号的模式。

15.7.2 自定义结构指令

在本节编写名叫 RepeatDirective 的结构指令,该指令的作用是根据提供的次数重复模板内容。

指令是使用注解@Directive 声明的类。建立指令的基本结构,代码如下:

```
import 'package:angular/angular.dart';

@Directive(selector: '[myRepeat]')
class RepeatDirective{
}
```

指令的选择器通常是方括号包裹的属性选择器[myRepeat],指令属性名字采用小驼峰命名法,并且带有前缀。避免使用 ng 前缀,该前缀是 Angular 占有的。为了编码规范指令类名称以 Directive 结尾。

1. 嵌入式模板

TemplateRef 表示能够用于实例化嵌入式视图的嵌入式模板。

可通过两种方式访问 TemplateRef:第一种,将指令放置在<template>元素上或者使用指令时以 * 开头,并通过 TemplateRef 令牌将此嵌入式视图的 TemplateRef 对象注入指令的构造函数中。第二种,通过 Query 从组件或指令中查找 TemplateRef。

2. 视图容器

ViewContainerRef 表示可以附加一个或多个视图的容器。该容器可以包含两种视图:第一种,通过 createComponent 方法实例化一个组件创建的宿主视图。第二种,通过 createEmbeddedView 方法实例化嵌入式模板 TemplateRef 创建的嵌入式视图。

要访问元素的 ViewContainerRef,可以将 ViewContainerRef 注入指令的构造函数中,或者通过 ViewChild 查询得到。

ViewContainerRef 类包含如下一些重要方法:

(1) clear():销毁该容器中的所有视图。

(2) createEmbeddedView(TemplateRef templateRef) → EmbeddedViewRef:基于 TemplateRef 实例化一个嵌入式视图,并将其附加到此容器中。

(3) insertEmbeddedView(TemplateRef templateRef, int index) → EmbeddedViewRef:基

于 TemplateRef 实例化一个嵌入式视图,并将其插入此容器中的指定索引 index 处。

使用 TemplateRef 可以获取 Angular 生成的< template >元素中的内容,使用 ViewContainerRef 可以访问宿主元素的视图容器。

通过构造函数将 TemplateRef 和 ViewContainerRef 注入指令类的私有变量中。代码如下:

```
//嵌入式模板
TemplateRef _templateRef;
//视图容器
ViewContainerRef _viewContainer;
//注入视图容器和嵌入式模板
RepeatDirective(this._templateRef, this._viewContainer);
```

该指令还需要接收一个值,声明一个输入属性 myRepeat。因为不需要获取它的值,所以只提供了该属性的 setter 方法。代码如下:

```
@Input()
set myRepeat(int times){
  if(times > 0){
    for(int i = 0; i < times; i++){
      //向视图容器插入嵌入式模板
      _viewContainer.insertEmbeddedView(templateRef,i);
    }
  }
}
```

times 表示需要重复嵌入式模板的次数,在循环体中将嵌入式模板 TemplateRef 不断附加到视图容器 ViewContainerRef 中。

完整的指令代码如下:

```
//chapter15/structural_directive/lib/src/repeat_directive.dart
import 'package:angular/angular.dart';

@Directive(selector: '[myRepeat]')
class RepeatDirective{
  //嵌入式模板
  TemplateRef _templateRef;
  //视图容器
  ViewContainerRef _viewContainer;
  //注入视图容器和嵌入式模板
  RepeatDirective(this._templateRef, this._viewContainer);
  //声明输入属性并接收一个 int 类型的参数
  //该参数表示在视图容器中重复嵌入式模板的次数
```

```
  @Input()
  set myRepeat(int times){
    if(times > 0){
      for(int i = 0; i < times; i++){
        //向视图容器插入嵌入式模板
        _viewContainer.insertEmbeddedView(_templateRef, i);
      }
    }
  }
}
```

将此指令添加到 AppComponent 的指令列表中。代码如下：

```
//chapter15/structural_directive/lib/app_component.dart
import 'package:angular/angular.dart';
import 'src/repeat_directive.dart';
@Component(
  selector: 'my-app',
  styleURLs: ['app_component.css'],
  templateURL: 'app_component.html',
  directives: [RepeatDirective],
)
class AppComponent {
}
```

然后在模板中使用指令，代码如下：

```
//chapter15/structural_directive/lib/app_component.html
<p *myRepeat="9">Repeat(重复指令)</p>
```

运行结果如图 15-15 所示。

Repeat（重复指令）
Repeat（重复指令）
Repeat（重复指令）
Repeat（重复指令）
Repeat（重复指令）
Repeat（重复指令）
Repeat（重复指令）
Repeat（重复指令）
Repeat（重复指令）

图 15-15　结构指令

15.8 HTTP 连接

38min

43min

前端应用通常需要通过 HTTP 与服务端通信。基础请求可以使用 dart:html 包中的 HttpRequest 类。若需要做复杂操作,推荐使用 http 包中的方法和类。

15.8.1 http 包

http 包包含一组高级函数和类,可轻松使用 HTTP 资源。它是多平台的,并且支持移动设备、桌面和浏览器。

常用函数:

(1) delete(dynamic URL, {Map < String, String > headers}):将具有给定标头的 HTTP DELETE 请求发送到给定 URL,该 URL 可以是 Uri 或 String。

(2) get(dynamic URL, {Map < String, String > headers}):将具有给定标头的 HTTP GET 请求发送到给定 URL,该 URL 可以是 Uri 或 String。

(3) post(dynamic URL, {Map < String, String > headers, dynamic body, Encoding encoding}):将具有给定标头和正文的 HTTP POST 请求发送到给定的 URL,该 URL 可以是 Uri 或 String。

(4) put(dynamic URL, {Map < String, String > headers, dynamic body, Encoding encoding}):将具有给定标头和正文的 HTTP PUT 请求发送到给定的 URL,该 URL 可以是 Uri 或 String。

对于 put 和 post 方法请求的正文可以是 String、List < int >或 Map < String,String >:①如果是 String,将使用 encoding 对请求进行编码,请求的 content-type 默认是 text/plain;②如果是 List,它将用作请求正文的字节列表;③如果是 Map,将使用 encoding 为表单字段编码,请求的 content-type 被设置为 application/x-www-form-URLencoded,并且不可更改。encoding 默认为 UTF-8。

上述函数会返回一个由 Future 封装的 Response 对象,http 包中 Response 类常用属性如下:

(1) body:将响应正文作为字符串返回。

(2) bodyBytes:将响应正文作为字节数组返回。

15.8.2 数据转换

在客户端和服务端传输数据时,通常会对模型数据进行转码和解码。最常用的数据传输格式是 JSON,json_string 包使得数据在 JSON 和 Dart 对象间相互转换变得容易。该包并不是 JSON 数据和 Dart 对象间相互转换最好的包,但它最能够展现 JSON 数据和 Dart 对象相互转换的原理,并且具有极佳的自定义特征。

json_string 包定义了一个 Mixin。

Jsonable：使用它并实现 toJson 方法可以将 Dart 对象编码为某些有效的 JSON 对象。除此之外还需要为 Dart 对象定义类方法 fromJson，该方法作为将 JSON 数据转为 Dart 对象的解码器。

json_string 包定义了一个类 JsonString，它具有以下构造函数：

(1) JsonString(String source, {bool enableCache: false})：如果 source 是有效的 JSON，则构造一个 JsonString 对象。如果可选的 enableCache 参数设置为 true，那么将提供解码值的缓存。

(2) JsonString. encode(Object value, {dynamic encoder(Object object)})：创建一个将 value 转换为有效 JSON 的 JsonString。

(3) JsonString. orNull(String source, {bool enableCache: false})：如果 source 是有效的 JSON，则构造一个 JsonString 对象；如果不是，则返回 null。

类 JsonString 具有以下类方法：

(1) encodeObject < T extends Object >(T value, {JsonObjectEncoder < T > encoder})：创建一个将值转换为有效 JSON 对象的 JsonString。如果 T 实现 Jsonable 接口，则在转换期间使用 toJson 方法返回的结果。如果没有，则必须提供编码器函数。

(2) encodeObjectList < T extends Object >(List < T > list, {JsonObjectEncoder < T > encoder})：创建一个将 list 转换为有效 JSON 列表的 JsonString。如果 T 实现 Jsonable 接口，则在转换期间使用 toJson 方法返回的结果。如果没有，则必须提供编码器函数。

类 JsonString 具有以下实例方法：

(1) decodeAsObject < T extends Object >(JsonObjectDecoder < T > decoder)：将 JSON 数据解码为 T 的实例。

(2) decodeAsObjectList < T extends Object >(JsonObjectDecoder < T > decoder)：将 JSON 数据解码为 List < T >的实例。

15.8.3 服务端

创建命令行应用程序，项目名为 http_server。

更新依赖项并执行 pub get 命令，代码如下：

```
//chapter15/http_server/pubspec.yaml
name: http_server
description: A simple command - line application.
#version: 1.0.0
#homepage: https://www.example.com

environment:
  sdk: '>= 2.7.0 < 3.0.0'

dependencies:
```

```
  shelf: ^0.7.7
  shelf_router: ^0.7.2
  json_string: ^2.0.1

dev_dependencies:
  shelf_router_generator: ^0.7.0+1
  build_runner: ^1.3.1
  pedantic: ^1.8.0
```

首先建立模型 Employee，它包含字段 id、name 和 salary。使用 mixin 类 Jsonable，并实现 toJson 方法用于将 Employee 实例编码为 JSON。提供 fromJson 方法用于将 JSON 对象解码为 Employee 实例。代码如下：

```
//chapter15/http_server/bin/employee.dart
//也用作 chapter15/http_client/lib/src/employee.dart
import 'package:json_string/json_string.dart';
//使用 mixin 类 Jsonable
class Employee with Jsonable{
  int id;
  String name;
  num salary;
  Employee({this.id,this.name,this.salary});

  //实现 toJson 方法,将模型转换为 json 编码器
  @override
  Map<String,dynamic> toJson(){
    //返回 json 对象
    return {
      'id':id,
      'name':name,
      'salary':salary
    };
  }

  //类方法 fromJson 将 json 转换为模型解码器
  static Employee fromJson(Map<String,dynamic> json){
    //返回 Employee 实例
    return Employee(
        id:json['id'],
        name:json['name'],
        salary:json['salary']
    );
  }
}
```

因为目前没有连接数据库，所以以硬编码的方式实例化一个 Employee 类型的列表。提供 getAll 方法用于返回整个列表，提供 getById 方法用于根据 id 返回对应 Employee 实例。代码如下：

```
//chapter15/http_server/bin/employee_service.dart
import 'employee.dart';
class EmployeeService{
  var emps = [
    Employee(id:5,name:'Magneta',salary:9971),
    Employee(id:6,name:'RubberMan',salary:4533),
    Employee(id:7,name:'Dynama',salary:6720),
    Employee(id:8,name:'Dr IQ',salary:4907),
    Employee(id:9,name:'Magma',salary:5278),
    Employee(id:10,name:'Tornado',salary:7800)
  ];
  List<Employee> getAll(){
    //返回整个 emps 列表
    return emps;
  }
  Employee getById(int id){
    //根据 id 从 emps 列表中返回单个 Employee 实例
    return emps.firstWhere((emp) => emp.id == id,orElse: (){
      //当 emps 列表中不存在 id 相匹配的实例时 返回 null
      return null;
    });
  }
}
```

定义路由器，此部分内容一方面是回顾路由器方面的知识，另一方面是使用 json_string 包提供的功能转换 Dart 对象和 JSON 数据。代码如下：

```
//chapter15/http_server/bin/routers.dart
import 'dart:async' show Future;
import 'dart:convert';
import 'package:shelf/shelf.dart';
import 'package:shelf_router/shelf_router.dart';
import 'package:json_string/json_string.dart';
import 'employee_service.dart';
import 'employee.dart';
//将生成的文件作为本库的一部分
//该文件名作为路由生成器生成文件的依据
//当未使用 library 指令时，单个文件默认为一个库
part 'routers.g.dart';

class Emp{
```

```
    //响应路由'/emp/'下带路径参数的 GET 请求
    @Route.get('/emp/<empId>')
    Future<Response> _getById(Request request)async{
      //解析查询参数 empId
      var empId = int.parse(params(request, 'empId'));
      //根据 id 返回对应实例
      var emp = EmployeeService().getById(empId);
      //状态码为 200 成功响应并返回 Json 数据
      //使用 jsonEncode 方法将 Employee 实例编码为 Json 数据并返回
      //jsonEncode 方法是由库 dart:convert 提供的
      return Response.ok(jsonEncode(emp));
    }
    //响应路由'/emps'下的 GET 请求
    @Route.get('/emps')
    Future<Response> _getAll(Request request)async{
      //返回整个列表
      var emps = EmployeeService().getAll();
      //状态码为 200 成功响应
      //使用 jsonEncode 将 emps 列表编码为 Json 数据并返回
      return Response.ok(jsonEncode(emps));
    }
    //响应路由'/emp'下的 POST 请求
    @Route.post('/emp')
    Future<Response> _create(Request request)async{
      var emp;
      //request.read()方法将请求主体转化为 Stream<List<int>>
      //utf8.decoder.bind()方法对 Stream<List<int>>进行解码
      //join()方法用于将可能的多个数据块组合在一起
      await utf8.decoder.bind(request.read()).join().then((content){
        //将 Json 内容解码为 Employee 实例
        emp = JsonString(content).decodeAsObject(Employee.fromJson);
      });
      return Response.ok(jsonEncode(emp));
    }

    @Route.all('/<ignored|.*>')
    Future<Response> _notFound(Request request) async{
      return Response.notFound('页面未找到');
    }
    //生成的_$AppRouter 函数可用于获取此对象的 handler
    //用于返回此路由器的处理程序
    Handler get handler => _$EmpRouter(this).handler;
  }
```

然后在编辑器中的命令行输入并执行以下命令：

```
pub run build_runner build
```

成功生成 routers.g.dart 文件后，添加跨域中间件，用于响应跨域请求。代码如下：

```
//chapter15/http_server/bin/main.dart
//创建中间件并提供 corsHeaders 配置
final cors = createCorsHeadersMiddleware(
    corsHeaders:{
      'Access-Control-Allow-Origin': '*',
      'Access-Control-Expose-Headers': 'Authorization, Content-Type',
      'Access-Control-Allow-Headers': 'Authorization, Origin, X-Requested-With, Content
-Type, Accept',
      'Access-Control-Allow-Methods': 'GET, POST, PUT, PATCH, DELETE'
    }
);

Middleware createCorsHeadersMiddleware({Map<String, String> corsHeaders}) {
  //未提供 corsHeaders 时，使用默认配置
  corsHeaders ??= {'Access-Control-Allow-Origin': '*'};

  //请求处理程序
  Response handleOptionsRequest(Request request) {
    if (request.method == 'OPTIONS') {
      return Response.ok(null, headers: corsHeaders);
    } else {
      return null;
    }
  }

  //响应处理程序，将 corsHeaders 配置应用于响应
  Response addCorsHeaders(Response response) => response.change(headers: corsHeaders);

  //创建中间件并返回
  return createMiddleware(requestHandler: handleOptionsRequest, responseHandler: addCorsHeaders);
}
```

实例化路由器对象；通过 Pipeline 对象组合中间件 logRequests 和 cors 及路由器实例提供的处理程序；使用适配器 serve 启动服务。代码如下：

```
//chapter15/http_server/bin/main.dart
import 'package:shelf/shelf.dart';
import 'package:shelf/shelf_io.dart' as io;
import 'routers.dart';
```

```
void main()async{
  //路由器实例
  final service = Emp();

  //通过 Pipeline 对象组合中间件和单个处理程序
  var handler = Pipeline()
      //添加默认中间件,请求日志
      .addMiddleware(logRequests())
      //添加跨域中间件
      .addMiddleware(cors)
      //添加路由器实例提供的处理程序
      .addHandler(service.handler);

  //通过 shelf_io.serve 方法启动一个 HttpServer
  var server = await io.serve(handler, 'localhost', 1024);
  print('服务地址:http://${server.address.host}:${server.port}');
}
```

启动脚本,控制台输出以下信息表示服务启动成功。

```
服务地址: http://localhost:1024
```

15.8.4　客户端

创建 AngularDartWebApp 模板生成的项目,项目命为 http_client。删除项目 lib/src 目录下的所有文件,并删除根组件中的导入信息和指令。

更新依赖项并执行 pub get 命令,代码如下:

```
//chapter15/http_client/pubspec.yaml
name: http_client
description: A web app that uses AngularDart Components
#version: 1.0.0
#homepage: https://www.example.com

environment:
  sdk: '>=2.7.0 <3.0.0'

dependencies:
  angular: ^5.3.0
  angular_components: ^0.13.0
  json_string: ^2.0.1
  http: ^0.12.2

dev_dependencies:
```

```
angular_test: ^2.3.0
build_runner: ^1.6.0
build_test: ^0.10.8
build_web_compilers: ^2.3.0
pedantic: ^1.8.0
test: ^1.6.0
```

将服务端的 employee.dart 文件复制到当前项目的 lib/src 目录下。创建服务，使用 http 包下的方法向服务端发出请求，并通过 Response 对象的 body 属性获取响应正文的字符串。使用 json_string 中的 JsonString 构造函数构建 JsonString 实例，使用 JsonString 实例的 decodeAsObject 或 decodeAsObjectList 方法将 Json 数据解码为 Dart 模型类或模型类组成的列表。模型类有时会被称为实体类，在本项目中指类 Employee。代码如下：

```
//chapter15/http_client/lib/src/service.dart
import 'dart:convert';
import 'package:angular/angular.dart';
import 'package:http/http.dart' as http;
import 'package:json_string/json_string.dart';

import 'employee.dart';
//服务的地址和端口
const _URL = 'http://localhost:1024';

@Injectable()
class EmployeeService{
  //根据 id 获取雇员信息
  Future<Employee> getById(int id)async{
    var emp;
    //使用 http 包下的 get 方法发起 GET 请求
    await http.get('$_URL/emp/$id').then((response){
      if(response.statusCode == 200){
        //response.body 将响应正文作为 String 返回
        //服务端返回的是 Json 数据
        //可以使用 JsonString 构造函数构建 JsonString 对象并解码为 Employee 实例
        emp = JsonString(response.body).decodeAsObject(Employee.fromJson);
      }
    });
    return emp;
  }
  //获取所有雇员信息
  Future<List<Employee>> getAll()async{
    var emps;
```

```
    await http.get('$_URL/emps').then((response){
      if(response.statusCode == 200){
        emps = JsonString(response.body).decodeAsObjectList(Employee.fromJson);
      }
    });
    return emps;
  }
  //添加雇员
  Future<Employee> post(Employee emp)async{
    var new_emp;
    //使用 http 包下的 post 方法发起 POST 请求
    //通过 body 参数传递 json 数据
    await http.post('$_URL/emp',body: jsonEncode(emp)).then((response){
      if(response.statusCode == 200){
        new_emp = JsonString(response.body).decodeAsObject(Employee.fromJson);
      }
    });
    return new_emp;
  }
}
```

在根组件中注入服务，并将服务类 EmployeeService 添加到 providers 参数列表中。在参数 directives 中添加常用指令和表单指令列表 formDirectives。所有需要向服务端发出请求的操作都通过注入的服务代理完成。代码如下：

```
//chapter15/http_client/lib/app_component.dart
import 'package:angular/angular.dart';
import 'package:angular_forms/angular_forms.dart';
import 'src/service.dart';
import 'src/employee.dart';

@Component(
  selector: 'my-app',
  styleURLs: ['app_component.css'],
  templateURL: 'app_component.html',
  providers: [EmployeeService],
  directives: [NgFor,NgIf,formDirectives],
)
class AppComponent implements OnInit{
  //服务实例
  final EmployeeService _employeeService;
  //缓存雇员列表
  List<Employee> emps = List<Employee>();
  //缓存通过 id 获取的雇员信息
```

```
  Employee emp_get;
  //缓存添加雇员的信息
  Employee emp_add = Employee();
  //注入服务
  AppComponent(this._employeeService);
  @override
  void ngOnInit()async{
    //从服务端获取雇员列表
    var all = await _employeeService.getAll();
    //向雇员列表 emps 中添加服务端获取的雇员信息
    emps.addAll(all);
  }
  void getById(String id)async{
    //将通过 id 获取的雇员赋值给 emp_get
    emp_get = await _employeeService.getById(int.parse(id));
  }
  void add()async{
    //向服务端发出新增请求
    var emp = await _employeeService.post(emp_add);
    //将新增雇员添加到雇员列表 emps 中
    emps.add(emp);
  }
}
```

在根组件模板中添加迭代雇员列表、新增雇员信息表单及通过 id 获取雇员信息的元素块。代码如下：

```
//chapter15/http_client/lib/app_component.html
<h1>HTTP 客户端</h1>
<h6>雇员列表</h6>
<table *ngIf="emps != null">
    <tr>
        <th>编号</th>
        <th>名字</th>
        <th>薪资</th>
    </tr>
    <tr *ngFor="let emp of emps">
        <td>{{emp.id}}</td>
        <td>{{emp.name}}</td>
        <td>{{emp.salary}}</td>
    </tr>
</table>

<h6>通过 id 获取单个雇员信息</h6>
<input type="number" #input placeholder="输入列表中存在的单个 id">
```

```
<button (click) = "getById(input.value)">获取单个雇员信息</button>
<div *ngIf = "emp_get != null">
    <span>{{emp_get.id}}</span>
    <span>{{emp_get.name}}</span>
    <span>{{emp_get.salary}}</span>
</div>

<h6>添加雇员</h6>
<form action = "#">
    <input type = "number" [(ngModel)] = "emp_add.id" placeholder = "雇员编号">
    <input type = "text" [(ngModel)] = "emp_add.name" placeholder = "雇员名字">
    <input type = "number" [(ngModel)] = "emp_add.salary" placeholder = "雇员薪资">
    <button (click) = "add()">添加雇员</button>
</form>
```

运行项目 web 目录下的 index.html 文件。运行成功后添加雇员和获取当前雇员信息的结果如图 15-16 所示。

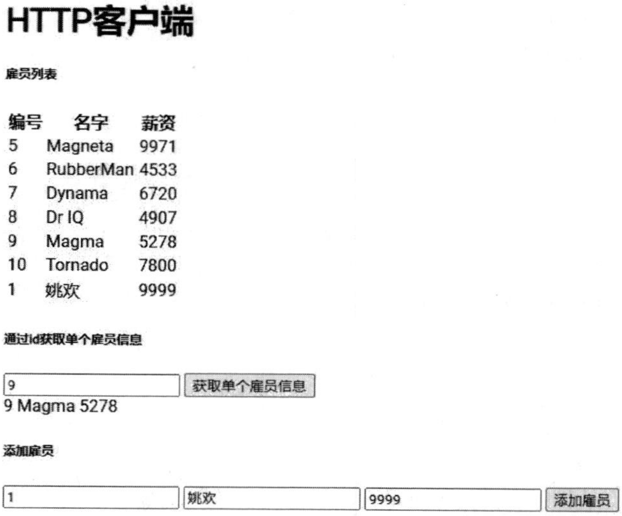

HTTP客户端

雇员列表

编号	名字	薪资
5	Magneta	9971
6	RubberMan	4533
7	Dynama	6720
8	Dr IQ	4907
9	Magma	5278
10	Tornado	7800
1	姚欢	9999

通过id获取单个雇员信息

9 获取单个雇员信息

9 Magma 5278

添加雇员

1 姚欢 9999 添加雇员

图 15-16　客户端请求页

15.9 部署项目

15min

部署 AngularDart 项目，首先需要将应用程序编译为 JavaScript。使用 webdev 工具构建应用程序，将应用程序编译为 JavaScript 并生成部署所有资源。dart2js 编译器支持 tree-shaking，它只导入所需的库并忽略未使用的类、函数、方法等。因此在使用 dart2js 构建应用时，可以得到一个相当小的 JavaScript 文件。

通过一些额外的工作，可以使可部署的应用程序更小、更快、更可靠。

15.9.1 webdev 工具

在使用 webdev 工具前需要全局安装它，以便在任何地方都可以直接使用该工具。编辑器会在合适的时候自动安装，若是手动安装则可在命令行中执行如下命令：

```
pub global activate webdev
```

使用 webdev 时，它还依赖于 build_runner 和 build_web_compilers 软件包。使用模板创建的项目将自动添加依赖，否则可在开发依赖中添加依赖项并执行 pub get 命令：

```
dev_dependencies:
  build_runner: ^1.6.0
  build_web_compilers: ^2.3.0
```

在开发应用程序时在命令行中使用 webdev serve 命令，该命令会在开发过程中持续构建 Web 应用程序。它默认使用 webdevc 编译器编译应用程序并发布在地址 localhost:8080 上，如果使用编辑器运行项目，端口将由编辑器随机指定。

```
pub global run webdev serve
```

开发编译器 webdevc 仅支持 Chrome 浏览器，如果需要在其他浏览器中也能正常运行，则可使用 dart2js 编译器。添加--release 标志以便使用 dart2js：

```
pub global run webdev serve - - release
```

也可以为项目指定端口为 80，使用示例如下：

```
pub global run webdev serve web:80
```

使用 webdev build 命令创建应用程序的可部署版本，默认使用 dart2js 编译器。

```
pub global run webdev build
```

命令执行完毕后，会在项目根目录下生成一个 build 文件夹，其中包含所有编译后的文件。dart2js 编译器将应用程序编译为 JavaScript，并将结果保存在 build/web/main. dart. js 文件中，通过 build 文件下的 index. html 主页文件即可访问部署完成的应用程序。

15.9.2 dart2js 选项

在编译为 JavaScript 文件时，可以向编译器指定选项以便控制编译优化程度。

(1) --minify：生成压缩的输出，默认压缩。

(2) -O0：禁用所有优化。禁用内联、禁用类型推断、禁用 rti 优化。

(3) -O1：启用默认优化。

(4) -O2：启用遵循语义的优化并且对所有程序适用，它会更改类型的字符串表示形式。在 minify 的基础上对运行时类型返回的字符串表示形式要求降低。

(5) -O3：仅在不抛出 Error 子类型的程序上启用优化并遵循语言语义。在-O2 的基础上省略隐式检查。

(6) -O4：在-O3 的基础上优化程度更进一步，但容易受用户输入中边界测试的影响。

需要使用配置文件以便应用 dart2js 选项，在项目根目录下创建 build.yaml 文件。以 15.9.1 节的项目 http_client 为例，添加代码如下：

```
//chapter15/http_client/build.yaml
targets:
  $default:
    builders:
      build_web_compilers|entrypoint:
        #这是编译的入口点的位置
        generate_for:
          - test/**.browser_test.dart
          - web/**.dart
        options:
          compiler: dart2js
          #dart2js 指定参数列表,也可以忽略
          dart2js_args:
            - -O2
```

需遵循以下做法，以便帮助 dart2js 更好地进行类型推断，从而可以生成更小和更快的 JavaScript 代码：

(1) 不要使用 Function.apply()方法。

(2) 不要覆写 noSuchMethod()方法。

(3) 避免将变量设置为 null。

(4) 传递给每个函数或方法的参数类型应保持一致。

然后在命令行执行如下命令：

```
pub global run webdev build
```

项目构建完成后的目录结构如图 15-17 所示。

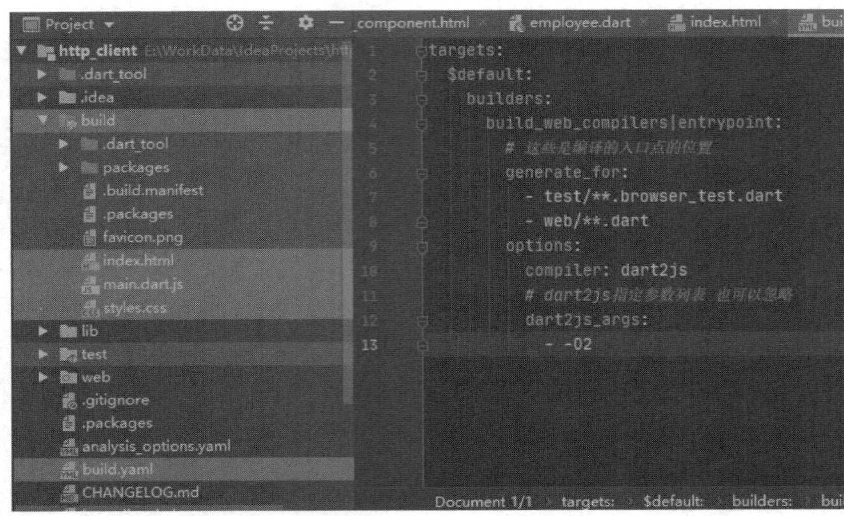

图 15-17　部署项目结构

第 16 章

材质化组件

本章介绍基于 Angular 框架的材质化组件库，该库中定义了常用的网页组件。可以直接将它们应用于项目，它们支持复杂的任务逻辑、高效且稳定。

在传统组件所使用的 Dart、HTML 和 CSS 语言的基础上，该库还使用了一种名叫 Sass 的语言。Sass 是一款强化 CSS 的辅助工具，它在 CSS 语法的基础上增加了变量(variables)、嵌套(nested rules)、混合(mixins)、导入(inline imports)等高级功能，这些拓展令 CSS 更加强大与优雅。使用 Sass 有助于更好地组织管理样式文件，以及更高效地开发项目。

在材质化组件库中 Sass 采用的语法格式是 SCSS(Sassy CSS)，这种格式仅在 CSS3 语法的基础上进行拓展，所有 CSS3 语法在 SCSS 中都是通用的，同时加入 Sass 的特色功能。这种格式以.scss 作为文件扩展名。

在项目中使用材质化组件库前，首先需要添加依赖项并执行 pub get 命令：

```
name: project_name
description: A web app that uses AngularDart Components
#version: 1.0.0
#homepage: https://www.example.com

environment:
  sdk: '>=2.7.0 <3.0.0'

dependencies:
  angular: ^5.3.0
  angular_components: ^0.13.0

dev_dependencies:
  sass_builder: ^2.1.3
  angular_test: ^2.3.0
  build_runner: ^1.6.0
  build_test: ^0.10.8
  build_web_compilers: ^2.3.0
```

```
pedantic: ^1.8.0
test: ^1.6.0
```

依赖项 angular_components 就是材质化组件库，可以在需要的地方导入其中的一个或多个组件。开发依赖项 sass_builder 使用 build 包和 Sass 的 Dart 实现转换 Sass 文件，该工具会将 *.scss 文件编译为 *.css 文件。本章创建的所有项目都需要添加这两个库，后续将不再赘述。

组件都具有默认样式，它们适用于大部分使用场景。如果需要修改组件的默认样式，最好具备 Sass 语言基础。本章与 Sass 相关的知识仅涉及样式表中 mixin 的使用，因此跟随本章后续对 Sass 的介绍也可以满足常规使用。

在 Sass 中使用指令@mixin 定义一个可以在整个样式表中重复使用的样式，其定义格式代码如下：

```
@mixin mixin-name($arg1, $arg2:value2, $arg3:value3){
  //样式内容
}
```

从代码中可以看出其定义格式类似于方法。mixin 名是普通标识符，括号中是需要向 mixin 传递的参数。标识符前加 $ 表示变量，在花括号（{}）中可以引用。可以像 Dart 中的函数一样为参数提供默认值，例如：变量 $arg2 和 $arg3 都有默认值。

在 Sass 中使用@include 指令将 mixin 引入文档中，定义和使用 mixin 的示例代码如下：

```
//定义 mixin bordered
//它带有两个参数
@mixin bordered($width, $color) {
  //$width 表示边框的宽度
  //$color 表示边框的颜色
  border: $width solid $color;
}

//使用 mixin 并传入参数
.myArticle {
  @include bordered(1px, blue);
}
.myNotes {
  @include bordered(2px, green );
}
```

上述内容存放在.scss 文件中，它会被转换成同名.css 文件，并且其内容将会被转换为如下 CSS 内容：

```
.myArticle {
  border: 1px solid blue;
}

.myNotes {
  border: 2px solid green;
}
```

组件库中几乎每个组件都提供 mixin 集合，它们包含在对应组件目录下的_mixins.scss 文件中。导入该文件，并通过提供的 mixin 可以控制组件的样式细节。

对于组件的定义与各个细节在前两章中已经介绍完毕，本章将不再详细介绍。本章会先列出组件可用的 mixin、属性、输入和输出属性，并将重点放在组件的使用上。

16.1 图标

24min

材质化风格的图标，它的组件名是 MaterialIconComponent，它在模板中的选择器是 <material-icon>元素。可用图标参考网址：https://material.io/resources/icons。

1. 属性

(1) size String：图标的大小。值包括 x-small、small、medium、large 和 x-large，分别对应 12px、13px、16px、18px 和 20px。如果未指定 size，则默认为 24px。

(2) flip：是否应该翻转图标以适应 RTL 的语言。方向为 RTL 时，并非所有图标都应翻转。通常，处理空间关系的图标应翻转。代表时间关系、物理对象或产品徽标的图标不应翻转。

(3) light：是否应减少图标的不透明度。

(4) baseline：图标是否需要与基线对齐。

2. 输入属性

icon dynamic：该组件应显示的 Icon 模型或图标标识符。

3. Sass mixin

(1) material-icon-size($size)：更改图标的大小。

(2) svg-icon($svg-icon)：将 SVG 图片或数据用作图标而不是字体图标。

(3) svg-icon-size($size)：设置 material-icon 的内部<i>元素的大小，进而控制 SVG 图标大小。

创建 Angular 项目 icon_demo，添加依赖项并启动项目。在根组件中导入材质化组件库，将 MaterialIconComponent 添加到指令列表，属性 iconColor 用于样式绑定以便控制图标的颜色，属性 iconModel 用于与图标组件的输入属性 icon 绑定。示例代码如下：

```
//chapter16/icon_demo/lib/app_component.dart
import 'package:angular/angular.dart';
//导入组件库
import 'package:angular_components/angular_components.dart';
//导入模型类
import 'package:angular_components/model/ui/icon.dart';

@Component(
  selector: 'my-app',
  styleURLs: ['app_component.css'],
  templateURL: 'app_component.html',
  directives: [MaterialIconComponent],
)
class AppComponent {
  //自定义图标颜色
  String iconColor = 'blue';
  //新建图标模型
  Icon iconModel = Icon('edit');
}
```

模板代码如下：

```
//chapter16/icon_demo/lib/app_component.html
<h6>普通图标</h6>
<p><material-icon icon="eco"></material-icon> 默认 24px</p>

<h6>图标大小</h6>
<p><material-icon size="x-small" icon="eco"></material-icon> x-small 12px</p>
<p><material-icon size="small" icon="eco"></material-icon> small 13px</p>
<p><material-icon size="medium" icon="eco"></material-icon> medium 16px</p>
<p><material-icon size="large" icon="eco"></material-icon> large 18px</p>
<p><material-icon size="x-large" icon="eco"></material-icon> x-large 20px</p>

<h6>图标 light</h6>
<p><material-icon light icon="eco"></material-icon>减少图标的不透明度</p>

<h6>图标 baseline</h6>
<p>图标<material-icon baseline icon="eco"></material-icon>与基线对齐</p>
<p>图标<material-icon icon="eco"></material-icon>未与基线对齐</p>

<h6>自定义图标颜色</h6>
<p><material-icon [style.color]="'green'" icon="eco"></material-icon></p>
<p><material-icon [style.color]="'orange'" icon="eco"></material-icon></p>
<p><material-icon [style.color]="iconColor" icon="eco"></material-icon></p>
```

```
<h6>Icon 模型</h6>
<p><material-icon [icon]="iconModel"></material-icon></p>

<h6>mixin 自定义图标大小</h6>
<p><material-icon class="custom-size" icon="eco"></material-icon></p>

<h6>mixin 自定义图标(SVG)</h6>
<material-icon icon="" class="svg-icon"></material-icon>

<h6>mixin 自定义图标(SVG)大小</h6>
<material-icon icon="" class="svg-icon svg-size"></material-icon>
```

样式代码如下：

```
//chapter16/icon_demo/lib/app_component.scss
//导入图标组件提供的 mixins
@import 'package:angular_components/material_icon/mixins';

.custom-size {
  //使用 material-icon-size mixin 设置图标大小
  @include material-icon-size(36px);
}

.svg-icon {
  //使用 mixin svg-icon 将 SVG 图像作为图标
  //这里将项目 web 目录下名为 bilibili.svg 的文件作为图标
  @include svg-icon(URL('/bilibili.svg'));
}

.svg-size{
  //使用 mixin svg-icon-size 设置 SVG 图标大小
  @include svg-icon-size(16px);
}
```

16.2 滑动条

20min

适用于整数值的材质化滑动条。组件名是 MaterialSliderComponent，在模板中使用元素<material-slider>创建滑动条组件的实例，可以通过使用鼠标或键盘拖动来控制滑块。使用双值浮点数学运算时，可能导致值不精确，如果向用户显示该值，需考虑格式化结果。

1. 输入属性

(1) disabled bool：如果禁用滑动条，则其值为 true。

(2) max num：最大进度值。默认值为 100，必须严格大于最小进度值。

(3) min num：最小进度值。默认值为 0，必须严格小于最大进度值。

(4) step num：步长。必须为正数。

(5) value num：当前值。

2. 输出属性

valueChange Stream < num >：当用户更改输入值时，发布事件。

3. Sass mixin

(1) slider-thumb-color($selector, $color)：用于设置滑块的颜色。第一个参数是指向滑动条元素的选择器，第二个参数是要设定的颜色。

(2) slider-track-color($selector, $left-color, $right-color: $mat-grey)：用于设置滑动条轨道的颜色。第 1 个参数是指向滑动条元素的选择器。第 2 个参数是滑块左侧轨道的颜色。第 3 个参数是滑块右侧轨道的颜色。

创建 Angular 项目 slider_demo，添加依赖项并启动项目。在根组件中导入材质化组件库，将 MaterialSliderComponent 添加到指令列表，将 DomService 和 windowBindings 添加到提供者列表。示例代码如下：

```
//chapter16/slider_demo/lib/app_component.dart
import 'package:angular/angular.dart';
import 'package:angular_components/angular_components.dart';
//导入 DomService,用于跨组件同步 DOM 操作
//例如:在 UI 更新或应用程序事件后检查更改
import 'package:angular_components/utils/browser/dom_service/angular_2.dart';
//导入 windowBindings
//提供在 Angular 中绑定使用的 Document、HtmlDocument 和 Window
import 'package:angular_components/utils/browser/window/module.dart';
@Component(
  selector: 'my-app',
  styleURLs: ['app_component.css'],
  templateURL: 'app_component.html',
  providers: [DomService,windowBindings],
  directives: [MaterialSliderComponent],
)
class AppComponent{
  //用于绑定到滑动条组件的 value 属性
  int value1 = 10;
  int value2 = 30;
  int value3 = 70;
}
```

模板代码如下：

```
//chapter16/slider_demo/lib/app_component.html
<h1>滑块</h1>
<h6>默认</h6>
<material-slider></material-slider>

<h6>绑定到组件属性</h6>
<p>值:{{value1}}</p>
<material-slider [(value)]="value1"></material-slider>

<h6>禁用</h6>
<material-slider [(value)]="value2"
                 [disabled]="true"></material-slider>

<h6>mixin 自定义滑块颜色</h6>
<material-slider class="thumb-colors"
                 [(value)]="value3"></material-slider>

<h6>mixin 自定义滑轨颜色</h6>
<material-slider class="track-colors"
                 [(value)]="value3"></material-slider>
```

样式代码如下：

```
//chapter16/slider_demo/lib/app_component.scss
//导入预定义材质化样式
@import 'package:angular_components/css/material/material';
//导入滑动条组件的 mixins
@import 'package:angular_components/material_slider/mixins';

:host {
  //限定滑动条的宽度和显示框类型
  material-slider {
    display: inline-block;
    width: 400px;
  }
  //$mat-green, $mat-green-400, $mat-green-100 均是预定义颜色变量
  //自定义滑块的颜色,第一项是选择器,第二项参数是滑块的颜色
  @include slider-thumb-color('material-slider.thumb-colors', $mat-green);
  //自定义轨道的颜色,第一项是选择器,第二项参数是滑块左侧轨道的颜色,第三项参数是滑块右
//侧轨道的颜色
  @include slider-track-color('material-slider.track-colors', $mat-green-400, $mat
-green-100);
}
```

16.3 旋转器

10min

当表示进度和活动在不确定的时间内完成时可用圆形旋转器显示，此组件名为MaterialSpinnerComponent，在模板中使用元素< material-spinner >创建滑动条组件的实例。

Sass mixin

material-spinner-thickness($stroke-width)：更改旋转器的边框宽度，使其显得更粗或更细。

修改 material-spinner 的 border-color 属性可以更改旋转器的颜色：

```
material - spinner {
  border - color: $ mat - red;
}
```

创建 Angular 项目 spinner_demo，将 MaterialSpinnerComponent 添加到指令列表。示例代码如下：

```
//chapter16/spinner_demo/lib/app_component.dart
import 'package:angular/angular.dart';
import 'package:angular_components/angular_components.dart';
@Component(
  selector: 'my - app',
  styleURLs: ['app_component.css'],
  templateURL: 'app_component.html',
  directives: [MaterialSpinnerComponent],
)
class AppComponent {}
```

模板代码如下：

```
//chapter16/slider_demo/lib/app_component.html
< h1 >旋转器</h1 >
< h6 >默认</h6 >
< material - spinner ></material - spinner >

< h6 >自定义旋转器颜色</h6 >
< material - spinner class = "orange"></material - spinner >
< material - spinner class = "green"></material - spinner >

< h6 >配合文字使用</h6 >
< div >
```

```
    <material-spinner></material-spinner>
    <span>处理中...</span>
</div>

<h6>mixin 自定义旋转器的边框宽度</h6>
<material-spinner class="custom-width"></material-spinner>
```

样式代码如下：

```
//chapter16/spinner_demo/lib/app_component.scss
//导入旋转器组件的 mixins
@import 'package:angular_components/material_spinner/mixins';

//自定义旋转器的颜色
.green{
  border-color:green;
}
.orange{
  border-color:orange;
}

//自定义旋转器的边框宽度
.custom-width{
  @include material-spinner-thickness(6px);
}
```

16.4 切换按钮

17min

用户可以单击切换按钮来更改状态。通常，只有一个打开或关闭选项时，才使用切换按钮。组件名是 MaterialToggleComponent，在模板中使用元素< material-toggle >创建切换按钮的实例。

1. 输入属性

(1) checked bool：切换按钮的当前状态。true 为打开，false 为关闭。

(2) disabled bool：启用或禁用切换按钮。禁用为 true，启用为 false。

(3) label String：切换按钮的标签。

2. 输出属性

checkedChange Stream < bool >：切换按钮被选中后触发的事件。

3. Sass mixin

(1) material-toggle-theme($primary-color, $off-btn-color: null)：第一个参数表示切换按钮打开时按钮的颜色，第二个参数表示切换按钮关闭时按钮的颜色，轨道会根据按钮颜色自适应。

(2) flip-toggle-label-position()：在切换框的右侧显示切换标签。

创建 Angular 项目 toggle_demo，将 MaterialToggleComponent 添加到指令列表。示例代码如下：

```
//chapter16/toggle_demo/lib/app_component.dart
import 'package:angular/angular.dart';
import 'package:angular_components/angular_components.dart';
@Component(
  selector: 'my-app',
  styleURLs: ['app_component.css'],
  templateURL: 'app_component.html',
  directives: [MaterialToggleComponent],
)
class AppComponent {
  //切换按钮状态控制变量
  bool toggle1 = true;
  bool bluetooth = false;
  bool finddevice = false;
}
```

模板代码如下：

```
//chapter16/toggle_demo/lib/app_component.html
<h1>切换按钮</h1>
<h6>默认</h6>
<material-toggle></material-toggle>

<h6>带有标签</h6>
<material-toggle label="自动提醒"></material-toggle>

<h6>禁用</h6>
<material-toggle label="显示电量"
                 [disabled]="true"></material-toggle>
<br>
<material-toggle label="显示网速"
                 [checked]="true"
                 [disabled]="true"></material-toggle>

<h6>绑定到组件属性</h6>
<p>状态:{{toggle1 ? '打开':'关闭'}}</p>
<material-toggle label="自动提醒"
                 [(checked)]="toggle1"></material-toggle>

<h6>切换按钮交互</h6>
```

```
<material-toggle label="蓝牙:{{bluetooth ? '打开':'关闭'}}"
                 [(checked)]="bluetooth"></material-toggle>
<br>
<material-toggle label="发现设备:{{finddevice ? '打开':'关闭'}}"
                 [disabled]="!bluetooth"
                 [(checked)]="finddevice"></material-toggle>

<h6>自定义宽度</h6>
<material-toggle class="theme-width"
                 label="自动提醒"></material-toggle>

<h6>自定义打开时的颜色</h6>
<div class="theme-orange">
    <material-toggle label="自动提醒"
                     [checked]="true"></material-toggle>
</div>

<h6>自定义关闭时的颜色</h6>
<div class="theme-double">
    <material-toggle label="自动提醒"></material-toggle>
</div>

<h6>标签显示在右边</h6>
<div class="theme-fliplabel">
    <material-toggle label="自动提醒"></material-toggle>
</div>
```

样式代码如下：

```
//chapter16/toggle_demo/lib/app_component.scss
@import 'package:angular_components/css/material/material';
//导入切换按钮提供的 mixins
@import 'package:angular_components/material_toggle/mixins';

//自定义切换按钮宽度
.theme-width{
  padding: 12px;
  width: 320px;
}

//自定义打开时的按钮颜色
.theme-orange{
   @include material-toggle-theme($primary-color:orange);
}
```

```
//自定义打开和关闭时的按钮颜色
.theme-double{
  @include material-toggle-theme(
    $primary-color:orange,
    $off-btn-color:gray);
}

//翻转标签
.theme-fliplabel{
    @include flip-toggle-label-position();
}
```

13min

16.5 选项卡

16.5.1 固定选项条

固定选项条是具有选项卡样式的按钮和活动的选项卡指示器的选项条组件。固定选项条具有相同大小的选项卡按钮，并且没有滚动。组件名是 FixedMaterialTabStripComponent，在模板中使用元素<material-tab-strip>可以创建选项条的实例。

1. 输入属性

(1) activeTabIndex int：活动面板的索引，从 0 开始。默认值为 0。

(2) tabIds List<String>：选项卡按钮 ID 的列表。

(3) tabLabels List<String>：选项卡按钮标签的列表。

2. 输出属性

(1) activeTabIndexChange Stream<int>：在触发 tabChange 事件后发布的 activeTabIndex 更新流。

(2) beforeTabChange Stream<TabChangeEvent>：TabChangeEvent 实例的流，在选项卡更改之前发布。调用 TabChangeEvent#preventDefault 将阻止更改选项卡。

(3) tabChange Stream<TabChangeEvent>：TabChangeEvent 实例的流，在选项卡已更改时发布。

3. Sass mixin

(1) tab-panel-tab-strip-width($selector, $tab-width)：设置选项卡面板中选项条的宽度。

(2) tab-strip-color($selector, $color, $accent-color)：设置选项卡的默认颜色和指示器的颜色。

(3) tab-strip-accent-color($selector, $accent-color)：设置选择状态下选项卡和指示器的颜色。

(4) tab-strip-tab-color($selector, $color)：设置选项卡的默认颜色。

(5) tab-strip-selected-tab-color($selector, $color):设置选择状态下选项卡的颜色。

(6) tab-strip-indicator-color($selector, $accent-color):设置指示器的颜色。

(7) tab-strip-show-bottom-shadow($dp: 2):设置选项卡条底部阴影的宽度。

16.5.2 选项卡面板

带有顶部导航栏的选项卡面板,组件名是 MaterialTabPanelComponent,在模板中使用元素< material-tab-pane >可以创建选项卡面板的实例。此组件实例是对< material-tab-strip >的扩展,因此 FixedMaterialTabStripComponent 的 Sass mixin 适用于此组件实例。

1. 输入属性

(1) activeTabIndex dynamic:活动面板的索引,从 0 开始。默认值为 0。

(2) centerTabs bool:是否将标签按钮居中对齐。否则,按钮从左端(LTR)对齐。

2. 输出属性

(1) beforeTabChange Stream < TabChangeEvent >:TabChangeEvent 实例的流,在选项卡更改之前发布。调用 TabChangeEvent # preventDefault 将阻止更改选项卡。

(2) tabChange Stream < TabChangeEvent >:TabChangeEvent 实例的流,在选项卡已更改时发布。

16.5.3 材质化选项卡

材质化风格的选项卡,作为 MaterialTabPanelComponent 的一部分显示或隐藏。组件名是 MaterialTabComponent,在模板中使用元素< material-tab >可以创建选项卡的实例。此组件通过 label 属性设置按钮的文本。使用 * deferredContent 模板指令可以延迟实例化选项卡的内容。此组件需要与 MaterialTabPanelComponent 配合使用。

输入属性

label String:此选项卡的标签。

创建 Angular 项目 tab_demo,将上述组件添加到指令列表。示例代码如下:

```
//chapter16/tab_demo/lib/app_component.dart
import 'package:angular/angular.dart';
import 'package:angular_components/angular_components.dart';
@Component(
  selector: 'my - app',
  styleURLs: ['app_component.css'],
  templateURL: 'app_component.html',
  directives: [
    FixedMaterialTabStripComponent,
    MaterialTabPanelComponent,
    MaterialTabComponent,
    DeferredContentDirective],
)
class AppComponent {
```

```
  //选项卡默认索引
  var tabIndex = 0;
  //选项标签列表
  var tabLabels = <String>['选项 1','选项 2','选项 3'];
  //用于响应选项更改事件
  void onTabChange(TabChangeEvent event) {
    tabIndex = event.newIndex;
  }
}
```

模板代码如下：

```
//chapter16/tab_demo/lib/app_component.html
<h1>选项卡</h1>
<h6>选项条</h6>
<p>选项卡与内容无直接关联</p>
<material-tab-strip (tabChange) = "onTabChange($event)"
                    [tabLabels] = "tabLabels">
</material-tab-strip>
<p>当前活跃选项卡索引:{{tabIndex}}</p>

<h6>自定义选项条宽度</h6>
<material-tab-strip style = "width:300px;"
                    (tabChange) = "onTabChange($event)"
                    [tabLabels] = "tabLabels">
</material-tab-strip>
<p>当前活跃选项卡索引:{{tabIndex}}</p>

<h6>选项面板</h6>
<p>选项卡与内容相关联</p>
<material-tab-panel [activeTabIndex] = "1">
    <material-tab label = "选项 1">选项 1 的内容</material-tab>
    <material-tab label = "选项 2">选项 2 的内容</material-tab>
    <material-tab label = "选项 3">
        <template deferredContent>延迟加载内容</template>
    </material-tab>
</material-tab-panel>

<h6>选项标签居中且自定义选项条宽度</h6>
<material-tab-panel centerStrip [activeTabIndex] = "0">
    <material-tab label = "选项 1">选项 1 的内容</material-tab>
    <material-tab label = "选项 2">选项 2 的内容</material-tab>
    <material-tab label = "选项 3">
        <template deferredContent>延迟加载内容</template>
    </material-tab>
```

```
</material-tab-panel>

<h6>mixin 自定义选项条宽度、默认选项卡颜色、选择状态下选项卡和指示器颜色、选项卡条底部阴影的宽度</h6>
<material-tab-panel class="tab-panel" [activeTabIndex]="0">
    <material-tab label="选项 1">选项 1 的内容</material-tab>
    <material-tab label="选项 2">选项 2 的内容</material-tab>
    <material-tab label="选项 3">
        <template deferredContent>延迟加载内容</template>
    </material-tab>
</material-tab-panel>
```

样式代码如下：

```
//chapter16/tab_demo/lib/app_component.scss
@import 'package:angular_components/css/material/material';
@import 'package:angular_components/material_tab/mixins';

//自定义选项面板选项条宽度
@include tab-panel-tab-strip-width('material-tab-panel[centerStrip]', 300px);

//自定义选项面板选项条宽度
@include tab-panel-tab-strip-width('material-tab-panel.tab-panel', 400px);

//自定义选项文字颜色
@include tab-strip-tab-color('.tab-panel', blue);

//自定义当前选项文字颜色
@include tab-strip-accent-color('.tab-panel', $mat-red);

//自定义选择项指示器颜色
@include tab-strip-indicator-color('.tab-panel', orange);

//自定义选项卡条底部阴影的宽度
.tab-panel{
  @include tab-strip-show-bottom-shadow($dp: 6);
}
```

16.6 计数卡与计数板

11min

16.6.1 计数卡

计数卡为独立组件，该组件可在更大的组件中重用或嵌入。组件名是 ScorecardComponent，在模板中使用元素< acx-scorecard >创建计分卡组件实例。

1. 内容元素

（1）name：标签区域中的自定义内容。

（2）value：值区域中的自定义内容。

（3）description：描述区域中的自定义内容。为了显示该部分内容，即使只是将其设置为空字符串，也需要设置 description 属性。

2. 输入属性

（1）changeGlyph bool：是否在描述区域显示变化箭头，可选。

（2）changeType String：设置计分卡描述的变化类型。用于确定描述的样式。可能的值为 POSITIVE 表示正，NEGATIVE 表示负，默认值为 NEUTRAL 表示无符号。

（3）description String：计分卡上的简短说明，可选。

（4）extraBig bool：是否对计分卡使用 CSS 类 big 定义的样式，可选。

（5）label String：计分卡的标题。

（6）selectable bool：单击是否可以更改计分卡的选择状态。

（7）selected bool：是否选择了计分卡。

（8）selectedColor Color：选中计分卡时背景要应用的颜色。

（9）suggestionAfter String：描述后的一条建议文本，可选。

（10）suggestionBefore String：描述前的一条建议文本，可选。

（11）tooltip String：用户将鼠标悬停在值上时，该值将显示在工具提示中。

（12）value String：向用户显示的值。

3. 输出属性

selectedChange Stream < bool >：selectedChange Stream < bool >。

16.6.2 计数板

计数板管理一行计数卡，组件名是 ScoreboardComponent，在明模板中使用元素< acx-scoreboard >实例化计分板组件。

1. 属性

enableUniformWidths bool：计分板上的计分卡是否应具有统一的宽度。

2. 输入属性

（1）isVertical bool：计分卡是否垂直显示。默认为 false。

（2）resetOnCardChanges bool：更改卡片时是否重置卡片选择。如果添加或删除卡，并且此卡设置为 true，则将取消选择所有卡。对于 ScoreboardType. radio，将选择第一张卡。

（3）scrollable bool：是否允许通过滚动按钮滚动计分板。滚动属性可以在应用运行时动态设置，将根据可滚动状态添加或删除窗口大小调整侦听器。

（4）type ScoreboardType：计分板的类型，例如 standard、selectable、radio、toggle。

创建 Angular 项目 scorecard_demo，将上述组件添加到指令列表。组件的使用需要

DomService 和 windowBindings 支持，因此需要在提供者列表中添加它们。示例代码如下：

```
//chapter16/scorecard_demo/lib/app_component.dart
import 'package:angular/angular.dart';
import 'package:angular_components/angular_components.dart';
import 'package:angular_components/utils/browser/dom_service/angular_2.dart';
import 'package:angular_components/utils/browser/window/module.dart';
@Component(
  selector: 'my-app',
  styleURLs: ['app_component.css'],
  templateURL: 'app_component.html',
  providers: [DomService, windowBindings],
  directives: [
    ScoreboardComponent,
    ScorecardComponent,
    NgFor],
)
class AppComponent {
  //预定义计分板类型变量
  final ScoreboardType selectable = ScoreboardType.selectable;
  final ScoreboardType toggle = ScoreboardType.toggle;
  final ScoreboardType radio = ScoreboardType.radio;
}
```

模板代码如下：

```
//chapter16/scorecard_demo/lib/app_component.html
<h1>计分卡</h1>
<h3>单独的计分卡</h3>
<h6>基础积分卡</h6>
<acx-scorecard label="余额 RMB" value="158.22" description="+24.20 (15%)">
</acx-scorecard>

<h6>正号并提供变化箭头</h6>
<acx-scorecard
        label="余额 RMB"
        value="158.22"
        description="+24.20 (15%)"
        [changeGlyph]="true"
        changeType="POSITIVE">
</acx-scorecard>

<h6>负号并提供变化箭头</h6>
<acx-scorecard
        label="余额 RMB"
        value="158.22"
```

```
        description = " - 24.20 (15 % )"
        [changeGlyph] = "true"
        changeType = "NEGATIVE">
</acx - scorecard>

< h6 >建议</h6 >
< acx - scorecard
        label = "余额 RMB"
        value = "158.22"
        description = " + 24.20 (15 % )"
        changeType = "POSITIVE"
        suggestionBefore = "值得关注"
        suggestionAfter = "增持">
</acx - scorecard>

< h6 >可选择</h6 >
< acx - scorecard
        label = "余额 RMB"
        value = "158.22"
        description = " - 24.20 (15 % )"
        [selectable] = "true">
</acx - scorecard>

< h6 >提示文本</h6 >
< acx - scorecard
        label = "余额 RMB"
        value = "158.22"
        description = " - 24.20 (15 % )"
        tooltip = "多考虑一下">
</acx - scorecard>

< h3 >计分板</h3 >
< h6 >基础</h6 >
< acx - scoreboard >
    < acx - scorecard label = "余额 RMB" value = "158.22" description = " + 24.20 (15 % )">
    </acx - scorecard>
    < acx - scorecard label = "余额 RMB" value = "158.22" description = " - 24.20 (15 % )">
    </acx - scorecard>
    < acx - scorecard label = "余额 RMB" value = "158.22" description = " + 24.20 (15 % )">
    </acx - scorecard>
</acx - scoreboard>

< h6 >可选择</h6 >
< acx - scoreboard [type] = "selectable">
    < acx - scorecard label = "余额 RMB" value = "158.22" description = " + 24.20 (15 % )">
    </acx - scorecard>
```

```
    <acx-scorecard label="余额 RMB" value="158.22" description="-24.20 (15%)">
    </acx-scorecard>
    <acx-scorecard label="余额 RMB" value="158.22" description="+24.20 (15%)">
    </acx-scorecard>
</acx-scoreboard>

<h6>可切换</h6>
<acx-scoreboard [type]="toggle">
    <acx-scorecard label="余额 RMB" value="158.22" description="+24.20 (15%)">
    </acx-scorecard>
    <acx-scorecard label="余额 RMB" value="158.22" description="-24.20 (15%)">
    </acx-scorecard>
    <acx-scorecard label="余额 RMB" value="158.22" description="+24.20 (15%)">
    </acx-scorecard>
</acx-scoreboard>

<h6>单选</h6>
<acx-scoreboard [type]="radio">
    <acx-scorecard label="余额 RMB" value="158.22" description="+24.20 (15%)">
    </acx-scorecard>
    <acx-scorecard label="余额 RMB" value="158.22" description="-24.20 (15%)">
    </acx-scorecard>
    <acx-scorecard label="余额 RMB" value="158.22" description="+24.20 (15%)">
    </acx-scorecard>
</acx-scoreboard>

<h6>垂直</h6>
<acx-scoreboard [isVertical]="true" [type]="selectable">
    <acx-scorecard label="余额 RMB" value="158.22" description="+24.20 (15%)">
    </acx-scorecard>
    <acx-scorecard label="余额 RMB" value="158.22" description="-24.20 (15%)">
    </acx-scorecard>
    <acx-scorecard label="余额 RMB" value="158.22" description="+24.20 (15%)">
    </acx-scorecard>
</acx-scoreboard>

<h6>可滚动</h6>
<acx-scoreboard scrollable [type]="selectable">
    <acx-scorecard
            *ngFor="let n of [1,2,3,4,5,6,7,8,9]"
            label="条目{{n}}"
            value="{{n}}"
            description="">
    </acx-scorecard>
</acx-scoreboard>
```

```
<h6>自定义内容</h6>
<acx-scorecard
        label="Estimated earnings"
        value="$158.22"
        description="+$24.20 (15%)"
        changeType="POSITIVE">
    <name>注入标签</name>
    <value>注入值</value>
    <description>注入描述信息</description>
    <div>注入的其他信息</div>
</acx-scorecard>
```

14min

16.7 按钮

16.7.1 按钮设置

扁平或突起的按钮，可以选择带有波纹效果。组件名是 MaterialButtonComponent，在模板中使用元素<material-button>创建按钮组件的实例。

1. 属性

（1）icon：删除按钮的最小宽度样式。需要在按钮中指定确定的图标，可以使用<glyph>、<material-icon>或<img>元素。

（2）no-ink：移除按钮上的波纹效果。

（3）clear-size：将按钮的最小宽度和边距置为0。

（4）dense：将字体大小调整为13px，将按钮高度调整为32px。

2. 输入属性

（1）disabled：是否禁用组件。

（2）raised：是否使按钮具有凸起的阴影。

（3）role：此组件的角色用于a11y。

（4）tabbable：组件是否可选卡片，适用于选项卡。

（5）tabindex：组件选项卡索引，如果tabbable为true，disabled为false，则使用该值，适用于选项卡。

3. 输出属性

trigger：通过单击或按键激活按钮时触发。

4. Sass mixin

（1）button-color($selector, $color)：将按钮字体颜色应用于与选择器匹配的按钮。给定的颜色不适用于按钮，否则它看起来像是启用的按钮。

（2）button-disabled-color($selector, $disabled-color)：将按钮在禁用状态下的字体

颜色应用于与选择器匹配的按钮。

(3) button-background-color($selector, $background-color):将按钮的背景颜色应用于与选择器匹配的按钮。

(4) icon-button-hover-color($selector, $color):当鼠标指针浮动在图标按钮上时的颜色。

(5) icon-button-color($ selector, $ color):图标按钮的默认颜色。

默认情况下,不透明度为 25%时,波纹与前景的颜色相同。要自定义颜色,需使用 material-ripple 选择器。

```
/*
  若无法正常使用,需将其添加到项目的 style.css 文件中
*/
#myButton5 material-ripple{
  color:blue;
}
```

实际上波纹的颜色并没有改变,可能是组件视图封装时不够精确。可以将样式添加到项目的 style. css 文件中。

16.7.2 浮动操作按钮

MaterialFab 是一个浮动操作按钮。它是一个大圆形的按钮,使用 mini 属性可使其变为小的按钮。它与 MaterialButton 有相同的输入属性和输出属性,故这里不再赘述。在 MaterialFab 中指定确定的图标,可以使用< glyph >、< material-icon >或< img >元素。组件名是 MaterialFabComponent,在模板中使用元素< material-fab >创建浮动操作按钮的实例。它包含 MaterialButton 所有的 mixin,且使用规则是一致的。

创建 Angular 项目 button_demo,将上述组件添加到指令列表。示例代码如下:

```
//chapter16/button_demo/lib/app_component.dart
import 'package:angular/angular.dart';
import 'package:angular_components/angular_components.dart';

@Component(
  selector: 'my-app',
  styleURLs: ['app_component.css'],
  templateURL: 'app_component.html',
  directives: [
    MaterialButtonComponent,
    MaterialIconComponent,
    MaterialFabComponent],
)
class AppComponent{
```

```
  //记录事件触发次数
  int i = 0;
  //响应按钮 trigger 事件
  trigger(){
    ++i;
  }
}
```

模板代码如下：

```
//chapter16/button_demo/lib/app_component.html
<h1>按钮</h1>
<h6>扁平按钮</h6>
<material-button>默认</material-button>
<material-button no-ink>no-ink</material-button>
<material-button clear-size>clear-size</material-button>
<material-button dense>dense</material-button>
<material-button disabled>disabled</material-button>
<material-button icon>
    <material-icon icon="add_alert"></material-icon>
</material-button>

<h6>突起按钮</h6>
<material-button raised>raised</material-button>
<material-button no-ink raised>no-ink</material-button>
<material-button clear-size raised>clear-size</material-button>
<material-button dense raised>dense</material-button>
<material-button disabled raised>disabled</material-button>
<material-button icon raised>
    <material-icon icon="add_alert"></material-icon>
</material-button>

<h6>按钮事件</h6>
<material-button raised (trigger)="trigger()">trigger</material-button>
<p>事件触发次数{{i}}</p>

<h6>mixin 自定义按钮</h6>
<material-button raised class="myButton1">蓝色背景</material-button>
<material-button raised class="myButton2">蓝底白字</material-button>
<material-button raised class="myButton2" disabled>禁用按钮</material-button>

<h6>mixin 自定义图标按钮 </h6>
<material-button icon raised class="myButton3">
    <material-icon icon="add_alert"></material-icon>
</material-button>
```

```
<material-button icon raised class="myButton4">
    <material-icon icon="add_alert"></material-icon>
</material-button>


<h6>自定义波纹颜色</h6>
<material-button raised class="myButton5">蓝色波纹</material-button>


<h1>浮动操作按钮</h1>
<h6>扁平按钮</h6>
<material-fab>
    <material-icon icon="add"></material-icon>
</material-fab>

<h6>突起按钮</h6>
<material-fab raised>
    <material-icon icon="add"></material-icon>
</material-fab>

<h6>mini 按钮</h6>
<material-fab mini>
    <material-icon icon="add"></material-icon>
</material-fab>
<material-fab mini raised>
    <material-icon icon="add"></material-icon>
</material-fab>
```

样式代码如下：

```
//chapter16/button_demo/lib/app_component.scss
@import 'package:angular_components/material_button/mixins';

//自定义按钮背景颜色
@include button-background-color('.myButton1', #4285f4);

//自定义按钮背景颜色、文字颜色、禁用状态下的文字颜色
@include button-background-color('.myButton2', #4285f4);
@include button-color('.myButton2', white);
@include button-disabled-color('.myButton2', white);

//自定义鼠标指针悬浮时图标按钮的颜色和图标按钮颜色
@include icon-button-hover-color('.myButton3', #4285f4);
@include icon-button-color('.myButton4', orange);
```

```
//自定义按钮波纹颜色,默认与按钮前景色一致
//若无法正常使用,需将其添加到项目的 style.css 文件中
.myButton5 material - ripple{
  color:blue;
}
```

9min

16.8 进度条

进度条用于可以确定完成百分比的情况，它使用户快速了解一次操作将花费多长时间。组件名是 MaterialProgressComponent，在模板中可以使用元素<material-progress>创建进度条的实例。

1. 输入属性

（1）activeProgress int：当前进度值。

（2）indeterminate bool：进度条是否是确定性的布尔值。默认值为 false。

（3）max int：最大进度值。默认值为 100。

（4）min int：最小进度值。预设值为 0。

（5）secondaryProgress int：次要进度。次要进度以较浅的颜色显示在主进度的后面。

2. Sass mixin

（1）material-progress-theme($indeterminate-color， $active-color， $secondary-color)：设置不确定进度条、主进度条、副进度条的颜色。

（2）wide-rounded-progress-bar($height)：设置进度条轨道的高度。

创建 Angular 项目 progress_demo，将组件 MaterialProgressComponent 和 MaterialButtonComponent 添加到指令列表。示例代码如下：

```
//chapter16/progress_demo/lib/app_component.dart
import 'package:angular/angular.dart';
import 'package:angular_components/angular_components.dart';
@Component(
  selector: 'my-app',
  styleURLs: ['app_component.css'],
  templateURL: 'app_component.html',
  directives: [
    MaterialProgressComponent,
    MaterialButtonComponent],
)
class AppComponent {
   //设置默认进度值
   int progress = 32;
   //追加进度
```

```
    void addProgress(){
      progress++;
    }
}
```

模板代码如下：

```
//chapter16/progress_demo/lib/app_component.html
<h1>进度条</h1>
<h6>主进度条</h6>
<material-progress [activeProgress]="25"></material-progress>

<h6>不确定进度条</h6>
<material-progress [indeterminate]="true"></material-progress>

<h6>副进度条</h6>
<material-progress [activeProgress]="25"
                   [secondaryProgress]="75"></material-progress>

<h6>数据绑定</h6>
<section class="custom-theme">
    <material-button raised (trigger)="addProgress()">追加进度</material-button>
    <material-progress [activeProgress]="progress"></material-progress>
</section>

<h6>mixin 自定义进度条主题</h6>
<material-progress class="custom-theme" [indeterminate]="true"></material-progress>

<h6>mixin 自定义进度条轨道高度</h6>
<material-progress class="custom-height"
                   [activeProgress]="25"
                   [secondaryProgress]="75"></material-progress>
```

样式代码如下：

```
//chapter16/progress_demo/lib/app_component.scss
@import 'package:angular_components/css/material/material';
//导入进度条组件定义的 mixins
@import 'package:angular_components/material_progress/mixins';

//自定义进度条主题
.custom-theme {
  @include material-progress-theme(
      //不确定进度条颜色
      $indeterminate-color: $mat-green-100,
```

```
        //主进度条颜色
        $active-color: $mat-green-500,
        //副进度条颜色
        $secondary-color: $mat-green-200);
}

//自定义进度条轨道高度
.custom-height{
  @include wide-rounded-progress-bar(6px);
}
```

22min

16.9 单选按钮

16.9.1 材质化单选按钮

材质化风格的单选按钮，组件名是 MaterialRadioComponent，在模板中使用元素<material-radio>创建单选按钮的实例。通常与元素<material-radio-group>一起使用。一旦选中，则不能通过用户操作取消选中同一单选按钮。

1. 属性

no-ink：设置此属性可禁用芯片上的波纹效应。

2. 输入属性

(1) checked bool：单选按钮是否被选择。

(2) disabled bool：单选按钮是否不应响应事件，并具有不允许进行交互的样式。

(3) value dynamic：此单选按钮代表的值，用于带有单选按钮群组的选择模型。

3. 输出属性

checkedChange Stream<bool>：选择状态更改时发布事件。

4. Sass mixin

material-radio-color($primary-color, $focus-indicator-color: $primary-color, $modifier:"): 第 1 个参数是按钮选择状态下的颜色，第 2 个参数表示波纹颜色，第 3 个参数是类选择器。

16.9.2 单选按钮组

包含多个材料单选按钮的组，强制在该组中仅选择一个值。组件名是 MaterialRadioGroupComponent，在模板中使用元素<material-radio-group>创建按钮组实例。

可以通过 selected 和 ngModel 获取值，但应避免同时使用两者，因为 ngModel 也通过监听 onChange 获取值，所以这些值可能看起来不同步。

1. 输入属性

(1) selected dynamic：当前所选单选按钮的值，首选 ngModel。

(2) selectionModel SelectionModel < dynamic >：包含值对象的选择模型。

2. 输出属性

selectedChange Stream < dynamic >：选择更改时发布事件，首选 ngModelChange。

类 SelectionModel 用于定义选择模型，提供管理选定值集合的模式。该类包含如下构造函数：

(1) SelectionModel. multi({List < T > selectedValues, KeyProvider < T > keyProvider})：创建多选模型。keyProvider 用于相等性检查。

(2) SelectionModel. single({T selected, KeyProvider < T > keyProvider})：创建单选模型。

Angular 项目 radio_demo，将组件 MaterialRadioComponent 和 MaterialRadioGroupComponent 添加到指令列表。示例代码如下：

```
//chapter16/radio_demo/lib/app_component.dart
import 'package:angular/angular.dart';
import 'package:angular_components/angular_components.dart';
@Component(
  selector: 'my-app',
  styleURLs: ['app_component.css'],
  templateURL: 'app_component.html',
  directives: [
    MaterialRadioComponent,
    MaterialRadioGroupComponent,
    NgFor],
)
class AppComponent {
  //设置单选按钮状态
  bool radio1 = false;
  bool radio2 = true;
  bool radio3 = false;
  bool radio4 = true;

  //已选选项
  String selectedOption;

  //定义单选模型
  final SelectionModel selectionModel = SelectionModel.single();

  //获取单选模型的值
  dynamic get selectedValue =>
      selectionModel.selectedValues.isEmpty ? '未知':
```

```
          selectionModel.selectedValues.first;

  //构建选项
  List<Option> options = [
    Option('梨','pear'),
    Option('香蕉','banana'),
    Option('桃子','peach'),
    Option('其他','others')
  ];
}
//定义选项类
class Option{
  String label;
  String value;
  Option(this.label,this.value);
}
```

模板代码如下：

```
//chapter16/radio_demo/lib/app_component.html
<h1>单选按钮</h1>

<h3>独立单选按钮</h3>
<h6>可选择的单选按钮</h6>
<material-radio [(checked)]="radio1">可选择按钮</material-radio>

<h6>预选择的单选按钮</h6>
<material-radio [(checked)]="radio2">预选择按钮</material-radio>

<h6>禁用的单选按钮</h6>
<material-radio [disabled]="true" [(checked)]="radio3">不可操作的按钮</material-
radio>

<h6>预选择且禁用的单选按钮</h6>
<material-radio [disabled]="true" [(checked)]="radio4">预选择且不可操作的按钮
</material-radio>

<h3>分组按钮</h3>
<h6>手动提供选择项</h6>
<p>已选择项:{{selectedOption}}</p>
<material-radio-group [(selected)]="selectedOption">
    <material-radio [checked]="true" value="option1">选项 1</material-radio>
    <material-radio value="option2">选项 2</material-radio>
</material-radio-group>
```

```
<h6>选择模型</h6>
<p>已选择项:{{selectedValue}}</p>
<material-radio-group [selectionModel]="selectionModel">
    <material-radio
            *ngFor="let option of options"
            [value]="option.value"
            no-ink>{{option.label}}</material-radio>
</material-radio-group>

<h6>mixin 自定义按钮在选择状态下的颜色、波纹颜色</h6>
<material-radio class="custom">可选择按钮</material-radio>
```

样式代码如下：

```
//chapter16/radio_demo/lib/app_component.scss
@import 'package:angular_components/material_radio/mixins';

//自定义按钮在选择状态下的颜色、波纹颜色
@include material-radio-color(green, orange, '.custom');
```

16.10 复选框

7min

复选框是可以选中或不选中的按钮。用户可以单击复选框以选中或取消选中它。通常，使用复选框允许用户从集合中选择多个选项。如果只有一个打开或关闭选项，应避免使用单个复选框，而应使用切换按钮。组件名是 MaterialCheckboxComponent，在模板中使用元素<material-checkbox>创建复选框实例。

1. 属性

no-ink：禁用波纹效果。

2. 输入属性

(1) checked bool：用户设置复选框的当前状态，选中则为 true，否则为 false。

(2) disabled bool：复选框处于禁用状态，不应响应事件，并具有不允许互动的样式。

(3) indeterminate bool：复选框的二选一状态，非用户可设置的状态。在 checked 和 indeterminate 之间，只有一个可以为 true，也可以都为 false。如果是 indeterminate 则为 true，选中或未选中都为 false。

(4) indeterminateToChecked bool：判断 indeterminate 状态切换时复选框要进入的状态。为 true 时将被选中，为 false 时将被取消选中。

(5) label String：复选框的标签。

(6) readOnly bool：该复选框是否可以通过用户交互来更改。

(7) themeColor String：复选框的颜色和选中时的波纹，默认情况下为 $mat-blue-500。

注意，即使未选中复选框，themeColor 也会应用到该复选框，这与标准材质规范有所不同。除非需要此行为，否则使用 mixin 设置 themeColor。

3. 输出属性

(1) change Stream < String >：复选框状态更改时触发，发送 checkedStr，即 ARIA 状态。

(2) checkedChange Stream < bool >：选中或取消选中复选框时触发，但设置为 indeterminate 时则不触发。发送值是 checked 的状态。

(3) indeterminateChange Stream < bool >：当复选框进入和退出 indeterminate 状态时触发，但设置为 checked 时则不触发。

4. Sass mixin

material-checkbox-color($color, $modifier: '')：第一个参数是复选框在选择状态下的颜色，第二个参数是类选择器。

Angular 项目 checkbox_demo，将组件 MaterialCheckboxComponent 添加到指令列表。示例代码如下：

```
//chapter16/checkbox_demo/lib/app_component.dart
import 'package:angular/angular.dart';
import 'package:angular_components/angular_components.dart';

@Component(
  selector: 'my-app',
  styleURLs: ['app_component.css'],
  templateURL: 'app_component.html',
  directives: [
    MaterialCheckboxComponent,
    MaterialIconComponent,
    MaterialIconComponent,
    MaterialToggleComponent],
)
class AppComponent {
  //初始化各个状态值
  //禁用状态
  bool disabledState = false;
  //选择状态
  bool checkedState = false;
  //不确定状态
  bool indeterminateState = false;
  //不确定行为
  bool indeterminateBehavior = false;
  //状态字符串
  String statusStr;
}
```

模板代码如下：

```
//chapter16/checkbox_demo/lib/app_component.dart
<h1>复选框</h1>
<section>
    <h5>所有状态的复选框:</h5>
    <div>
        <h6>没标签</h6>
        <material-checkbox></material-checkbox>

        <h6>未选中状态</h6>
        <material-checkbox label="未选中 unchecked"></material-checkbox>

        <h6>选中状态</h6>
        <material-checkbox [checked]="true" label="选中 checked"></material-checkbox>

        <h6>不确定</h6>
        <material-checkbox [indeterminate]="true" label="不确定状态 indeterminate">
</material-checkbox>

        <h6>不可用</h6>
        <material-checkbox [disabled]="true" label="不可用 disabled"></material-checkbox>

        <h6>选中且不可用</h6>
        <material-checkbox [checked]="true" [disabled]="true" label="checked 和
disabled">
        </material-checkbox>

        <h6>不确定且不可用</h6>
        <material-checkbox [indeterminate]="true" [disabled]="true" label="indeterminate 和
disabled">
        </material-checkbox>

        <h6>选中且不确定 = 不确定</h6>
        <material-checkbox [indeterminate]="true" [checked]="true" label="indeterminate 和
checked">
        </material-checkbox>

        <h6>自定义复选框颜色</h6>
        <material-checkbox
                label="红色复选框"
                themeColor="#FF0000">
        </material-checkbox>

        <h6>自定义复选框内容; 顶部对齐</h6>
```

```
        <material-checkbox class="top">
            <div class="custom">
                建议
                <material-icon icon="help" baseline class="help-icon"></material-icon><br/>
                <textarea cols="40" rows="2"></textarea>
            </div>
        </material-checkbox>
    </div>
</section>

<section>
    <h5>事件和属性控制复选框的状态:</h5>
    <p>仔细观察各个状态的变化</p>
    <div>
        <material-checkbox
                            [disabled]="disabledState"
                            [(checked)]="checkedState"
                            [(indeterminate)]="indeterminateState"
                            [indeterminateToChecked]="indeterminateBehavior"
                            (change)="statusStr=$event"
                            label="事件和属性">
        </material-checkbox>
    </div>
    <div>
        <material-toggle
                [(checked)]="disabledState"
                label="切换到 {{disabledState ? '可用' : '禁用'}}">
        </material-toggle><br/>
        <material-toggle
                [(checked)]="indeterminateState"
                  label="切换到 {{indeterminateState ? '未设置 indeterminate' : '设置 indeterminate'}}">
        </material-toggle><br/>
        <material-toggle
                [(checked)]="indeterminateBehavior"
                 label="切换到 {{indeterminateBehavior ? '从 indeterminate 到 unchecked' : '从 indeterminate 到 checked'}}">
        </material-toggle><br/>
    </div>
    <div class="debug-info">
        status = {{statusStr}}<br/>
        checked = {{checkedState}}<br/>
        disabled = {{disabledState}}<br/>
        indeterminate = {{indeterminateState}}<br/>
        indeterminateToChecked = {{indeterminateBehavior}}<br/>
```

```
    </div>
</section>

<h6>mixin 自定义选择状态下的复选框颜色</h6>
<material-checkbox class="custom">自定义颜色</material-checkbox>
```

样式代码如下：

```
//chapter16/checkbox_demo/lib/app_component.scss
@import 'package:angular_components/css/material/material';
@import 'package:angular_components/material_checkbox/mixins';

//内容与复选框顶部对齐
material-checkbox.top {
  align-items: flex-start;
}

.debug-info {
  background: $mat-grey-200;
}

//自定义复选框颜色
@include material-checkbox-color(orange, '.custom');
```

16.11　输入框

12min

输入框可以是单行或多行文本字段，它可以有一个标签。组件名是MaterialInputComponent，可以在模板中使用元素<material-input>创建输入框的实例。使用时必须在指令列表中声明materialInputDirectives而不是MaterialInputComponent。

1. 属性

（1）type：输入的类型，默认是text。也可以支持email、password、URL、number、tel和search。

（2）multiple：用户可以输入通过逗号分隔的多个值。此属性仅在type为email时适用，否则忽略。

2. 输入属性

（1）label String：此输入的标签。如果未在文本框中输入任何内容，则默认显示文本。用户输入文字时消失。

（2）rightAlign bool：输入内容是否应始终右对齐。默认值为false。

（3）required bool：是否必须输入。如果没有输入文本，则在第一次失去焦点时显示验证错误。

(4) requiredErrorMsg String：自定义错误消息，当字段必填且为空时显示。

(5) disabled bool：是否禁用输入。

(6) leadingText String：需要显示在输入框前方的文本，例如货币符号或类似符号。

(7) leadingGlyph String：在输入框前方显示的任何符号，例如链接图标或类似图标。

(8) trailingText String：在输入框的后方显示的任何文本，例如货币符号或类似符号。

(9) trailingGlyph String：在输入框的后边显示的任何符号，例如链接图标或类似图标。

(10) hintText dynamic：要在输入框下方显示的提示。如果输入中有错误消息，则不会显示此文本。

(11) showHintOnlyOnFocus bool：输入框未聚焦时是否显示提示文本。

(12) characterCounter int Function(String)：自定义字符计数器函数。接收输入文本，返回应将文本视为多少个字符。

3. 输出属性

(1) blur Stream<FocusEvent>：触发失去焦点事件时发布事件。

(2) change Stream<String>：触发更改事件时发布事件，例如回车或失去焦点。

(3) focus Stream<FocusEvent>：元素聚焦时触发的事件。

(4) inputKeyPress Stream<String>：每当输入文本更改时发布事件，每次按键都会触发。

拥有 multiline 属性的输入框是一个多行文本框，它需要提供指令 MaterialMultilineInputComponent。还需要在提供者列表中添加 DomService 和 windowBindings。它还具有两个特殊属性：rows 表示多行输入默认显示的行数，maxRows 表示显示的最大行数，超过 maxRows 的所有内容则会出现滚动条。

创建 Angular 项目 input_demo，将 materialInputDirectives 和 MaterialMultilineInputComponent 添加到指令列表。示例代码如下：

```
//chapter16/input_demo/lib/app_component.dart
import 'package:angular/angular.dart';
import 'package:angular_components/angular_components.dart';
import 'package:angular_components/utils/browser/dom_service/angular_2.dart';
import 'package:angular_components/utils/browser/window/module.dart';
import 'package:angular_forms/angular_forms.dart';

@Component(
  selector: 'my-app',
  styleURLs: ['app_component.css'],
  templateURL: 'app_component.html',
  providers: [DomService,windowBindings],
  directives: [
    formDirectives,
```

```
    materialInputDirectives,
    MaterialMultilineInputComponent],
)
class AppComponent {
  //初始化输入框文本
  String textValue = 'Text value';
}
```

模板代码如下：

```
//chapter16/input_demo/lib/app_component.html
<h6>单行输入框</h6>
<p><material-input></material-input></p>
<p><material-input label="标签"></material-input></p>
<p><material-input label="右对齐" [rightAlign]="true"></material-input></p>
<p><material-input label="必填" required
                   requiredErrorMsg="此输入框必填"></material-input></p>
<p><material-input [disabled]="true" label="此输入框被禁用"></material-input></p>

<h6>前导和后导</h6>
<p><material-input label="前导文本" leadingText="$"></material-input></p>
<p><material-input label="后导文本" trailingText=".00"
                   [rightAlign]="true"></material-input></p>
<p><material-input label="前导图标" leadingGlyph="link"></material-input></p>
<p><material-input label="后导图标" trailingGlyph="email"></material-input></p>

<h6>浮动标签</h6>
<p><material-input floatingLabel label="浮动标签"></material-input></p>
<p><material-input floatingLabel label="浮动标签和双向数据绑定"
                   [ngModel]="textValue"></material-input></p>

<h6>失去焦点时更新</h6>
<p><material-input blurUpdate [(ngModel)]="textValue"></material-input></p>
<div>值:{{textValue}}</div>

<h6>更改时更新</h6>
<p><material-input changeUpdate [(ngModel)]="textValue"></material-input></p>
<div>值:{{textValue}}</div>

<h6>提示文本</h6>
<p><material-input hintText="输入提示信息"></material-input></p>

<p>仅在聚焦时才显示提示信息</p>
<p><material-input hintText="使用 showHintOnlyOnFocus"
                   showHintOnlyOnFocus required></material-input></p>
```

```
<h6>字符计数</h6>
<p><material-input [maxCount]="10" label="计数"></material-input></p>

<h6>邮箱</h6>
<p><material-input floatingLabel label="email" type="email"></material-input></p>

<h6>密码</h6>
<p><material-input floatingLabel label="password" type="password"></material-input>
</p>

<h6>URL 地址</h6>
<p><material-input floatingLabel label="URL" type="URL"></material-input></p>

<h6>多行输入框</h6>
<p><material-input multiline floatingLabel
                    label="默认多行输入框,随着字符增多而增加行"></material-input></p>
<p><material-input multiline floatingLabel rows="2" maxRows="4"
                    label="显示 2 行,最多显示 4 行,随着字符增多而增加行"></material-
input></p>
<p><material-input multiline floatingLabel rows="2" [maxCount]="90"
                    label="最多 90 个字符,否则显示错误信息"></material-input></p>
```

12min

16.12 列表

16.12.1 材质化列表

材质化列表是用于与用户进行交互的一组列表条目的容器组件，它构成了选择和菜单组件的基础。组件名是 MaterialListComponent，在模板中使用元素< material-list >可以创建列表的实例。

MaterialListComponent 类充当列表的根节点，提供样式和收集列表条目事件的能力。如果需要对列表条目进行分组，则可以使用带有 group 属性的元素包裹相应的列表条目。如果需要在组中提供标签，则可以将 label 属性置于组内的块元素上。

1. 属性

(1) size String：列表大小，预置大小包括 x-small、small、medium、large 和 x-large，它们的实际宽度分别是 64px * {1.5,3,5,6,7}。其默认大小根据内容自动调整。

(2) min-size String：列表的最小大小，列表宽度至少会达到该指定大小。

2. 输入属性

size String：预设宽度 1～5，分别与属性中的预设宽度对应。默认情况下其值为 0，材质列表将扩展到其父级的完整宽度。

16.12.2 列表条目

材质化列表条目是与用户交互的块级元素，当用户单击或按 Enter 键时发出并触发事件。组件名是 MaterialListItemComponent，在模板中使用元素< material-list-item >创建列表条目的实例。

1. 输入属性

(1) disabled bool：禁用触发器并为列表条目设置禁用样式。

(2) tabbable bool：组件是否可用作选项卡。

(3) tabindex String：组件的选项卡索引。与 tabbable 配合使用，当 tabbable 为 true 时有效。

2. 输出属性

trigger Stream < UIEvent >：通过单击，单击或按键激活按钮时触发。

3. 样式

(1) material-list-item-primary：该样式常应用于显示列表条目中的图标。

(2) material-list-item-secondary：该样式常用于显示列表条目的辅助信息。

创建 Angular 项目 list_demo，将 MaterialListComponent 和 MaterialListItemComponent 添加到指令列表。示例代码如下：

```
//chapter16/list_demo/lib/app_component.dart
import 'package:angular/angular.dart';
import 'package:angular_components/angular_components.dart';

@Component(
  selector: 'my-app',
  styleURLs: ['app_component.css'],
  templateURL: 'app_component.html',
  directives: [
    MaterialIconComponent,
    MaterialListComponent,
    MaterialListItemComponent, ],
)
class AppComponent {
  //自定义列表背景颜色
  String bgColor = '#f0c9cf';
  //响应列表条目的 trigger 事件
  void toggleColor(String color){
    bgColor = color;
  }
}
```

模板代码如下：

```
//chapter16/list_demo/lib/app_component.html
<h1>列表</h1>
<h6>普通列表,默认宽度为父容器的宽度</h6>
<material-list>
    <material-list-item>条目 1</material-list-item>
    <material-list-item>条目 2</material-list-item>
    <material-list-item>条目 3</material-list-item>
</material-list>

<h6>宽度为 large 并带有禁用条目的列表</h6>
<material-list size="large">
    <material-list-item disabled>条目 1</material-list-item>
    <material-list-item>条目 2</material-list-item>
    <material-list-item>条目 3</material-list-item>
</material-list>

<h6>分组且带标签的列表</h6>
<material-list size="large">
    <div group>
        <div label>分组 1</div>
        <material-list-item>条目 1</material-list-item>
        <material-list-item>条目 2</material-list-item>
    </div>
    <div group>
        <div label>分组 2</div>
        <material-list-item>条目 1</material-list-item>
        <material-list-item>条目 2</material-list-item>
    </div>
</material-list>

<h6>带图标和辅助信息的列表</h6>
<material-list size="large">
    <material-list-item>
        <material-icon icon="today" class="material-list-item-primary">
        </material-icon>
        条目 1
    </material-list-item>
    <material-list-item disabled>
        条目 2
        <span class="material-list-item-secondary">不可选</span>
    </material-list-item>
    <material-list-item>条目 3</material-list-item>
    <material-list-item>条目 4</material-list-item>
</material-list>

<h6>通过 trigger 事件切换列表背景</h6>
```

```
<material - list size = "large" [style.background - color] = "bgColor">
    <material - list - item (trigger) = "toggleColor('#f07c82')">香叶红</material - list -
item >
    <material - list - item (trigger) = "toggleColor('#eea2a4')">牡丹粉红</material - list -
item >
    <material - list - item (trigger) = "toggleColor('#f03752')">海棠红</material - list -
item >
</material - list >
```

16.13 片记与片集

28min

16.13.1 片记

片记小部件呈现为带阴影的圆形框，通常水平排列，它用于呈现简要信息。任何对象都可以通过实现 HasUIDisplayName 接口来使用片记。组件名是 MaterialChipComponent，在模板中使用元素< material-chip >创建片记的实例。

当 removable 属性为 true 或在片集实例上设置了 selectionModel 时，删除按钮才会显示。

当 hasLeftIcon 为 true 时，应将左图标内容设置为 MaterialIconComponent 组件实例或 SVG 图像。

1. 输入属性

（1）deleteButtonAriaMessage String：移除按钮的 Aria 标签。

（2）hasLeftIcon bool：片记是否应显示自定义图标，默认值为 false。

（3）itemRenderer String Function(T)：一个 ItemRenderer 函数，获取一个对象并返回一个字符串。如果 ItemRenderer 不是无状态的，并且可能为同一输入项返回不同的值，则需要更新 ItemRenderer 引用，否则该更改将不会得到体现。提供时，将用于生成片记的标签。

（4）removable bool：片记是否应显示移除按钮，默认值为 true。

（5）value dynamic：要渲染的数据模型。在片记的内容中提供标签，或提供一个 ItemRenderer。

2. 输出属性

remove Stream < dynamic >：移除片记时触发事件，该事件返回片记的值。

16.13.2 片集

片记的集合，将列表中的所有对象以片的形式展示。组件名是 MaterialChipsComponent，在模板中使用元素< material-chips >创建片集的实例。

输入属性

(1) itemRenderer String Function(T)：将列表中的条目呈现为字符串的函数。注意：仅当提供 SelectionModel 时，才使用此 ItemRenderer。如果片集是手动渲染的，itemRenderer 属性也需要手动设置。关于 OnPush 的注意事项：如果 ItemRenderer 不是纯函数，并且具有可能以不同方式呈现同一项目的内部状态，则引用必须更改才能生效。

(2) removable bool：是否可以移除片记。

(3) selectionModel SelectionModel < T >：此组件控制的选择模型。

创建 Angular 项目 chips_demo，将 MaterialChipsComponent、displayNameRendererDirective 和 MaterialChipComponent 添加到指令列表。示例代码如下：

```
//chapter16/chips_demo/lib/app_component.dart
import 'package:angular/angular.dart';
import 'package:angular_components/angular_components.dart';

@Component(
  selector: 'my-app',
  styleURLs: ['app_component.css'],
  templateURL: 'app_component.html',
  directives: [
    MaterialChipsComponent,
    MaterialChipComponent,
    displayNameRendererDirective,
    MaterialIconComponent,
    NgFor],
)
class AppComponent {
  //Label 实现了 HasUIDisplayName 接口
  //选择模型的泛型为< HasUIDisplayName >
  //需要同指令 displayNameRenderer 一起使用
  //SelectionModel.multi()方法返回选择模型 SelectionModel
  //selectedValues 参数用于传递一个任意类型的列表
  SelectionModel < HasUIDisplayName > labelSelection = _labelSelectionModel();
  static SelectionModel < HasUIDisplayName > _labelSelectionModel() =>
      SelectionModel.multi(selectedValues: [
        Label('书法'),
        Label('国画'),
        Label('诗歌'),
        Label('元曲')
      ]);

  //Subject 未实现 HasUIDisplayName 接口
  //选择模型的泛型为< Subject >
  //需要同渲染函数 renderSubjectChip 一起使用
  SelectionModel < Subject > subjectSelection = _subjectSelectionModel();
```

```
  static SelectionModel<Subject> _subjectSelectionModel() =>
      SelectionModel.multi(selectedValues: [
        Subject(1,'语文'),
        Subject(1,'数学'),
        Subject(1,'物理'),
        Subject(1,'美术')
      ]);
  //渲染函数
  ItemRenderer<dynamic> renderSubjectChip =
      (dynamic protoChip) {
      return protoChip.name;
  };
  //Subject 未实现 HasUIDisplayName 接口
  //需要同渲染函数 renderSubjectChip 一起使用
  //通过响应 remove 事件移除 subjects 列表条目
  List<Subject> subjects = [
    Subject(1,'语文'),
    Subject(1,'数学'),
    Subject(1,'物理'),
    Subject(1,'美术')
  ];
  //Label 实现了 HasUIDisplayName 接口
  //需要同指令 displayNameRenderer 一起使用
  //通过响应 remove 事件移除 labels 列表条目
  List<Label> labels = [
    Label('书法'),
    Label('国画'),
    Label('诗歌'),
    Label('元曲')
  ];
  //向 labels 列表添加新条目
  void add(String val){
    labels.add(Label(val));
  }
}
//定义实现 HasUIDisplayName 接口的 Label 类
class Label implements HasUIDisplayName{
  @override
  //TODO: 实现 uiDisplayName
  String get uiDisplayName => name;
  String name;
  Label(this.name);
}
//定义一个普通类 Subject
class Subject{
  int id;
```

```
  String name;
  Subject(this.id,this.name);
}
```

模板代码如下：

```
//chapter16/chips_demo/lib/app_component.html
<h1>片集</h1>
<h5>片集</h5>
<p>选择模型 labelSelection</p>
<material-chips [selectionModel]="labelSelection"
                [removable]="true"
                displayNameRenderer>
</material-chips>
<p>选择模型 subjectSelection</p>
<material-chips [selectionModel]="subjectSelection"
                [removable]="true"
                [itemRenderer]="renderSubjectChip">
</material-chips>

<h5>片记</h5>
<p>手动指定内容,移除按钮不起作用</p>
<material-chips>
    <material-chip>默认</material-chip>
    <material-chip [removable]="false">不可移除</material-chip>
    <material-chip [hasLeftIcon]="true">
        <material-icon left-icon icon="link" size="large"></material-icon>
        带有左图标
    </material-chip>
</material-chips>
<h5>使用数据模型生成片记</h5>

<p>数据模型 subjects 列表</p>
<material-chips>
    <material-chip *ngFor="let subject of subjects"
                   [removable]="true"
                   [itemRenderer]="renderSubjectChip"
                   (remove)="subjects.remove(subject)"
                   [value]="subject"></material-chip>
</material-chips>

<p>数据模型 labels 列表</p>
<material-chips>
    <material-chip *ngFor="let c of labels"
                   [removable]="true"
```

```
                    (remove) = "labels.remove(c)"
                    [value] = "c"
                    displayNameRenderer ></material - chip >
</material - chips >

< input #label type = "text" placeholder = "填入新 label"/>
< button (click) = "add(label.value)">添加 Label </button >
```

min

16.14 按钮组

两个水平相邻的按钮组件,例如:是或否、保存或取消、同意或不同意等。按钮上的文本可以更改,也可以突出显示。组件名是 MaterialYesNoButtonsComponent,可以使用元素 < material-yes-no-buttons >创建按钮组实例。

可以使用 MaterialSaveCancelButtonsDirective 之类的指令提供自定义文本,该指令用 Save 或 Cancel 替换 yes 或 no。

要以相反的顺序显示按钮,需添加 reverse 属性。

1. 输入属性

(1) disabled bool:是否应禁用按钮。默认值为 false。

(2) noAutoFocus bool:no 按钮是否应自动对焦。默认值为 false。

(3) noDisabled bool:是否应禁用 no 按钮。默认值为 false。

(4) noDisplayed bool:是否显示 no 按钮。默认值为 true。

(5) noText String:要在取消按钮上显示的文本。例如,关闭、Not now 等。默认值为 No。

(6) pending bool:表示待定状态,如果值为 true,将隐藏是和否按钮,并显示一个旋转器。这应该用于指示异步操作,例如:保存或验证输入。默认值为 false。

(7) raised bool:是否应该凸出按钮。默认值为 false。

(8) yesAutoFocus bool:Yes 按钮是否应该自动对焦。默认值为 false。

(9) yesDisabled bool:是否应禁用 Yes 按钮。默认值为 false。

(10)yesDisplayed bool:是否显示 Yes 按钮。默认值为 true。

(11) yesHighlighted bool:是否应突出显示 Yes 按钮。默认值为 false。

(12) yesRaised bool:是否应凸起 Yes 按钮。默认值为 false。

(13) yesText String:要在 Yes 按钮上显示的文本。例如,确定、应用等。默认值为 Yes。

2. 输出属性

(1) no Stream < UIEvent >:当 No 按钮被按下时要调用的回调。发布的事件是 KeyboardEvent 或 MouseEvent。

（2）yes Stream < UIEvent >：按下 Yes 按钮时要调用的回调。发布的事件是 KeyboardEvent 或 MouseEvent。

3. 相关指令

保存取消按钮组指令 MaterialSaveCancelButtonsDirective，对应元素是< material-yes-no-buttons[saveCancel]>，它会将 Yes 和 No 替换为 Save 和 Cancel。

提交取消按钮组指令 MaterialSubmitCancelButtonsDirective，对应元素是< material-yes-no-buttons[submitCancel]>，它会将 Yes 和 No 更改为 Submit 和 Cancel。

4. Sass mixin

（1）material-yes-button-color($value)：更改 yes 按钮的颜色。

（2）material-yes-button-text-color($value)：更改 yes 按钮上的文字颜色。

（3）material-no-button-color($value)：更改 no 按钮的颜色。

（4）material-no-button-text-color($value)：更改 no 按钮上的文字颜色。

创建 Angular 项目 yes_no_buttons_demo，添加 MaterialYesNoButtonsComponent、MaterialSaveCancelButtonsDirective、MaterialSubmitCancelButtonsDirective 到指令列表，将 DomService 和 windowBindings 添加到提供者列表。示例代码如下：

```
//chapter16/yes_no_buttons_demo/lib/app_component.dart
import 'package:angular/angular.dart';
import 'package:angular_components/angular_components.dart';
import 'package:angular_components/utils/browser/dom_service/angular_2.dart';
import 'package:angular_components/utils/browser/window/module.dart';
@Component(
  selector: 'my-app',
  styleURLs: ['app_component.css'],
  templateURL: 'app_component.html',
  providers: [DomService,windowBindings],
  directives: [
    MaterialYesNoButtonsComponent,
    MaterialSaveCancelButtonsDirective,
    MaterialSubmitCancelButtonsDirective],
)
class AppComponent {
  //初始化待定状态
  bool pending = false;
  //响应 yes 事件
  void save(){
    pending = true;
    Future.delayed(Duration(seconds: 2),() => pending = false);
  }
}
```

模板代码如下：

```
//chapter16/yes_no_buttons_demo/lib/app_component.html
<h1>确认和取消按钮组</h1>
<h6>默认</h6>
<material-yes-no-buttons></material-yes-no-buttons>

<h6>凸起</h6>
<material-yes-no-buttons raised></material-yes-no-buttons>

<h6>仅 Yes 按钮凸起</h6>
<material-yes-no-buttons yesRaised></material-yes-no-buttons>

<h6>Yes 按钮高亮</h6>
<material-yes-no-buttons yesHighlighted></material-yes-no-buttons>

<h6>自动聚焦</h6>
<material-yes-no-buttons yesAutoFocus></material-yes-no-buttons>

<h6>禁用按钮</h6>
<material-yes-no-buttons [disabled]="true"></material-yes-no-buttons>
<material-yes-no-buttons [yesDisabled]="true"></material-yes-no-buttons>
<material-yes-no-buttons [noDisabled]="true"></material-yes-no-buttons>

<h6>不显示其中某个按钮</h6>
<material-yes-no-buttons [noDisplayed]="false"></material-yes-no-buttons>
<material-yes-no-buttons [yesDisplayed]="false"></material-yes-no-buttons>

<h6>自定义文本</h6>
<material-yes-no-buttons noText="取消"
                         yesText="保存"></material-yes-no-buttons>

<h6>翻转</h6>
<material-yes-no-buttons reverse
                         noText="取消"
                         yesText="保存"></material-yes-no-buttons>

<h6>触发 yes 事件可观察到待定状态</h6>
<material-yes-no-buttons noText="取消"
                         yesText="保存"
                         (yes)="save()"
                         [pending]="pending"></material-yes-no-buttons>

<h6>保存和取消</h6>
<material-yes-no-buttons saveCancel></material-yes-no-buttons>

<h6>提交和取消</h6>
<material-yes-no-buttons submitCancel></material-yes-no-buttons>
```

```
<h6>自定义主题</h6>
<material-yes-no-buttons raised class="green-yes-no"></material-yes-no-buttons>
<br>
<material-yes-no-buttons class="green-yes-no"></material-yes-no-buttons>
<br>
<material-yes-no-buttons class="blue-text-yes-no"></material-yes-no-buttons>
<br>
<material-yes-no-buttons raised class="yes-no"></material-yes-no-buttons>
```

样式代码如下：

```
//chapter16/yes_no_buttons_demo/lib/app_component.scss
@import 'package:angular_components/css/material/material';
@import 'package:angular_components/material_yes_no_buttons/mixins';

.green-yes-no{
  //设置 yes 和 no 按钮背景色，当设置按钮背景色时必须提供 raised 属性，否则无效
  @include material-no-button-color(green);
  @include material-yes-button-color(green);
}

.blue-text-yes-no{
  //设置 yes 和 no 按钮文本颜色，波纹颜色依赖于文本颜色
  @include material-yes-button-text-color(blue);
  @include material-no-button-text-color(blue);
}

.yes-no{
   //yes 按钮背景色
  @include material-yes-button-color($mat-red-500);
  //yes 按钮文本颜色
  @include material-yes-button-text-color(white);
  //no 按钮背景色
  @include material-no-button-color($mat-red-500);
  //no 按钮文本颜色
  @include material-no-button-text-color(white);
}
```

21min

16.15 日期、时间选择器

16.15.1 日期范围选择器

一种采用材料设计风格的日期范围选择器。组件名是 MaterialDateRangePickerComponent，

在模板中使用元素< material-date-range-picker >创建日期范围选择器的实例。

1. 输入属性

(1) minDate Date：无法选择早于 minDate 的日期。

(2) maxDate Date：无法选择晚于 maxDate 的日期。

(3) showNextPrevButtons bool：是否显示“下一个”和“上一个”按钮。默认值为 true。

(4) range DatepickerComparison：所选日期范围和比较。此日期选择器使用 DatepickerComparison 代替普通的 DateRangeComparison 对象，在内部实现添加其他所需的功能，例如“名称”和“下一个”或“上一个”。

(5) presets List < DatepickerPreset >：用户可以选择的预设日期范围的列表。这些预设项目受 minDate 和 maxDate 的限制，如果它们的终点在 minDate 之前或起点在 maxDate 之后，则将其完全排除。

2. 输出属性

rangeChange Stream < DatepickerComparison >：所选日期范围或比较范围更改时发布。

创建 Angular 项目 date_time_picker_demo，在 src 目录下创建日期范围选择器。示例代码如下：

```
//chapter16/date_time_picker_demo/lib/src/date_range_component.dart
import 'package:angular/angular.dart';
import 'package:quiver/time.dart';
import 'package:angular_components/angular_components.dart';
import 'package:angular_components/utils/browser/window/module.dart';
@Component(
  selector: 'date-range',
  templateURL: 'date_range_component.html',
  providers: [windowBindings, datepickerBindings],
  directives: [
    MaterialDateRangePickerComponent],
)
class DateRangeComponent {
  //日期选择器预设列表
  List<DatepickerPreset> presets;

  //用于绑定日期范围 range
  DatepickerComparison range;
  DatepickerComparison emptyRange;

  //日期范围对象与输入属性[minDate]和[maxDate]配合使用
  DateRange limitRange = DateRange(Date.today().add(months: -3), Date.today());

  DateRangeComponent() {
```

```
    //创建一个本地时钟
    var clock = Clock();
    //默认日期选择器预设列表接收一个时钟对象
    presets = defaultPresets(clock);
    //预设 range 的日期选择范围为本周
    range = DatepickerComparison.noComparison(presets
        .singleWhere((preset) => preset.range.title == 'This week')
        .range);
  }
}
```

模板代码如下：

```
//chapter16/date_time_picker_demo/lib/src/date_range_component.html
<h3>时间范围选择器</h3>
<h6>默认</h6>
<material-date-range-picker>
</material-date-range-picker>

<h6>限制日期选择范围</h6>
<material-date-range-picker [minDate]="limitRange.start"
                            [maxDate]="limitRange.end"
                            [showNextPrevButtons]="false"
                            [(range)]="emptyRange">
</material-date-range-picker>

<h6>使用预设列表的日期选择器</h6>
<material-date-range-picker
                            [presets]="presets"
                            [(range)]="range"
                            [showNextPrevButtons]="false">
</material-date-range-picker>
```

16.15.2 日期选择器

材料设计风格的单个日期选择器。用户可以输入自定义日期，或单击日历以选择日期。组件名是 MaterialDatepickerComponent，在模板中使用元素< material-datepicker >创建日期选择器实例。

1. 输入属性

（1）date Date：所选的日期。常用于绑定到组件属性。

（2）minDate Date：无法选择早于 minDate 的日期。

（3）maxDate Date：无法选择晚于 maxDate 的日期。

2. 输出属性

dateChange Stream < Date >：所选日期更改时发布事件。

16.15.3 时间选择器

一种采用材料设计风格的单一时间选择器。组件名是 MaterialTimePickerComponent，在模板中使用元素< material-time-picker >创建时间选择器实例。

1. 输入属性

（1）time DateTime：所选的时间。

（2）utc bool：是否以 UTC 时区返回时间，默认返回本地时区的时间。

2. 输出属性

timeChange Stream < DateTime >：所选时间更改时发布事件。

16.15.4 日期和时间选择器

一种采用材料设计风格的单一日期和时间选择器。组件名是 MaterialDateTimePickerComponent，在模板中使用元素< material-date-time-picker >创建日期时间选择器实例。

1. 输入属性

（1）dateTime DateTime：所选日期时间。

（2）utc bool：是否使用 UTC 时区中的 dateTime，默认返回本地时区的 dateTime。

2. 输出属性

dateTimeChange Stream < DateTime >：选定的 dateTime 更改时发布事件。

在 src 目录下创建日期时间选择器的示例代码如下：

```
//chapter16/date_time_picker_demo/lib/src/date_time_component.dart
import 'package:angular/angular.dart';
import 'package:angular_components/angular_components.dart';
import 'package:angular_components/utils/browser/window/module.dart';
@Component(
  selector: 'date - time',
  templateURL: 'date_time_component.html',
  providers: [windowBindings, datepickerBindings],
  directives: [
    MaterialDatepickerComponent,
    MaterialTimePickerComponent,
    MaterialDateTimePickerComponent,],
)
class DateTimeComponent {
  //用于绑定到日期选择器
  Date date = Date.today();
```

```
  //用于绑定到时间选择器
  DateTime time = DateTime.now();

  //用于绑定到日期时间选择器
  DateTime dateTime = DateTime.now();
}
```

模板代码如下：

```
//chapter16/date_time_picker_demo/lib/src/date_time_component.html
<h3>日期选择器</h3>
<material-datepicker [(date)]="date">
</material-datepicker>

<h3>时间选择器</h3>
<material-time-picker [(time)]="time">
</material-time-picker>

<h6>使用 UTC 时区</h6>
<material-time-picker [utc]="true"
                      [(time)]="time">
</material-time-picker>

<h3>日期时间选择器</h3>
<material-date-time-picker [(dateTime)]="dateTime">
</material-date-time-picker>

<h6>使用 UTC 时区</h6>
<material-date-time-picker [utc]="true"
                           [(dateTime)]="dateTime">
</material-date-time-picker>
```

在根组件中将 DateRangeComponent 和 DateTimeComponent 添加到指令列表。代码如下：

```
//chapter16/date_time_picker_demo/lib/app_component.dart
import 'package:angular/angular.dart';
import 'src/date_range_component.dart';
import 'src/date_time_component.dart';

@Component(
  selector: 'my-app',
  styleURLs: ['app_component.css'],
  templateURL: 'app_component.html',
```

```
    directives: [DateRangeComponent,DateTimeComponent],
)
class AppComponent {
}
```

在模板中添加元素代码如下：

```
//chapter16/date_time_picker_demo/lib/app_component.html
<h1>日期时间</h1>
<date-range></date-range>
<date-time></date-time>
```

16.16 步骤指示器

11min

16.16.1 材质化步骤指示器

材质化风格的步骤指示器，它是带编号的指示器，用于传达进度或用作导航工具。组件名是 MaterialStepperComponent，在模板中使用元素<material-stepper>创建步骤指示器实例。

1. 输入属性

(1) keepInactiveStepsInDom bool：如果值为 true，则在非活动状态下不会从 DOM 卸载步骤，而是通过 CSS 隐藏这些步骤。这样可以在加载复杂的 DOM 之间快速切换步骤。

(2) legalJumps String：合法的跳转。定义为非继续或取消按钮触发的步进开关。可能的值：none：默认，不允许跳转；backwards：允许跳转至已完成的步骤；all：允许任意跳转，与步骤状态无关。

(3) noText String：返回上一步按钮上显示的文本，默认为 Cancel。

(4) orientation String：步骤的布局方向。包含两个值 horizontal 和 vertical，分别代表水平和垂直。

(5) size String：设置大小，用于确定各个步骤标头元素的大小，包括步骤编号、步骤名称等。可能的值为 default 和 mini。

(6) stickyHeader bool：指示列出了可用步骤的标头，是否应该停留在页面顶部。仅适用于带有水平标头的步骤指示器。

(7) yesText String：在下一步按钮上显示的文本。默认情况下显示 Continue，如果源码中未提供该输入属性，则无法使用。

2. 输出属性

activeStepChanged Stream<StepDirective>：当前活动的步骤更改时触发的事件。

3. Sass mixin

(1) material-stepper-theme($selector:'',$step-color: $mat-blue-500,$disabled-color:

$mat-grey-500,$button-color：$mat-blue-500)：第 1 个参数是选择器，第 2 个参数是步骤的索引颜色，第 3 个参数是处于非活跃状态的步骤的索引颜色，第 4 个参数是 yes 按钮的颜色。

(2) material-stepper-step-name-disabled-color($color：$mat-grey-500)：设置不可跳转步骤名的颜色。

(3) material-stepper-step-name-selectable-color($color：$mat-grey-500)：设置可跳转步骤名的颜色。

16.16.2 步骤指令

用来标记步骤指示器中的一步。指令名是 StepDirective，在模板中使用带有 step 属性的元素创建单个步骤。

1. 输入属性

(1) canContinue bool：该步骤是否可以继续。这可以用来防止继续执行某个步骤，直到当前步骤的所有部分都满足验证要求为止。

(2) cancelHidden bool：在此步骤中是否应隐藏取消按钮。

(3) complete bool：该步骤是否完成。当进入下一步时设置此值。

(4) completeSummary String：在垂直默认大小的步骤指示器中步骤完成时显示的摘要文本。对于其他步骤指示器不适用。

(5) hideButtons bool：在此步骤中是否应隐藏按钮。

(6) name String：显示为标题。

(7) optional bool：该步骤是否可选。可选步骤带有一个额外的标签，表示它们是可选的，应该可以跳过。默认值为 false。

2. 输出属性

(1) cancel Stream<AsyncAction<bool>>：单击取消按钮时调用。如果事件处理程序调用 $event.cancel()，则不会取消该步骤。

(2) continue Stream<AsyncAction<bool>>：单击继续按钮时调用。如果事件处理程序调用 $event.cancel()，则该步骤将不会继续。

(3) jumpHere Stream<AsyncAction<bool>>：当用户想要跳至此步骤时调用。如果事件处理程序调用 $event.cancel()，则该步骤将不会继续。

创建 Angular 项目 stepper_demo，将 MaterialStepperComponent、StepDirective 和 SummaryDirective 添加到指令列表，在提供者列表中添加 scrollHostProviders、DomService 和 windowBindings。示例代码如下：

```
//chapter16/stepper_demo/lib/app_component.dart
import 'package:angular/angular.dart';
import 'package:angular_components/angular_components.dart';
```

```
import 'package:angular_components/model/action/async_action.dart';
import 'package:angular_components/utils/angular/scroll_host/angular_2.dart';
import 'package:angular_components/utils/browser/dom_service/angular_2.dart';
import 'package:angular_components/utils/browser/window/module.dart';
@Component(
  selector: 'my-app',
  styleURLs: ['app_component.css'],
  templateURL: 'app_component.html',
  providers: [scrollHostProviders,DomService,windowBindings],
  directives: [
    MaterialStepperComponent,
    StepDirective,
    SummaryDirective,
    MaterialButtonComponent],
)
class AppComponent{
  //该方法用于响应 continue 和 cancel 事件
  void validDelayedCheck(AsyncAction<bool> action) {
    //在这里没有进行实际的验证操作,而是延迟 1s 后允许 continue 和 cancel 操作继续执行
    action.cancelIf(Future.delayed(const Duration(seconds: 1), () {
      //不取消
      return false;
    }));
  }
}
```

模板代码如下：

```
//chapter16/stepper_demo/lib/app_component.html
<h1>步骤指示器</h1>
<h6>水平步骤指示器,可往回调转,大小取值为 default</h6>
<material-stepper legalJumps="backwards"
                  orientation="horizontal"
                  size="default">
    <template step name="第 1 步">
        <p>操作内容</p>
    </template>
    <template step name="第 2 步">
        <p>操作内容</p>
    </template>
    <template step name="第 3 步">
        <p>操作内容</p>
    </template></material-stepper>

<h6>垂直步骤指示器,可任意调转,大小取值为 mini</h6>
```

```
<material-stepper legalJumps="all"
                  orientation="vertical"
                  size="mini">
    <template step name="第 1 步">
        <p>操作内容</p>
    </template>
    <template step name="第 2 步">
        <p>操作内容</p>
    </template>
    <template step name="第 3 步">
        <p>操作内容</p>
    </template>
</material-stepper>

<h6>带验证和可选步骤</h6>
<material-stepper legalJumps="all"
                  orientation="vertical"
                  size="default">
    <template step name="第 1 步"
              (continue)="validDelayedCheck($event)">
        <p>操作内容</p>
    </template>
    <template step name="第 2 步" [optional]="true"
              (cancel)="validDelayedCheck($event)">
        <p>操作内容</p>
    </template>
    <template step name="第 3 步"
              (continue)="validDelayedCheck($event)">
        <p>操作内容</p>
    </template>
    <template step name="第 4 步"
              (continue)="validDelayedCheck($event)"
              (cancel)="validDelayedCheck($event)">
        <p>操作内容</p>
    </template>
</material-stepper>

<h6>mixin 步骤的摘要文本和主题及自定义按钮文本</h6>
<section class="themed">
    <material-stepper legalJumps="all"
                      orientation="vertical"
                      size="default"
                      yesText="继续"
                      noText="返回">
        <template step name="第 1 步">
```

```
            <p>操作内容</p>
        </template>
        <template step name="第 2 步" #step2="step">
            <p>操作内容</p>
        </template>
        <template [summary]="step2" >
            第 2 步的摘要文本
        </template>
        <template step name="第 3 步">
            <p>操作内容</p>
        </template>
    </material-stepper>
</section>

<h6>mixin 自定义不可跳转步骤名的颜色 </h6>
<material-stepper class="custom-disabled"
                  orientation="vertical"
                  size="default"
                  yesText="继续"
                  noText="返回">
    <template step name="第 1 步">
        <p>操作内容</p>
    </template>
    <template step name="第 2 步">
        <p>操作内容</p>
    </template>
    <template step name="第 3 步">
        <p>操作内容</p>
    </template>
</material-stepper>

<h6>mixin 自定义可跳转步骤名的颜色 </h6>
<material-stepper class="custom-selectable"
                  legalJumps="all"
                  orientation="vertical"
                  size="default"
                  yesText="继续"
                  noText="返回">
    <template step name="第 1 步">
        <p>操作内容</p>
    </template>
    <template step name="第 2 步">
        <p>操作内容</p>
    </template>
    <template step name="第 3 步">
```

```
        <p>操作内容</p>
    </template>
</material-stepper>
```

样式代码如下：

```
//chapter16/stepper_demo/lib/app_component.scss
@import 'package:angular_components/css/material/material';
@import 'package:angular_components/material_stepper/mixins';
@import 'package:angular_components/material_yes_no_buttons/mixins';
//用于设置主题
//第 1 个参数表示选择器
//第 2 个参数表示步骤标头的颜色
//第 3 个参数表示 yes 按钮的颜色
@include material-stepper-theme($selector: '.themed',
                                $step-color: $mat-teal-500,
                                $button-color: $mat-teal-500);
//自定义不可跳转步骤名的颜色
.custom-disabled{
    @include material-stepper-step-name-disabled-color($color: green);
}
//自定义可跳转步骤名的颜色
.custom-selectable{
    @include material-stepper-step-name-selectable-color($color: orange);
}
```

18min

16.17 对话框

遵循材质化风格设计的对话框。组件名是 MaterialDialogComponent，在模板中使用元素<material-dialog>创建对话框的实例。

1. 内容元素

(1) [header]：对话框的标题内容，需在元素上添加该属性。

(2) [footer]：对话框的页脚内容，需在元素上添加该属性。

2. 属性

(1) headered：在对话框标题上添加灰色背景。

(2) info：将对话框设置为信息对话框。

3. 输入属性

(1) error String：将错误显示在对话框的 error 部分。

(2) hideFooter bool：是否隐藏对话框页脚。

(3) hideHeader bool：是否隐藏对话框标题。

(4) listenForFullscreenChanges bool：在对话框进入或退出全屏模式时是否监听。

(5) shouldShowScrollStrokes bool：当用户通过滚动访问更多内容时，是否显示内容部分的上下边框。默认显示。

4. 输出属性

fullscreenMode Stream < bool >：对话框进入或退出全屏模式时的事件流。

5. Sass mixin

material-dialog-fullscreen($ width-threshold：100vw，$ height-threshold：100vh)：控制对话框进入全屏的条件，第一个参数是可视窗口的宽度，第二个参数是可视窗口的高度。

对话框需要与模态组件 ModalComponent 配合使用，模态组件的选择器是< modal >，使用该选择器作为对话框的父元素。需要将对话框的控制变量双向数据绑定到模态组件的 visible 变量上。在对话框中需要通过手动指定关闭按钮，按钮的 trigger 事件修改对话框的控制变量。

如果按钮需要自动聚焦，则需要与指令 AutoFocusDirective 配合使用。如果希望对话框能自动关闭，则需要与指令 AutoDismissDirective 配合使用。

创建 Angular 项目 dialog_demo，示例代码如下：

```
//chapter16/dialog_demo/lib/app_component.dart
import 'package:angular/angular.dart';
import 'package:angular_components/angular_components.dart';

@Component(
  selector: 'my-app',
  styleURLs: ['app_component.css'],
  templateURL: 'app_component.html',
  providers: [overlayBindings],
  directives: [
    MaterialDialogComponent,
    MaterialButtonComponent,
    MaterialIconComponent,
    ModalComponent,
    AutoFocusDirective,
    AutoDismissDirective,
    NgIf,],
)
class AppComponent {
  //基本对话框控制变量
  bool basicDialog = false;
  //滚动对话框控制变量
  bool scrollingDialog = false;
  //信息对话框控制变量
  bool infoDialog = false;
  //自动关闭对话框控制布尔量
```

```
  bool autoDismissDialog = false;
  //带标题背景对话框控制布尔量
  bool headeredDialog = false;
  //错误对话框控制变量
  bool erroDialog = false;
  //全屏对话框控制变量
  bool fullscreenDialog = false;
  //全屏状态
  bool isInFullscreenMode = false;
  //错误信息
  String dialogWithErrorErrorMessage;
  //在错误信息有和无间切换
  void toggleErrorMessage() {
    if (dialogWithErrorErrorMessage == null) {
      dialogWithErrorErrorMessage = '自定义错误信息';
    } else {
      dialogWithErrorErrorMessage = null;
    }
  }
}
```

模板代码如下：

```
//chapter16/dialog_demo/lib/app_component.html
<h1>对话框</h1>
<h6>基本对话框</h6>
<material-button (trigger)="basicDialog = true"
                 [disabled]="basicDialog"
                 raised>
    打开普通对话框
</material-button>

<modal [(visible)]="basicDialog">
    <material-dialog>
        <h1 header>对话框标题</h1>
        <p>普通对话框 </p>
        <div footer>
            <material-button autoFocus clear-size (trigger)="basicDialog = false">
                关闭
            </material-button>
        </div>
    </material-dialog>
</modal>

<material-button (trigger)="scrollingDialog = true"
```

```
                    [disabled] = "scrollingDialog"
                    raised>
    打开滚动对话框
</material-button>

<modal [(visible)] = "scrollingDialog">
    <material-dialog class = "scrolling">
        <h1 header>对话框标题</h1>
        <p class = "scrolling-content">内容高度超出对话框高度</p>
        <p class = "scrolling-content">在垂直方向可滚动</p>
        <p class = "scrolling-content">更多信息</p>
        <div footer>
            <material-button autoFocus clear-size (trigger) = "scrollingDialog = false">
                关闭
            </material-button>
        </div>
    </material-dialog>
</modal>

<h6>信息对话框</h6>
<material-button (trigger) = "infoDialog = true"
                    [disabled] = "infoDialog"
                    raised>
    打开普通对话框
</material-button>
<modal [(visible)] = "infoDialog">
    <material-dialog info class = "info-dialog">
        <div header>
            <material-button icon autoFocus (trigger) = "infoDialog = false">
                <material-icon icon = "close"></material-icon>
            </material-button>
            <h1>信息</h1>
        </div>
        <p>信息对话框 </p>
    </material-dialog>
</modal>

<h6>自动关闭对话框</h6>
<material-button (trigger) = "autoDismissDialog = true"
                    [disabled] = "autoDismissDialog"
                    raised>
    打开自动关闭对话框
</material-button>
<modal [(visible)] = "autoDismissDialog">
    <material-dialog info class = "info-dialog"
                        [autoDismissable] = "autoDismissDialog"
```

```
                  (dismiss) = "autoDismissDialog = false">
          < div header >
              < material - button icon autoFocus (trigger) = "autoDismissDialog = false">
                  < material - icon icon = "close"></material - icon >
              </material - button >
              < h1 >信息</h1 >
          </div >
          < p >单击对话框以外的地方,对话框将自动关闭.</p >
      </material - dialog >
</modal >

< h6 >带标题背景的对话框</h6 >
< material - button (trigger) = "headeredDialog = true"
                  [disabled] = "headeredDialog"
                  raised >
    打开带标题背景的对话框
</material - button >
< modal [(visible)] = "headeredDialog">
    < material - dialog headered class = "info - dialog">
        < div header >
            < h1 >信息</h1 >
        </div >
        < p >带标题背景的对话框</p >
        < div footer >
            < material - button autoFocus clear - size (trigger) = "headeredDialog = false">
                关闭
            </material - button >
        </div >
    </material - dialog >
</modal >

< h6 >带错误信息的对话框</h6 >
< material - button (trigger) = "erroDialog = true"
                  [disabled] = "erroDialog"
                  raised >
    打开带错误信息的对话框
</material - button >
< modal [(visible)] = "erroDialog">
    < material - dialog headered class = "info - dialog"
                      [error] = "dialogWithErrorErrorMessage">
        < div header >
            < h1 >信息</h1 >
        </div >
        < material - button raised (trigger) = "toggleErrorMessage()">
            {{dialogWithErrorErrorMessage == null ? '显示' : '隐藏'}} 错误信息
        </material - button >
```

```
        <div footer>
            <material-button autoFocus clear-size (trigger)="erroDialog = false">
                关闭
            </material-button>
        </div>
    </material-dialog>
</modal>

<h6>带错误信息的对话框</h6>
<material-button (trigger)="fullscreenDialog = true"
                 [disabled]="fullscreenDialog"
                 raised>
    打开带错误信息的对话框
</material-button>
<modal [(visible)]="fullscreenDialog">
    <material-dialog
            headered
            class="fullscreen-dialog"
            [class.fullscreen-mode]="isInFullscreenMode"
            [listenForFullscreenChanges]="true"
            (fullscreenMode)="isInFullscreenMode = $event">
        <div header>
            <h1>信息</h1>
        </div>
        <p *ngIf="isInFullscreenMode">
            对话框当前处于全屏模式,放大视窗以退出全屏模式.
        </p>
        <p *ngIf="!isInFullscreenMode">
            如果视窗足够小,则此对话框将全屏显示.
        </p>
        <div footer>
            <material-button autoFocus clear-size (trigger)="fullscreenDialog = false">
                关闭
            </material-button>
        </div>
    </material-dialog>
</modal>
```

样式代码如下：

```
//chapter16/dialog_demo/lib/app_component.scss
@import 'package:angular_components/css/material/material';
@import 'package:angular_components/material_dialog/mixins';

//设置对话框的高度和宽度,内容超出高度后将可滚动
```

```
.scrolling{
height: 300px;
width:70 %;
}

.scrolling-content{
height:160px;
}

//设置信息对话框的宽度
.info-dialog {
  width: 340px;
}

//通过 mixin 控制对话框进入全屏的条件
//第一个参数是高度,第二个参数是宽度,变量 $mat-grid 的值为 8px
.fullscreen-dialog{
  @include material-dialog-fullscreen($mat-grid * 100, $mat-grid * 80);
}
```

16.18 扩展面板

15min

一种材质化风格的扩展面板。组件名是 MaterialExpansionPanel，在模板中使用元素 <material-expansionpanel>创建扩展模板实例。

一个或多个面板组合在一个扩展面板集中。单击面板时，面板内容会展开。面板由名称、值、可选的辅助文本和展开的面板内容组成。

当面板内容处于折叠状态时，具有属性 value 的内容元素将被用作面板内容的值。

与面板的交互是通过扩展面板集完成的，该集合考虑了集合中其他面板的状态，并将适当的操作发布到每个单独的面板上。

1. 属性

（1）wide：指示面板展开后的宽度略宽于折叠状态。

（2）flat：指示面板展开时不应浮动或与其他面板分开。

（3）forceContentWhenClosed：关闭扩展面板时，将扩展面板内容保留在 DOM 中。尽量少使用，此属性会影响应用性能。

2. 内容引用

focusOnOpen：在内容中使用#focusOnOpen 标记一个 Focusable 或 DOM 元素，当扩展面板打开时，使该项目被聚焦。

3. 输入属性

（1）alwaysHideExpandIcon bool：如果值为 true，则展开图标永远不可见。

(2) alwaysShowExpandIcon bool：如果值为 true，则无论是否使用自定义图标，展开图标都应始终可见。

(3) cancelDisplayed bool：是否显示取消按钮，默认值为 true。

(4) cancelText String：在取消按钮上显示的文本。默认值为 Cancel。

(5) closeOnSave bool：如果值为 true，则保存成功后，面板将尝试关闭。

(6) disableHeaderExpansion bool：如果值为 true，则单击标题不会展开或折叠面板。

(7) disabled bool：如果值为 true，则面板将保持折叠状态而无法展开，或者如果默认情况下是展开的，则面板将保持展开状态。

(8) expandIcon String：可选的图标名称，用于使用自定义图标替换展开箭头。

(9) focusOnOpen dynamic：设置聚焦子元素，当面板打开时，可以聚焦在子元素上。

(10) hideExpandedHeader bool：如果值为 true，当面板展开时，则隐藏显示面板名称的标头。

(11) expanded bool：如果值为 true，则默认情况下会展开面板；如果值为 false，则将关闭面板。

(12) name String：扩展面板的简称标签。

(13) saveDisabled bool：是否禁用保存按钮。

(14) saveText String：要在保存按钮上显示的文本。默认值为 Save。

(15) secondaryText String：一些可选的辅助摘要文本，描述了面板内托管的小部件的状态。

(16) showSaveCancel bool：是否应显示保存和取消按钮，默认值为 true。

4. 输出属性

(1) cancel Stream<AsyncAction<bool>>：取消面板时触发事件。

(2) close Stream<AsyncAction<bool>>：面板尝试关闭时触发事件，此操作可能被取消。

(3) expandedChange Stream<bool>：当面板折叠或展开时触发事件。

(4) expandedChangeByUser Stream<bool>：当用户折叠或展开面板时触发事件。

(5) open Stream<AsyncAction<bool>>：面板尝试打开时触发事件，此操作可能被取消。

(6) save Stream<AsyncAction<bool>>：保存面板时触发事件。

扩展面板集合将一组扩展面板组合在一起，一次只能打开一个扩展面板。MaterialExpansionPanelSet 必须将 MaterialExpansionPanel 用作直接子代。组件名是 MaterialExpansionPanelSet，在模板中使用元素<material-expansionpanel-set>创建扩展模板实例。

创建 Angular 项目 expansionpanel_demo，示例代码如下：

```
//chapter16/expansionpanel_demo/lib/app_component.dart
import 'dart:async';
import 'package:angular/angular.dart';
import 'package:angular_components/angular_components.dart';
import 'package:angular_components/model/action/async_action.dart';

@Component(
  selector: 'my-app',
  styleURLs: ['app_component.css'],
  templateURL: 'app_component.html',
  providers: [overlayBindings],
  directives: [
    AutoFocusDirective,
    MaterialInputComponent,
    MaterialDialogComponent,
    MaterialYesNoButtonsComponent,
    MaterialIconComponent,
    MaterialExpansionPanelSet,
    MaterialExpansionPanel,
    MaterialExpansionPanelAutoDismiss,
    ModalComponent],
)
class AppComponent{
  //对话框显示控制属性
  bool showConfirmation = false;
  //完成对象,包含完成状态和信息
  Completer<bool> dialogFutureCompleter;
  //显示对话框函数
  void showConfirmationDialog(AsyncAction event) {
    showConfirmation = true;
    //初始完成对象
    dialogFutureCompleter = Completer();
    //事件对象根据对话框完成状态控制是否取消
    event.cancelIf(dialogFutureCompleter.future);
  }
  //关闭对话框函数
  void closeDialog(bool proceed) {
    showConfirmation = false;
    if (dialogFutureCompleter != null) {
      //设置完成状态
      dialogFutureCompleter.complete(!proceed);
    }
  }
}
```

模板代码如下：

```
//chapter16/expansionpanel_demo/lib/app_component.html
<h1>扩展面板</h1>
<h6>默认面板</h6>
<material-expansionpanel-set>
    <material-expansionpanel name="面板名">默认面板</material-expansionpanel>
    <material-expansionpanel name="面板 2"
                             secondaryText="摘要文本">
        <p>面板名和摘要文本</p>
    </material-expansionpanel>
    <material-expansionpanel name="面板 3">
        <p value>折叠状态面板的值</p>
        <p>面板 3 设置折叠状态面板的值</p>
    </material-expansionpanel>
</material-expansionpanel-set>

<h6>wide 面板</h6>
<material-expansionpanel-set>
    <material-expansionpanel wide name="隐藏标头"
                             [hideExpandedHeader]="true">
        <p>隐藏标头</p>
    </material-expansionpanel>
    <material-expansionpanel wide name="自定义按钮文本"
                             saveText="保存"
                             cancelText="关闭">
        <p>自定义 saveText 和 cancelText</p>
    </material-expansionpanel>
    <material-expansionpanel wide name="禁用面板" disabled>
    </material-expansionpanel>
</material-expansionpanel-set>

<h6>flat 面板</h6>
<material-expansionpanel-set>
    <material-expansionpanel flat>
        <div name>
            <material-icon icon="notifications" size="medium"></material-icon>
            提示
        </div>
        <p value>面板可以自定义 name</p>
        <p>可以将 name 作为元素的属性,该元素将作为面板的显示标签</p>
    </material-expansionpanel>
    <material-expansionpanel flat name="隐藏保持和取消按钮"
                             [showSaveCancel]="false">
        <p>隐藏保持和取消按钮</p>
    </material-expansionpanel>
    <material-expansionpanel flat name="自定义扩展按钮"
```

```
                                   expandIcon = "edit">
        自定义扩展按钮
    </material - expansionpanel >
</material - expansionpanel - set >

< section >
    < p >以下面板没有包含在面板集合中,可以独立打开</p >
    < material - expansionpanel name = "仅能通过扩展图标打开面板"
                                   [disableHeaderExpansion] = "true">
        单击面板标头将不会打开或折叠面板
    </material - expansionpanel >
    < material - expansionpanel name = "默认展开面板"
                                   [expanded] = "true">
        此面板默认展开
    </material - expansionpanel >
    < material - expansionpanel autoDismissable
                                   name = "自动折叠面板">
        < p >当展开面板后,单击面板之外的地方,面板自动折叠.</p >
    </material - expansionpanel >
    < material - expansionpanel name = "展开面板自动对焦">
        < material - input #focusOnOpen ></material - input >
    </material - expansionpanel >
    < material - expansionpanel shouldExpandOnLeft
                                   name = "在面板左侧展示扩展图标">
        < p > shouldExpandOnLeft 属性使得扩展图标显示在左侧</p >
    </material - expansionpanel >
    < material - expansionpanel name = "取消时确认按钮"
                                   (cancel) = "showConfirmationDialog( $ event)">
        < p >单击取消按钮时,如果取消状态一直持续没有完成,则在对话框中确定是否真正取消.
</p >
    </material - expansionpanel >
</section >

<!-- 确认对话框 -->
< modal [(visible)] = "showConfirmation">
    < material - dialog class = "confirmation - dialog" >
        < h3 >确定取消吗?</h3 >
        < div footer >
            < material - yes - no - buttons raised
                                   yesHighlighted
                                   (yes) = "closeDialog(true)"
                                   (no) = "closeDialog(false)">
            </material - yes - no - buttons >
        </div >
    </material - dialog >
</modal >
```

16.19 下拉菜单

19min

材质风格的下拉选择菜单，组件名是 MaterialDropdownSelectComponent，在模板中使用元素< material-dropdown-select >创建下拉菜单实例。

1. 输入属性

(1) options dynamic：设置选择组件的可用选项。接收 SelectionOptions 或 List。如果传递了 List，则将使用 StringSelectionOptions 类创建选择选项。如果需要更高级的功能，包括分组选项、自定义过滤或异步搜索，则可以传递 SelectionOptions 的实现。

(2) selection dynamic：设置选择组件的选择值或选择模型。默认使用 SingleSelectionModel，可以自定义 SelectionModel 或使用 MultiSelectionModel。

(3) itemRenderer String Function(T)：将选项对象转换为字符串的函数。

2. 输出属性

selectionChange Stream < dynamic >：只要更改选择，就会发出选定的值。对于单选，它将是所选值或为 null。对于多选，它将是所选值的列表或空列表。

创建 Angular 项目 dropdown_select_demo。示例代码如下：

```
//chapter16/dropdown_select_demo/lib/app_component.dart
import 'package:angular/angular.dart';
import 'package:angular_components/angular_components.dart';
import 'package:angular_forms/angular_forms.dart';
@Component(
  selector: 'my-app',
  styleURLs: ['app_component.css'],
  templateURL: 'app_component.html',
  providers: [popupBindings],
  directives: [
    MaterialDropdownSelectComponent,
    MaterialSelectDropdownItemComponent,
    DropdownSelectValueAccessor,
    MultiDropdownSelectValueAccessor,
    NgModel],
)
class AppComponent{
  //普通单选值
  String collegeValue;

  //简单 List 数据模型
  static List<String> colleges = <String>['文学院','哲学院','数学院','计算机学院'];

  //普通多选模型
```

```
final SelectionModel<String> multiSelectModel = SelectionModel<String>.multi();

//生成多选标签
String get multiSelectLabel {
  var selectedValues = multiSelectModel.selectedValues;
  if (selectedValues.isEmpty) {
    return '选择学院';
  } else if (selectedValues.length == 1) {
    return selectedValues.first;
  } else {
    return '${selectedValues.first} + ${selectedValues.length - 1} 更多';
  }
}

//单选与 Angular 表单
College selectionValue;

//复杂模型数据
static List<College> collegeList = <College>[
  College(1,'文学院'),
  College(2,'哲学院'),
  College(3,'数学院'),
  College(4,'计算机学院')
];

//通过 StringSelectionOptions 类创建选择选项
//其构造函数接收一个 List 对象
StringSelectionOptions<College> collegeOptions =
StringSelectionOptions<College>(collegeList);

 //多选与 Angular 表单
List<dynamic> selectionValues = [];

//显示条目生成函数
static final ItemRenderer itemRenderer =
    (item) => (item as HasUIDisplayName).uiDisplayName;

//生成多选标签
String get selectionValuesLabel {
  final size = selectionValues.length;
  if (size == 0) {
    return '选择学院';
  } else if (size == 1) {
    return itemRenderer(selectionValues.first);
  } else {
```

```
        return '${itemRenderer(selectionValues.first)} + ${size - 1} 更多';
      }
    }
  }

  //实现 HasUIDisplayName 接口
  //由该对象组成的选项或列表不需要提供 ItemRenderer 函数
  class College implements HasUIDisplayName{
    int id;
    String name;
    @override
    //TODO: 实现 uiDisplayName
    String get uiDisplayName => name;
    College(this.id,this.name);
    @override
    String toString() => uiDisplayName;
  }
```

模板代码如下：

```
//chapter16/dropdown_select_demo/lib/app_component.html
<h1>下拉列表</h1>
<h6>普通单选</h6>
<material-dropdown-select
        buttonText="{{collegeValue == null ? '选择学院' : collegeValue}}"
        [(selection)]="collegeValue"
        [options]="colleges">
</material-dropdown-select>

<h6>普通多选</h6>
<material-dropdown-select
        [buttonText]="multiSelectLabel"
        [options]="colleges"
        [selection]="multiSelectModel">
</material-dropdown-select>

<h6>单选和双向数据绑定</h6>
<material-dropdown-select
        [buttonText]="selectionValue == null ? '选择学院' : selectionValue.name"
        [options]="collegeList"
        [(ngModel)]="selectionValue">
</material-dropdown-select>

<h6>多选和双向数据绑定</h6>
<material-dropdown-select multi
```

```
                                        [buttonText] = "selectionValuesLabel"
                                        [options] = "collegeOptions"
                                        [itemRenderer] = "itemRenderer"
                                        [(ngModel)] = "selectionValues">
    </material - dropdown - select >
```

13min

16.20 弹出框

材质化风格的弹出组件，组件名是 MaterialPopupComponent，在模板中使用元素 < material-popup >创建弹出框实例。

注意事项：

(1) 关闭和打开弹出窗口会自动延迟以添加动画。

(2) 如果内容大小使页面滚动，则利用 PopupInterface 中定义的 forceSpaceSpaceConstraints 会有所帮助。

(3) 如果内容更改并且需要重新调整位置，需使用在 PopupInterface 中定义的 trackLayoutChanges。

材质弹出窗口还支持延迟加载的内容。

该组件将自身发布为 DropdownHandle，因此其子代可以通过注入它来控制其可见性。示例代码如下：

```
class MyComponent {
  final DropdownHandle _dropdownHandle;

  MyComponent(this._dropdownHandle);

  void onSomethingThatShouldCloseTheDropdown() {
    _dropdownHandle.close();
  }
}
```

1. 输入属性

(1) source PopupSource：设置弹出窗口应该相对创建的源。

(2) visible bool：设置是否显示弹出窗口，用于关闭或打开弹出窗口。

(3) enforceSpaceConstraints bool：设置弹出窗口是否应根据相对于视口的可用空间自动重新定位。

(4) preferredPositions Iterable < Object >：当提供了 enforceSpaceConstraints 属性时，设置弹出框弹出的位置，它接收一个 RelativePosition 类型的列表，它会依照给定列表依次选择合适的弹出位置，如果列表中没有合适的弹出位置，则它将自动选择弹出位置。

(5) offsetX int：弹出框相对源在 x 轴上的偏移量，参照点是源的左上角那个点。

(6) offsetY int：弹出框相对源在 y 轴上的偏移量，参照点是源的左上角那个点。

(7) ink bool：将弹出窗口的背景色设置为 ink($mat-grey-700)。

(8) matchMinSourceWidth bool：设置弹出窗口是否应将最小宽度设置为源的宽度。

2. 输出属性

visibleChange Stream<bool>：当弹出窗口的 visible 属性更改时触发同步事件。

创建 Angular 项目 popup_demo，示例代码如下：

```
//chapter16/popup_demo/lib/app_component.dart
import 'package:angular/angular.dart';
import 'package:angular_components/angular_components.dart';
@Component(
  selector: 'my-app',
  styleURLs: ['app_component.css'],
  templateURL: 'app_component.html',
  providers: [popupBindings],
  exports: [RelativePosition],
  directives: [
    MaterialButtonComponent,
    MaterialPopupComponent,
    PopupSourceDirective,
    MaterialDropdownSelectComponent,],
)
class AppComponent {
  //表示弹出框可见性的列表
  List<bool> visible = List.filled(11, false);

  //单选模型
  SelectionModel<RelativePosition> position =
  RadioGroupSingleSelectionModel(RelativePosition.OffsetBottomRight);

  //选择的方向
  RelativePosition get popupPosition => position.selectedValues.first;

  //手动指定弹出位置
  RelativePosition get customPosition => RelativePosition.OffsetBottomRight;

  //选项
  final SelectionOptions<RelativePosition> positions =
  SelectionOptions.fromList(positionMap.keys.toList());

  //标签生成函数
  ItemRenderer positionLabel =
    (position) => positionMap[position];
```

```
  //按钮标签
  String get buttonLabel => positionLabel(popupPosition);
}
//弹出位置集合
final positionMap = <RelativePosition, String>{
  //相对源的 4 个角
  RelativePosition.OffsetBottomRight: '右下角',
  RelativePosition.OffsetBottomLeft: '左下角',
  RelativePosition.OffsetTopRight: '右上角',
  RelativePosition.OffsetTopLeft: '左上角',
  //行内方向,沿 x 或 y 轴延伸方向
  RelativePosition.InlineBottom: '行内向下',
  RelativePosition.InlineBottomLeft: '行内向左和下',
  RelativePosition.InlineTop: '行内向上',
  RelativePosition.InlineTopLeft: '行内向左和上',
  //相对源的正方向
  RelativePosition.AdjacentTop: '上方',
  RelativePosition.AdjacentRight: '右方',
  RelativePosition.AdjacentLeft: '左方',
  RelativePosition.AdjacentBottom: '下方',
  //相对源的两个方向
  RelativePosition.AdjacentTopLeft: '左边与源对齐的上方',
  RelativePosition.AdjacentTopRight: '右边与源对齐的上方',
  RelativePosition.AdjacentLeftTop: '上边与源对齐的左方',
  RelativePosition.AdjacentRightTop: '上边与源对齐的右方',
  RelativePosition.AdjacentRightBottom: '下边与源对齐的右方',
  RelativePosition.AdjacentBottomRight: '右边与源对齐的下方',
  RelativePosition.AdjacentBottomLeft: '左边与源对齐的下方',
  RelativePosition.AdjacentLeftBottom: '下边与源对齐的左方',
};
```

模板代码如下：

```
//chapter16/popup_demo/lib/app_component.html
<h1>弹出框</h1>
<section>
    <p>由系统根据可视窗口大小自动选择弹出方向</p>
    <material-button
            raised
            popupSource
            #source = "popupSource"
            (trigger) = "visible[0] = !visible[0]">
        {{visible[0] ? '关闭' : '打开'}} 弹出框
    </material-button>
```

```
    <material-popup [source]="source" [(visible)]="visible[0]">
        <div style="height: 100px;width:200px;">弹出框内容</div>
    </material-popup>
</section>

<section>
    <p>带 ink 背景并手动指定弹出方向,在可视窗口能完全容纳弹出框时按指定方向弹出.否则由它自动选择弹出方向</p>
    <material-button
            raised
            popupSource
            #source1="popupSource"
            (trigger)="visible[2] = !visible[2]">
        {{visible[2] ? '关闭' : '打开'}} 弹出框
    </material-button>

    <material-popup [source]="source1" [(visible)]="visible[2]"
                    [enforceSpaceConstraints]="true"
                    ink
                    [preferredPositions]="[customPosition,popupPosition]">
        <div style="height: 100px;width:200px;">弹出框内容</div>
    </material-popup>
</section>

<section>
    <p>请选择方向,然后单击打开弹出框</p>
<material-dropdown-select
        [selection]="position"
        [options]="positions"
        [itemRenderer]="positionLabel"
        [buttonText]="buttonLabel">
</material-dropdown-select>
<div style="height:60px;"></div>
<material-button
        raised
        popupSource
        #customsrc="popupSource"
        class="trigger"
        (trigger)="visible[1] = !visible[1]">
    {{visible[1] ? '关闭' : '打开'}} 弹出框
</material-button>
<material-popup [source]="customsrc"
                [(visible)]="visible[1]"
                [enforceSpaceConstraints]="true"
                [preferredPositions]="[popupPosition]">
```

```
    <div style="height: 200px; padding: 24px;">
        <p>弹出框内容</p>
        <material-button
                raised
                (trigger)="visible[1] = false">
            关闭
        </material-button>
    </div>
</material-popup>
</section>


<section>
    <p>offsetX 为 30px,offsetY 为 60px</p>
    <material-button
            raised
            popupSource
            #source2="popupSource"
            (trigger)="visible[3] = !visible[3]">
        {{visible[3] ? '关闭' : '打开'}} 弹出框
    </material-button>

    <material-popup [source]="source2" [(visible)]="visible[3]"
                    [offsetX]="30"
                    [offsetY]="60">
        <div style="height: 100px;width:200px;">弹出框内容</div>
    </material-popup>
</section>

<section>
    <p>最小宽度等于源的宽度,通过导出的 RelativePosition 对象指定弹出方向</p>
    <material-button
            raised
            popupSource
            #source3="popupSource"
            (trigger)="visible[4] = !visible[4]">
        {{visible[4] ? '关闭' : '打开'}} 弹出框,最小宽度等于源的宽度
    </material-button>
    <material-popup [source]="source3" [(visible)]="visible[4]"
                    [enforceSpaceConstraints]="true"
                    [preferredPositions]="[RelativePosition.AdjacentTop]"
                    [matchMinSourceWidth]="true">
        <p>弹出框内容</p>
    </material-popup>
</section>
```

16.21 选项菜单

19min

选项菜单常用于在多个选项中选取一个或多个条目的使用场景。

16.21.1 选项容器

MaterialSelectComponent 是用于从集合中选择项目的容器，并标记所选的选项，在模板中使用元素< material-select >创建选项容器实例。

可以通过 SelectionOptions 实例或在模板中硬编码指定选项，通过模板或对照选择模型将选项标记为已选择项。

输入属性

(1) disabled bool：选择是否应显示为禁用。默认值为 false。

(2) itemRenderer String Function(T)：将选择选项给定的值转换为字符串，并呈现。

(3) options SelectionOptions < T >：SelectionOptions 实例提供的呈现选项。

(4) selection SelectionModel < T >：此容器的选择模型，可以是单选或多选模型。

(5) width dynamic：呈现列表的宽度，可用值从 1 到 5。

16.21.2 选择条目

MaterialSelectItemComponent 是可以被选择的一种特殊的列表项，在模板中使用元素< material-select-item >创建选择条目实例。

1. 输入属性

(1) isHidden bool：是否隐藏该选项，默认值为 false。

(2) itemRenderer String Function(T)：将选项呈现为 String 的函数。

(3) selected bool：手动将选项标记为已选。

(4) selection SelectionModel < T >：选择模型，可以是单选或多选模型。

(5) useCheckMarks bool：复选标记，如果值为 true，将替代复选框来指示多选中的已选选项。

(6) value T：当前选择项表示的值。如果对象实现了 HasUIDisplayName，则它将使用 uiDisplayName 字段作为项目的标签进行渲染。否则，必须提供 itemRenderer 属性，标签才由该组件生成。

2. 输出属性

trigger Stream < UIEvent >：通过单击，单击或按键激活按钮时触发。

创建 Angular 项目 select_demo，示例代码如下：

```
//chapter16/select_demo/lib/app_component.dart
import 'package:angular/angular.dart';
import 'package:angular_components/angular_components.dart';
```

```
@Component(
  selector: 'my-app',
  styleURLs: ['app_component.css'],
  templateURL: 'app_component.html',
  directives: [
    MaterialSelectComponent,
    MaterialSelectItemComponent,
    NgFor,
    displayNameRendererDirective],
)
class AppComponent {
  String season;

  //用作选项的列表
  static List<Season> seasonList = [
    Season('spring','春'),
    Season('summer','夏'),
    Season('autumn','秋'),
    Season('winter','冬'),
  ];

  //单选模型,接收的是 String
  final SelectionModel<String> singleSelection =
  SelectionModel.single();

  //响应选项的 trigger 事件,将单选模型的值赋值给变量 season
  void selected(){
    season = singleSelection.selectedValues.first;
  }

  //构建选择项,用于 options 参数
  final SelectionOptions<Season> seansonOptions =
  SelectionOptions.fromList(seasonList);

  //单选模型,接收单个 Season 对象,用于 selection 参数
  final SelectionModel<Season> singleSeasonSelection =
  SelectionModel.single();

  //多选模型,可接收多个 Season 对象,用于 selection 参数
  final SelectionModel<Season> multiSelection =
  SelectionModel.multi();

  //获取存储在多选模型中的已选项
  List<Season> get selectedList => multiSelection.selectedValues.toList();
}
```

```
//实现 HasUIDisplayName 接口的 Season 类
class Season implements HasUIDisplayName{
  String en;
  String cn;
  @override
  //实现 uiDisplayName
  String get uiDisplayName => cn;
  Season(this.en,this.cn);
}
```

模板代码如下：

```
//chapter16/select_demo/lib/app_component.html
<h1>选项</h1>
<h6>手动指定选项</h6>
<material-select [width]="2">
    <material-select-item (trigger)="season = 'spring'"
                          [selected]="season == 'spring'">春</material-select-item>
    <material-select-item (trigger)="season = 'summer'"
                          [selected]="season == 'summer'">夏</material-select-item>
    <material-select-item (trigger)="season = 'autumn'"
                          [selected]="season == 'autumn'">秋</material-select-item>
    <material-select-item (trigger)="season = 'winter'"
                          [selected]="season == 'winter'">冬</material-select-item>
</material-select>

<h6>使用 NgFor 生成选项</h6>
<material-select [width]="2">
    <material-select-item *ngFor="let s of seasonList"
                          (trigger)="season = s.en"
                            [selected]="season == s.en">{{s.cn}}</material-select-
item>
</material-select>

<h6>使用单选模型,并手动指定选项</h6>
<p>通过单选模型取值 season 的值:{{season}}</p>
<material-select [width]="2" [selection]="singleSelection">
    <material-select-item value="spring" (trigger)="selected()">春</material-select
-item>
    <material-select-item value="summer" (trigger)="selected()">夏</material-select
-item>
    <material-select-item value="autumn" (trigger)="selected()">秋</material-select
-item>
```

```
    <material-select-item value="winter" (trigger)="selected()">冬</material-select
-item>
</material-select>

<h6>使用单选模型,并提供选项</h6>
<material-select [selection]="singleSeasonSelection"
                 [options]="seansonOptions"
                 displayNameRenderer
                 width="2"></material-select>

<h6>使用多选模型,并提供选项 </h6>
<p>通过多选模型取值 已选项:<span *ngFor="let s of selectedList">{{s.cn }}</span></p>
<material-select [selection]="multiSelection"
                 [options]="seansonOptions"
                 displayNameRenderer
                 width="2"></material-select>
```

8min

16.22 工具提示

16.22.1 工具提示指令

工具提示可以附加到任何元素上，背景是水墨。指令名是 MaterialTooltipDirective，作用于模板中带有 materialTooltip 属性的元素。

输入属性

（1）alignPositionX String：弹出窗口在水平方向上的对齐方式。

（2）alignPositionY String：弹出窗口在垂直方向上的对齐方式。

（3）tooltipPositions List < RelativePosition >：工具提示应尝试显示的位置。

（4）materialTooltip String：要在工具提示中显示的文本。

16.22.2 工具提示卡片

工具提示卡片用于展示目标元素相关的信息，在提示信息较多时使用。组件名是 MaterialPaperTooltipComponent，在模板中使用元素< material-tooltip-card >创建工具提示卡片实例。

其目标可以是任何元素，此组件需与指令 ClickableTooltipTargetDirective 结合使用，且需将 focusContents 设置为 true。

如果工具提示内容是另一个组件，则需使用指令 DeferredContentDirective 加载组件。

支持 header 和 footer 元素作为内容，其他内容将采用工具提示正文样式。

输入属性

（1）focusContents bool：打开时工具提示内容是否应自动聚焦。

(2) offsetX int：工具提示最终定位位置的 x 偏移量。

(3) offsetY int：工具提示最终定位位置的 y 偏移量。

(4) preferredPositions List < RelativePosition >：尝试显示工具提示的相对位置。

(5) for TooltipTarget：此工具提示所针对的元素。

16.22.3 工具提示目标指令

工具提示目标指令将元素标记为工具提示的目标，它会立即打开工具提示，且工具提示不会聚焦。指令名是 ClickableTooltipTargetDirective，作用于带有 clickableTooltipTarget 属性的元素，导出为 tooltipTarget。该指令与工具提示组件一起使用，例如 MaterialPaper-TooltipComponent，它可以完全控制简单工具提示的内容。

1. 输入属性

(1) alignPositionX String：弹出窗口在水平方向上的对齐方式。

(2) alignPositionY String：弹出窗口在垂直方向上的对齐方式。

2. 输出属性

tooltipActivate Stream < bool >：激活工具提示时触发的事件。

16.22.4 图标提示

显示工具提示的图标。组件名是 MaterialIconTooltipComponent，在模板中使用元素 < material-icon-tooltip >创建图标提示实例。

与在 MaterialIconComponent 上显示 MaterialTooltipCard 基本相同，除此之外它在单击时也会显示工具提示，与没有单击触发器的 MaterialTooltipTarget 相反。

1. 属性

(1) icon：图标的名称。

(2) size：图标的大小。

(3) type：图标的类型。可能的值：help 显示 help_outline 图标。info 显示 info_outline 图标。error 显示 error_outline 图标。

2. 输入属性

(1) offsetX int：工具提示最终定位位置的 x 偏移量。

(2) offsetY int：工具提示最终定位位置的 y 偏移量。

(3) preferredPositions List < RelativePosition >：尝试显示工具提示的相对位置。

创建 Angular 项目 tooltip_demo，示例代码如下：

```
//chapter16/tooltip_demo/lib/app_component.dart
import 'package:angular/angular.dart';
import 'package:angular_components/angular_components.dart';
@Component(
  selector: 'my-app',
  styleURLs: ['app_component.css'],
```

```
  templateURL: 'app_component.html',
  providers: [popupBindings,materialTooltipBindings],
  directives: [
    MaterialTooltipDirective,
    MaterialPaperTooltipComponent,
    ClickableTooltipTargetDirective,
    MaterialIconTooltipComponent,
    MaterialIconComponent],
)
class AppComponent {}
```

模板代码如下：

```
//chapter16/tooltip_demo/lib/app_component.html
<h1>工具提示</h1>
<h6>工具提示指令</h6>
<material-icon
        icon="print"
        materialTooltip="打印">
</material-icon>
<p materialTooltip="段落">段落元素</p>

<h6>工具提示卡片</h6>
<p>
可单击的工具提示
    <material-icon
            icon="help_outline"
            clickableTooltipTarget
            size="medium"
            #clickableRef="tooltipTarget">
    </material-icon>
</p>
<material-tooltip-card [for]="clickableRef" focusContents>
    <header>标头</header>
    <p>提示内容</p>
    <footer>脚注</footer>
</material-tooltip-card>

<span clickableTooltipTarget #spanRef="tooltipTarget">悬浮标签</span>
<material-tooltip-card [for]="spanRef" focusContents>
    <header>标头</header>
    <p>普通文本</p>
    <footer>脚注</footer>
</material-tooltip-card>
```

```
<h6>图标工具提示</h6>
<p>附带图标工具提示的段落
    <material-icon-tooltip icon="help">
        帮助信息
    </material-icon-tooltip>
</p>
```

16.23 布局组件

27min

应用布局是由样式、指令和组件组成的系统，它根据材质化设计规范实现了应用栏、抽屉和导航样式。创建 Angular 项目 layout_demo。

样式由 package:angular_components/app_layout/layout.scss.css 文件提供。只需将该文件添加到@Component 注解的 styleURLs 列表中。最好的做法是放在所有样式文件前，这样可以根据需要进行微调。示例代码如下：

```
//chapter16/layout_demo/lib/app_component.dart
import 'package:angular/angular.dart';

@Component(
  selector: 'my-app',
  styleURLs: ['package:angular_components/app_layout/layout.scss.css','app_component.css'],
  templateURL: 'app_component.html',
)
class AppComponent {}
```

16.23.1 应用栏

应用栏是通过现有 HTML 元素与 CSS 样式配合使用的，导航栏样式如表 16-1 所示。

表 16-1 导航栏样式

class	说明
material-header	将元素作为标头的容器，即页眉
shadow	将阴影应用于标头元素
dense-header	使得标头内元素更加紧凑
material-header-row	标头中的行，只能包含一行
material-drawer-button	标头行中左侧的按钮，用于调出抽屉
material-header-title	标头的标题
material-spacer	放在标题和导航元素中间，用于填充它们之间的空间
material-navigation	导航元素，仅与锚标签<a>一起使用

标头就是常说的导航栏，示例代码如下：

```
//chapter16/layout_demo/lib/app_component.html
<header class = "material - header shadow dense - header">
    <div class = "material - header - row">
        <material - button icon
                        class = "material - drawer - button">
            <material - icon icon = "menu"></material - icon>
        </material - button>
        <span class = "material - header - title">应用栏</span>
        <div class = "material - spacer"></div>
        <nav class = "material - navigation">
            <a>链接 1</a>
        </nav>
        <nav class = "material - navigation">
            <a>链接 2</a>
        </nav>
        <nav class = "material - navigation">
            <a>链接 3</a>
        </nav>
    </div>
</header>
```

运行结果如图 16-1 所示。

图 16-1　导航栏

16.23.2　抽屉

共包含 3 种类型的抽屉：永久、持久和临时抽屉。所有抽屉都由元素<material-drawer>实例化，这些抽屉的实现方式略有差异，以保证各自的最佳性能和使用场景。对于抽屉之外的内容，可由<material-content>元素或带有 CSS 类 material-content 的元素包裹。

1. 永久抽屉

抽屉会固定在页面中，不能被关闭。使用时将 permanent 属性添加到<material-drawer>元素上即可。示例代码如下：

```
//chapter16/layout_demo/lib/src/permanent_component.dart
import 'package:angular/angular.dart';
import 'package:angular_components/angular_components.dart';
@Component(
    selector: 'permanent - drawer',
```

```
    styleURLs: ['package:angular_components/app_layout/layout.scss.css'],
    templateURL: 'permanent_component.html',
)
class PermanentComponent {

}
```

模板代码如下：

```
//chapter16/layout_demo/lib/src/permanent_component.html
<material-drawer permanent>
    <div>
        <!-- 抽屉内容 -->
        固定抽屉
    </div>
</material-drawer>
<material-content>
    <header class="material-header shadow">
        <div class="material-header-row">
            标头
        </div>
    </header>
    <div>
        内容
    </div>
</material-content>
```

2. 持久抽屉

持久抽屉可以通过触发器关闭或打开，例如：按钮。将 persistent 属性添加到 <material-drawer>元素，然后将指令 MaterialPersistentDrawerDirective 添加到组件的指令列表，以实例化持久抽屉。该指令将自身导出为 drawer，可以通过模板引用变量对其进行控制，例如：使用该指令的 toggle()方法关闭或打开抽屉。抽屉支持结构指令 deferredContent，当抽屉关闭时该指令允许动态添加或删除抽屉中的内容。抽屉中的内容通常是由导航组成的，实际上可以添加任何东西。示例代码如下：

```
//chapter16/layout_demo/lib/src/persistent_component.dart
import 'package:angular/angular.dart';
import 'package:angular_components/angular_components.dart';
@Component(
    selector: 'persistent-drawer',
    styleURLs: ['package:angular_components/app_layout/layout.scss.css'],
    templateURL: 'persistent_component.html',
    directives: [
```

```
        MaterialButtonComponent,
        MaterialIconComponent,
        MaterialPersistentDrawerDirective,
        DeferredContentDirective,
    ]
)
class PersistentComponent {

}
```

模板代码如下：

```
//chapter16/layout_demo/lib/src/persistent_component.html
<material-drawer persistent #drawer="drawer">
    <div *deferredContent>
        <!-- 抽屉内容 -->
        持久抽屉
    </div>
</material-drawer>
<material-content>
    <header class="material-header shadow">
        <div class="material-header-row">
            <!-- 通过抽屉的 toggle()方法关闭或打开抽屉 -->
            <material-button icon
                            class="material-drawer-button" (trigger)="drawer.toggle()">
                <material-icon icon="menu"></material-icon>
            </material-button>
            <!-- 其他标头信息 -->
        </div>
    </header>
    <div>
        可通过单击导航栏菜单按钮打开或关闭抽屉
    </div>
</material-content>
```

3. 临时抽屉

临时抽屉是位于所有页面内容上方的抽屉。使用时将 MaterialTemporaryDrawerComponent 添加到指令列表，并将 temporary 属性应用于<material-drawer>元素。

临时抽屉有一个可选的属性 overlay，当抽屉打开时，在非抽屉内容上方显示半透明的遮罩。示例代码如下：

```
//chapter16/layout_demo/lib/src/temporary_component.dart
import 'package:angular/angular.dart';
```

```
import 'package:angular_components/angular_components.dart';

@Component(
    selector: 'temporary-drawer',
    styleURLs: ['package:angular_components/app_layout/layout.scss.css'],
    templateURL: 'temporary_component.html',
    directives: [
      MaterialButtonComponent,
      MaterialIconComponent,
      MaterialTemporaryDrawerComponent
    ]
)
class TemporaryComponent {

}
```

模板代码如下：

```
//chapter16/layout_demo/lib/src/temporary_component.html
<material-drawer temporary #drawer = "drawer" overlay>
    <div>
        <!-- 抽屉内容 -->
        临时抽屉
    </div>
</material-drawer>
<material-content>
    <header class = "material-header shadow">
        <div class = "material-header-row">
            <!-- 通过 toggle 关闭或打开抽屉 -->
            <material-button icon
                                class = "material-drawer-button" (trigger) = "drawer.toggle()">
                <material-icon icon = "menu"></material-icon>
            </material-button>
            <!-- 其他标头信息 -->
        </div>
    </header>
    <div>
        内容
    </div>
</material-content>
```

所有抽屉都具有 HTML 属性 end，该属性将抽屉放置在另一侧。如果原来抽屉在左侧则放在右侧，如果原来在右侧则放在左侧。示例代码如下：

```
<material-drawer temporary end>
</material-drawer>
```

4. 应用栏与抽屉交互

应用栏常与抽屉协同工作，以满足应用程序整体布局。应用程序栏可以位于<material-content>元素的内部，也可以位于外部。如果位于<material-content>的内部，它将与内容布局在一起。如果位于<material-content>的外部，永久和持久抽屉及内容将位于应用栏下方。导航栏位于抽屉及内容上方的示例代码如下：

```
<header class="material-header shadow dense-header">
    <div class="material-header-row">
        <material-button icon
                            class="material-drawer-button" (trigger)="drawer.toggle()">
            <material-icon icon="menu"></material-icon>
        </material-button>
        <span class="material-header-title">应用栏</span>
        <div class="material-spacer"></div>
        <nav class="material-navigation">
            <a>链接 1</a>
        </nav>
    </div>
</header>

<material-drawer persistent #drawer="drawer" overlay>
    <!-- 抽屉内容 -->
</material-drawer>
<material-content>
    <!-- 内容 -->
</material-content>
```

5. 导航样式

布局样式表中还提供了抽屉内的导航元素样式，导航元素由列表组件和特殊的 CSS 类构成。将列表组件 MaterialListComponent 作为抽屉的子元素，在列表组件中通过组元素将内容分组，组元素是在元素上使用 group 属性指定的。

CSS 类 mat-drawer-spacer 是可选的，当应用栏位于元素<material-content>内部时，抽屉内容将留出与应用栏等高的空间，使用 dense-header 属性的导航栏不适用。

使用组件 MaterialListItemComponents 作为抽屉中的条目，条目通常放在组元素中，如果每个组需要标签，则可以在块元素上使用 label 属性。示例代码如下：

```
<material-drawer permanent>
  <material-list>
    <!-- 空出与应用栏等高的空间 -->
```

```
    <div group class = "mat - drawer - spacer"></div>
    <!-- 不带标签的组元素 -->
    <div group>
      <material - list - item>
        <material - icon icon = "inbox"></material - icon> Inbox
      </material - list - item>
      <material - list - item>
        <material - icon icon = "star"></material - icon> Star
      </material - list - item>
    </div>
    <!-- 带标签的组元素 -->
    <div group>
      <div label> Tags </div>
      <material - list - item>
        <material - icon icon = "star"></material - icon> Favorites
      </material - list - item>
    </div>
  </material - list>
</material - drawer>
```

6. 堆叠式抽屉

持久抽屉和临时抽屉具有输入属性 visible 和输出属性 visibleChange。因此可以使用双向数据绑定控制抽屉的打开与关闭，其效果与调用 toggle()方法一致。

堆叠式抽屉可以在抽屉中放置另一个抽屉，并且可以一直嵌套下去。为了好的交互体验，建议最多使用两个抽屉。堆叠式抽屉继承了临时抽屉，因此也具有输入属性 visible 和输出属性 visibleChange。堆叠式抽屉只能通过双向数据绑定到 visible 属性来控制抽屉的打开与关闭。

将 MaterialStackableDrawerComponent 添加到组件的指令列表，并定义两个布尔属性，用于初始两个抽屉的 visible 属性。示例代码如下：

```
//chapter16/layout_demo/lib/src/stackable_component.dart
import 'package:angular/angular.dart';
import 'package:angular_components/angular_components.dart';

@Component(
    selector: 'stackable - drawer',
    styleURLs: ['package:angular_components/app_layout/layout.scss.css'],
    templateURL: 'stackable_component.html',
    directives: [
      MaterialButtonComponent,
      MaterialIconComponent,
      MaterialStackableDrawerComponent,
      MaterialListComponent,
```

```
      MaterialListItemComponent
    ]
)
class StackableComponent {
  //用于控制抽屉 1 的打开或关闭
  bool drawer1Visable = false;
  //用于控制抽屉 2 的打开或关闭
  bool drawer2Visable = false;
}
```

模板代码如下：

```
//chapter16/layout_demo/lib/src/stackable_component.html
<header class="material-header shadow">
    <div class="material-header-row">
        <material-button icon
                          class="material-drawer-button" (trigger)="drawer1Visable = true">
            <material-icon icon="menu"></material-icon>
        </material-button>
        <span class="material-header-title">应用栏</span>
        <div class="material-spacer"></div>
        <nav class="material-navigation">
            <a>链接 1</a>
        </nav>
        <nav class="material-navigation">
            <a>链接 2</a>
        </nav>
        <nav class="material-navigation">
            <a>链接 3</a>
        </nav>
    </div>
</header>
<material-drawer stackable [(visible)]="drawer1Visable">
    <!-- 抽屉内容 -->
    <material-list>
        <!-- 空出与应用栏等高的空间 -->
        <div group class="mat-drawer-spacer"></div>
        <!-- 不带标签的组元素 -->
        <div group>
            <material-list-item>
                <material-icon icon="inbox"></material-icon> Inbox
            </material-list-item>
            <material-list-item>
                <material-icon icon="star"></material-icon> Star
            </material-list-item>
```

```
            </div>
            <!-- 带标签的组元素 -->
            <div group>
                <div label>Tags</div>
                <material-list-item>
                    <material-icon icon="star"></material-icon>Favorites
                </material-list-item>
            </div>
        </material-list>
        <div>第一个抽屉</div>
        <material-button (trigger)="drawer2Visable = true">显示第二个抽屉
        </material-button>
        <material-drawer stackable [(visible)]="drawer2Visable">
            <!-- 抽屉内容 -->
            <div>第二个抽屉</div>
        </material-drawer>
    </material-drawer>
    <material-content>
        <!-- 内容 -->
    </material-content>
```

将上述所有抽屉组件添加到根组件的指令列表，更新根组件。示例代码如下：

```
//chapter16/layout_demo/lib/app_component.dart
import 'package:angular/angular.dart';
import 'package:angular_components/angular_components.dart';

import 'src/permanent_component.dart';
import 'src/persistent_component.dart';
import 'src/stackable_component.dart';
import 'src/temporary_component.dart';

@Component(
    selector: 'my-app',
    styleURLs: ['package:angular_components/app_layout/layout.scss.css', 'app_component.css'],
    templateURL: 'app_component.html',
    directives: [
      PermanentComponent,
      PersistentComponent,
      TemporaryComponent,
      StackableComponent,
      MaterialButtonComponent,
      MaterialIconComponent,
      NgSwitch,
      NgSwitchWhen,
```

```
      NgSwitchDefault
    ]
)
class AppComponent {
  //缓存当前抽屉名
  String currentDrawer = '';
  //切换抽屉示例
  switchDrawer(String drawer){
    currentDrawer = drawer;
  }
}
```

模板代码如下：

```
//chapter16/layout_demo/lib/app_component.html
<div [ngSwitch] = "currentDrawer">
    <permanent - drawer * ngSwitchCase = "'permanent'"></permanent - drawer>
    <persistent - drawer * ngSwitchCase = "'persistent'"></persistent - drawer>
    <temporary - drawer * ngSwitchCase = "'temporary'"></temporary - drawer>
    <stackable - drawer * ngSwitchCase = "'stackable'"></stackable - drawer>
    <div * ngSwitchDefault>
        <header class = "material - header shadow dense - header">
        <div class = "material - header - row">
            <material - button icon
                            class = "material - drawer - button">
                <material - icon icon = "menu"></material - icon>
            </material - button>
            <span class = "material - header - title">应用栏</span>
            <div class = "material - spacer"></div>
            <nav class = "material - navigation">
                <a>链接 1</a>
            </nav>
            <nav class = "material - navigation">
                <a>链接 2</a>
            </nav>
            <nav class = "material - navigation">
                <a>链接 3</a>
            </nav>
        </div>
        </header>
    </div>
</div>
<div style = "width: 600px;margin:10px auto;">
    <material - button raised (trigger) = "switchDrawer('permanent')">切换到永久抽屉
</material - button>
```

```
    <material-button raised (trigger)="switchDrawer('persistent')">切换到持久抽屉
</material-button>
    <material-button raised (trigger)="switchDrawer('temporary')">切换到临时抽屉
</material-button>
    <material-button raised (trigger)="switchDrawer('stackable')">切换到堆叠式抽屉
</material-button>
    <material-button raised (trigger)="switchDrawer('')">切换到应用栏</material-button>
</div>
```

刷新浏览器，单击不同按钮切换到不同的应用程序布局，以观察不同抽屉的使用示例。

第四部分

第 17 章 项目实战 Deadline

前面学习了 Dart 基础、服务端编程和 Web 框架 Angular 的相关知识，本章通过项目实战将这些内容贯穿起来。

实战项目 Deadline 用于规划大小事务，并记录完成时间。

17.1 MySQL 数据库

19min

数据库负责对数据进行管理、维护和使用。主流的数据库有 Oracle、SQL Server、DB 2、Sysbase 和 MySQL 等，本节介绍 MySQL 数据库的安装与使用。

17.1.1 数据库安装

MySQL 数据库由 Oracle 公司负责提供技术支持和维护，并提供多个版本供选择，其中社区版 MySQL Community Edition 适合于个人开发者和中小型公司，本书也采用该版本用于项目实战中。

社区版提供多个平台版本，以满足在 Windows、Linux 和 macOS 等操作系统上安装和运行。本书以 Windows 操作系统为例，其他操作系统安装步骤类似，下载地址是 https://dev.MySQL.com/downloads/Windows/installer/8.0.html，如图 17-1 所示。

此版本下载的安装文件是 mysql-installer-community-8.0.21.0.msi，双击该文件启动安装过程，这里对重要步骤进行说明。

1. 安装类型

如图 17-2 所示是安装类型选择对话框。在此对话框中包含 5 种可选择的安装类型：Developer Default、Server only、Client only、Full 和 Custom。本书选择 Server only 并单击 Next 按钮。

进入步骤 Installation，单击 Execute 按钮。待状态处于 Complete 时，继续单击 Next 按钮。进入步骤 Product Configuration，继续单击 Next 按钮。

2. 配置

所需文件安装完成后，会进入 MySQL 的配置过程。如图 17-3 所示是数据库类型选择对话框，Standalone 是单个服务器，InnoDB Cluster 是数据库集群。

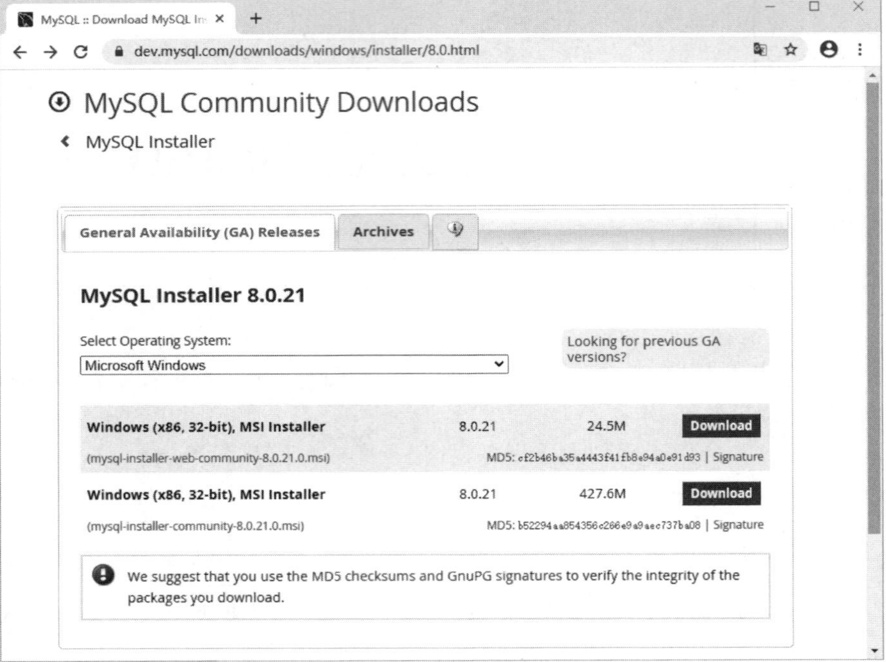

图 17-1　社区版 MySQL 下载

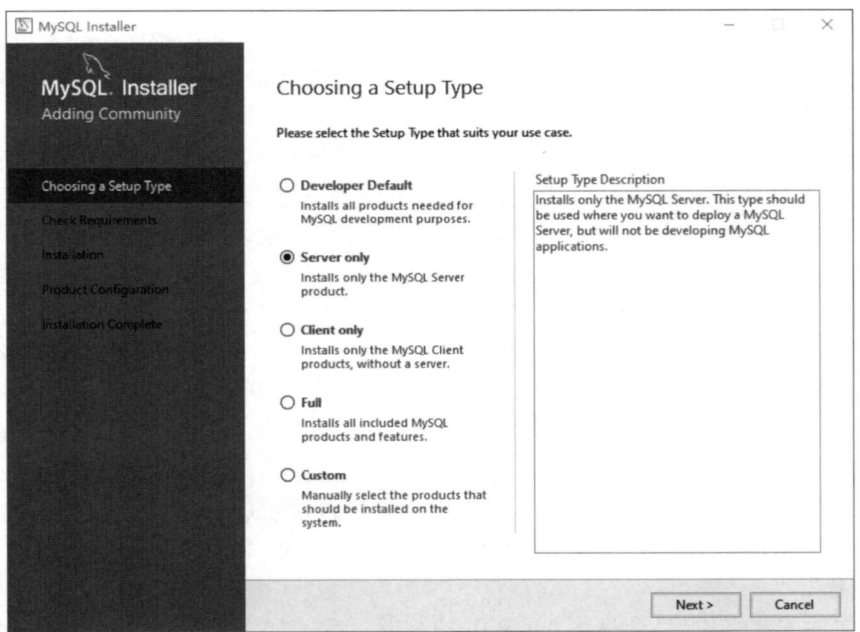

图 17-2　安装类型选择对话框

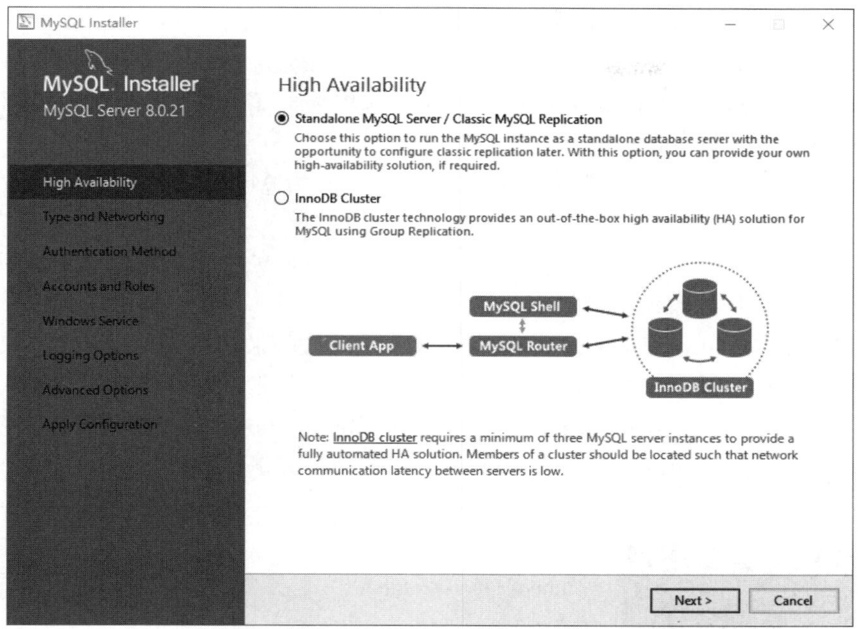

图 17-3　数据库类型选择对话框

选择 Standalone 即可，单击 Next 按钮进入如图 17-4 所示的服务器配置类型对话框。

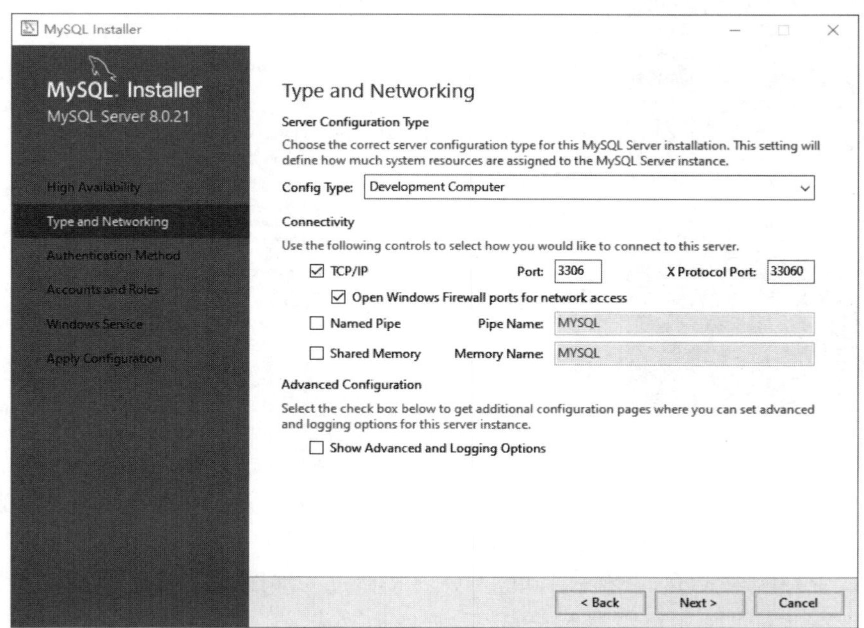

图 17-4　服务器配置类型对话框

在此对话框可以选择配置类型、通信协议和端口等。单击 Config Type 下拉列表可以选择如下配置类型：

(1) Development Computer：该选项代表个人用桌面工作站，该配置将使用最少的系统资源运行 MySQL 服务器。

(2) Server Computer：该选项代表服务器，该配置为 MySQL 服务器分配适当的系统资源。

(3) Dedicated Computer：该选项表示服务器只允许 MySQL 服务，该配置为 MySQL 服务器分配所有可用的系统资源。

这里选择 Development Computer，单击 Next 按钮进入如图 17-5 所示的授权方式对话框。它包含以下两种授权方式：

(1) Use Strong Password Encryption for Authentication：使用强密码加密授权。

(2) Use Legacy Authentication Method：传统授权方法，保留对 5.x 版本的兼容性。

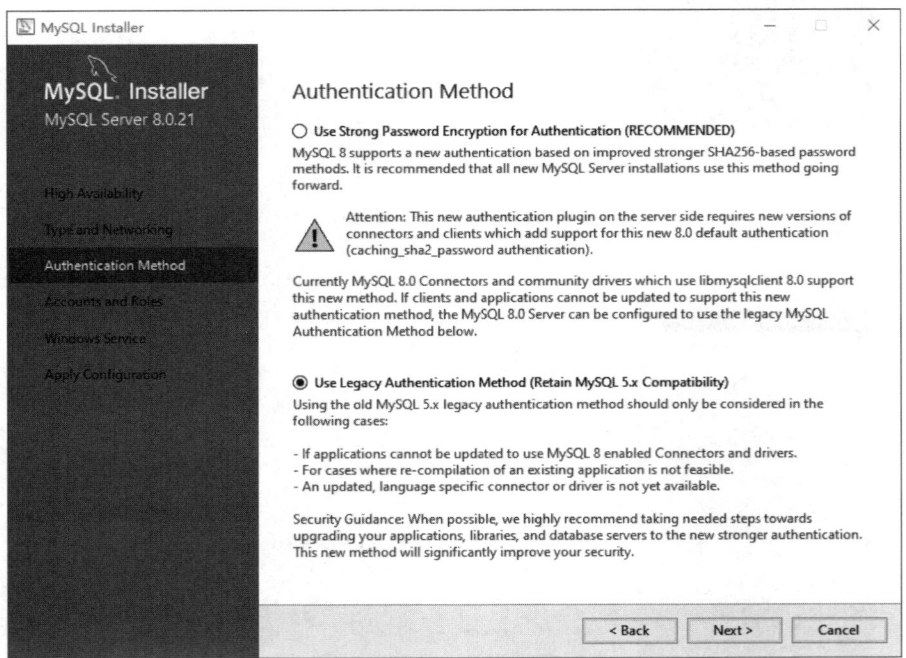

图 17-5 授权方式对话框

这里选择传统授权方法，单击 Next 按钮进入如图 17-6 所示的账号和角色设置对话框。在该对话框中设置账号 root 的密码，也可以添加其他账号。

这里设置好账号 root 的密码后，单击 Next 按钮进入如图 17-7 所示的配置 Windows 服务对话框。在此对话框中可以将 MySQL 数据库配置成为一个 Windows 服务，Windows 服务可以在后台随着 Windows 系统的启动而启动。当前版本默认的服务名是 MySQL80。

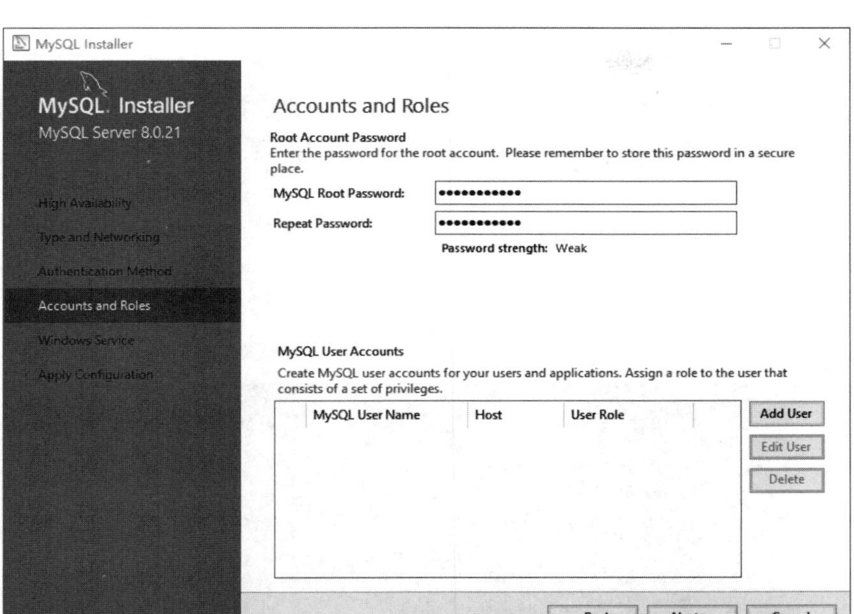

图 17-6　账号和角色设置对话框

完成如图 17-7 所示的配置后，就不需要进行手动配置了，只需单击 Next 按钮即可。

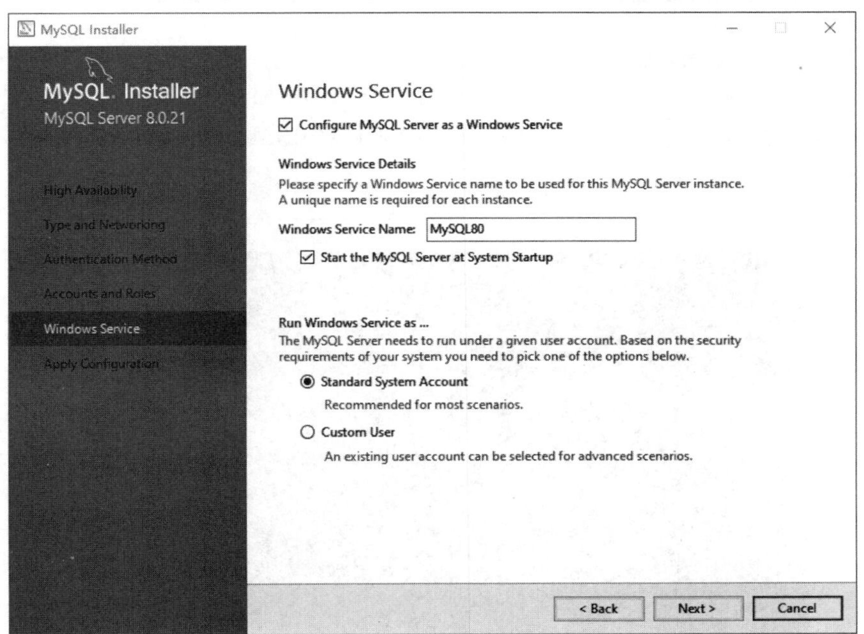

图 17-7　配置 Windows 服务对话框

17.1.2 数据库连接

MySQL 是客户端/服务器结构的，所以应用程序必须连接到服务器才能使用其服务功能。下面介绍使用 MySQL 自己的客户端和第三方客户端连接到服务器。

1. 自带客户端

MySQL for Windows 版本提供一个菜单项目可以快速连接服务器，打开步骤：右击屏幕左下角的 Windows 图标，在应用列表中找到 MySQL 8.0 Command Line Client，单击后会打开一个终端窗口，如图 17-8 所示。

图 17-8 MySQL 命令行客户端

这个工具就是 MySQL 命令行客户端工具，可以使用 MySQL 命令行客户端工具连接到 MySQL 服务器，要求输入 root 密码。输入密码后按 Enter 键，如果密码正确，则连接到 MySQL 服务器，如图 17-9 所示。

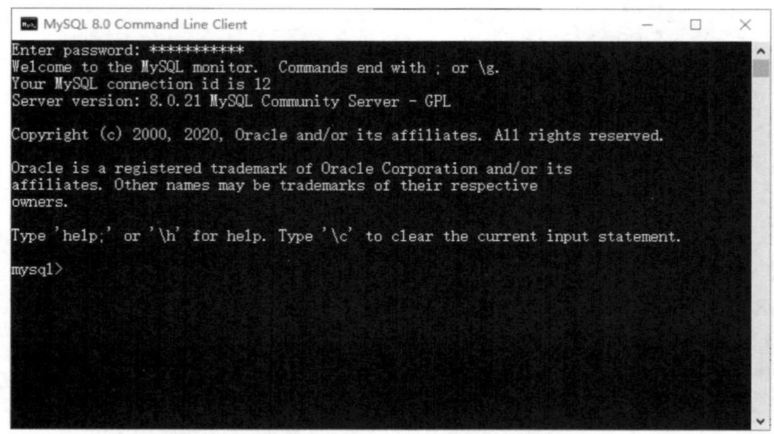

图 17-9 命令行客户端连接到服务器

2. Navicat

Navicat forMySQL 是管理和开发 MySQL 的理想解决方案，它是一套单一的应用程序。这套全面的前端工具为数据库管理、开发和维护提供了一款直观而强大的图形界面。

下载地址是 http://www.navicat.com.cn/download/navicat-for-MySQL，它提供了 Windows、Linux、macOS 3 个平台的客户端供选择。双击下载的文件即可安装应用，安装过程很简单，一直单击“下一步”按钮即可。

打开 Navicat 客户端，在界面中单击“连接”按钮，在下拉列表中选择 MySQL，如图 17-10 所示。单击 MySQL 选项后会弹出新建连接对话框，在对话框中输入自定义连接名和账号 root 的密码，最后单击“确定”按钮即可，如图 17-11 所示。

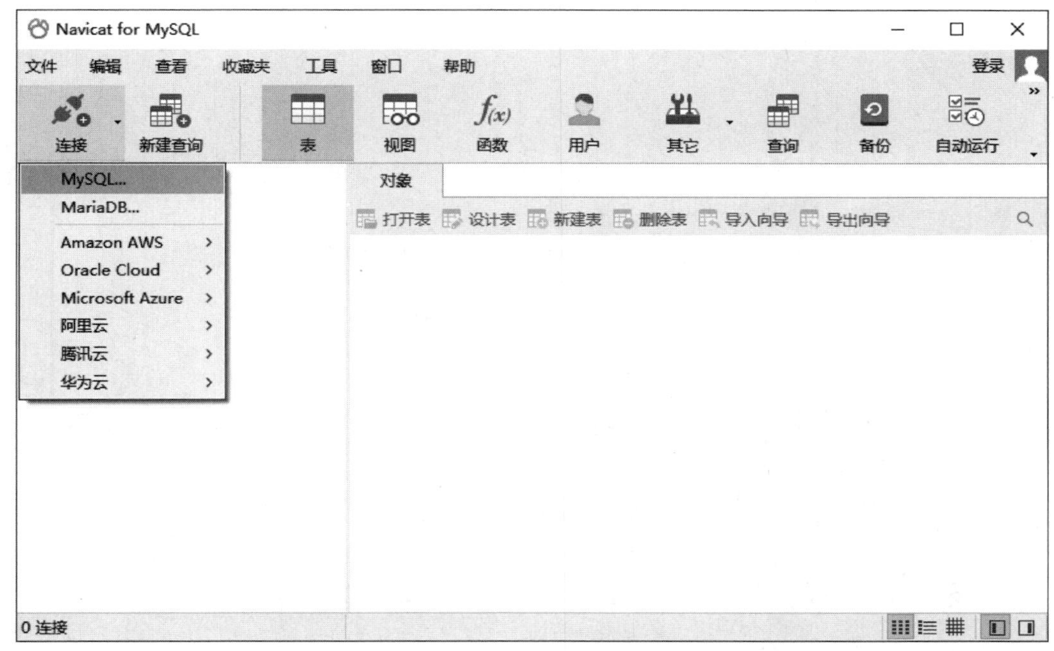

图 17-10 连接对话框

建立新连接后，双击连接名以建立连接，右击连接名在菜单中选择新建数据库，打开新建数据库对话框，如图 17-12 所示。在数据库名处输入 deadline，字符集是 utf8mb4，单击“确定”按钮完成数据库的建立。

双击数据库 deadline，在展开的表上右击，在选项菜单中选择新建表 plan，数据结构如图 17-13 所示。

计划列表表结构如表 17-1 所示。

图 17-11　新建 MySQL 连接对话框

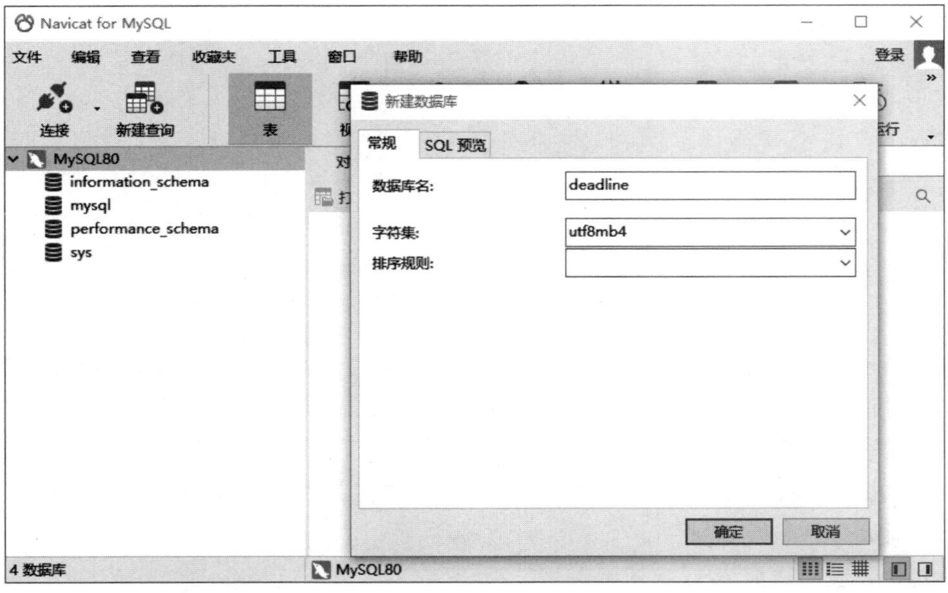

图 17-12　新建数据库对话框

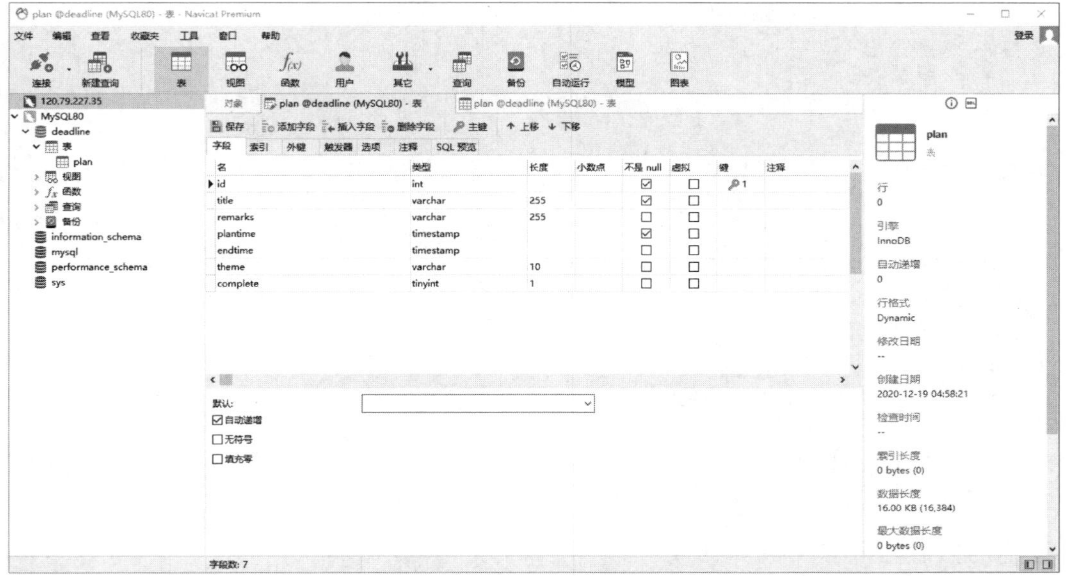

图 17-13　新建表 plan

表 17-1　计划表

字 段 名	数 据 类 型	长　度	主　键	备　注
id	int	0	是	计划 id
title	varchar	255	否	标题
remarks	varchar	255	否	备注
plantime	timestamp	0	否	计划完成时间
endtime	timestamp	0	否	实际完成时间
theme	varchar	10	否	主题
complete	tinyint	1	否	完成状态

17.2　数据库连接包

37min

包 MySQL1 提供了应用程序与 MySQL 数据库通信的能力，可以通过 SQL 语句查询、添加、删除和更新数据。

创建 shelf 应用程序，命名为 deadline_server。添加依赖并执行 pub get 命令：

```
//chapter17/deadline_server/pubspec.yaml
name: deadline_server
description: A web server built using the shelf package.
# version: 1.0.0
# homepage: https://www.example.com
```

```
environment:
  # sdk 版本依赖信息
  sdk: '>= 2.7.0 < 3.0.0'

dependencies:
  # shelf 框架包
  shelf: ^0.7.5
  # shelf 框架路由包
  shelf_router: ^0.7.2
  # shelf 框架异常处理包
  shelf_exception_handler: ^0.2.0
  # JSON 和 Dart 对象相互转换包
  json_string: ^2.0.1
  # MySQL 数据库驱动
  MySQL1: ^0.17.1

dev_dependencies:
  # 路由生成器
  shelf_router_generator: ^0.7.2 + 2
  build_runner: ^1.3.1
```

在项目的 test 目录下创建 MySQL_test. dart 文件，将在该文件中演示如何连接数据库、执行 SQL 语句，以及关闭连接。

17.2.1 连接配置

包 MySQL1 中的类 ConnectionSettings 用于配置连接信息，常用属性和方法如下：

(1) host：数据库所在主机地址，默认为 localhost。

(2) port：数据库对外开放的端口，MySQL 数据库默认是 3306 端口。

(3) user：可以访问数据库的用户名。

(4) password：用户名对应的密码。

(5) db：数据库名。

(6) ConnectionSettings({String host: 'localhost', int port: 3306, String user, String password, String db, bool useCompression: false, bool useSSL: false, int maxPacketSize: 16 * 1024 * 1024, Duration timeout: const Duration(seconds: 30), int characterSet: CharacterSet. UTF8MB4})：构造函数，通过指定参数创建 ConnectionSettings 实例。

示例代码如下：

```
//数据库连接配置
var settings = ConnectionSettings(
```

```
    host: 'localhost',
    port: 3306,
    user: 'root',
    password: 'rootflzx3QC',
    db: 'deadline'
);
```

17.2.2　连接与执行

包 MySQL1 中的类 MySQLConnection 用于与数据库连接。使用 connect 类方法打开连接，完成后，必须调用 close 方法关闭连接。以下是一些常用方法：

(1) connect(ConnectionSettings c)：类方法。接收 ConnectionSettings 类型的数据库配置信息连接到对应数据库，返回 MySQLConnection 的实例对象并包装在 Future 中。

(2) close()：关闭数据库连接。

(3) query(String sql, [List < Object > values])：在数据库上执行 sql 查询，并将 values 拼凑到对应 sql 参数上。返回由 Future 封装的 Results 对象。

(4) queryMulti(String sql, Iterable < List < Object >> values)：为 values 中每组对应 sql 参数执行一次 sql 查询。返回由 Future 封装的 List < Results >对象。

示例代码如下：

```
//返回数据库连接对象
var conn = await MySQLConnection.connect(settings);
//存储当前时间
var datetime = DateTime.now().toUtc();
//插入数据
var result1 = await conn.query('insert into plan(title,remarks,plantime) values(?,?,?)',
      ['新计划','备注信息',datetime]);
  //批量插入数据
var result2 = await conn.queryMulti('insert into plan(title,remarks,plantime) values(?,?,?)',
      [
        ['新计划 1','备注信息 1',datetime],
        ['新计划 2','备注信息 2',datetime],
        ['新计划 3','备注信息 3',datetime]
      ]
  );
```

17.2.3　结果集

Results 是执行 query 或 queryMulti 方法返回的结果行的可迭代对象。常用属性和方法如下：

(1) affectedRows：返回受影响的行数。

(2) fields：返回字段的列表。

(3) insertId：对于插入语句如果表中有一个自增长的列，则会返回该值。否则为空。

(4) isEmpty：如果返回结果中没有元素则返回 true。

(5) length：返回查询结果中元素的数量。

(6) single：检查此可迭代对象是否只有一个元素，然后返回该元素。如果为空或具有多个元素，则抛出 StateError。

(7) forEach(void f(Row element))：按迭代顺序将函数 f 应用于此集合的每个元素。

Row 表示一行数据，可以通过索引或名字来检索字段。按名字检索字段时，字段必须是有效的 Dart 标识符，且不能是 List 对象的字段。常用属性和方法如下：

(1) fields：返回字段和值组成的 Map 集合。

(2) values：返回由值组成的 List 集合。

示例代码如下：

```
//result1 是 Results 类型的数据
print('受影响的行: ${result1.affectedRows}');
print('自增长列返回的 id: ${result1.insertId}');

//查询数据
var result3 = await conn.query('select * from plan');

print('返回数据中包含表的字段: ${result3.fields}');
print('返回数据中包含的条数: ${result3.length}');

//result3 是 List<Results>类型的数据
result3.forEach((row){
    //打印由字段和值组成的 Map
    print(row.fields);
    //打印值组成的 List
    print(row.values);
  });
```

完整数据库测试代码如下：

```
//chapter17/deadline_server/bin/MySQL1_test.dart
import 'package:MySQL1/MySQL1.dart';

void main() async{
  //数据库连接配置
  var settings = ConnectionSettings(
      host: 'localhost',
      port: 3306,
```

```
        user: 'root',
        password: 'rootflzx3QC',
        db: 'deadline'
    );
    //返回数据库连接对象
    var conn = await MySQLConnection.connect(settings);
    //存储当前时间
    var datetime = DateTime.now().toUtc();
    //插入数据
    var result1 = await conn.query('insert into plan(title,remarks,plantime) values(?,?,?)',
        ['新计划','备注信息',datetime]);
    //批量插入数据
    var result2 = await conn.queryMulti('insert into plan(title, remarks, plantime) values
(?,?,?)',
        [
          ['新计划 1','备注信息 1',datetime],
          ['新计划 2','备注信息 2',datetime],
          ['新计划 3','备注信息 3',datetime]
        ]
    );

    //result1 是 Results 类型的数据
    print('受影响的行: ${result1.affectedRows}');
    print('自增长列返回的 id: ${result1.insertId}');

    //查询数据
    var result3 = await conn.query('select * from plan');

    print('返回数据中包含表的字段: ${result3.fields}');
    print('返回数据中包含的条数: ${result3.length}');

    //result3 是 List<Results>类型的数据
    result3.forEach((row){
      //打印由字段和值组成的 Map
      print(row.fields);
      //打印值组成的 List
      print(row.values);
    });

}
```

17.2.4 工具类

现在编写一个工具类,用于返回 MySQLConnection 的实例,这样就不需要每次连接数据库都创建配置信息。示例代码如下:

```
//chapter17/deadline_server/bin/utils/dbconn.dart
import 'package:MySQL1/MySQL1.dart';

class Db{
  //返回数据库连接对象 MySQLConnection
  static Future<MySQLConnection> conn()async{
     //数据库连接配置
     var settings = ConnectionSettings(
         host: 'localhost',
         port: 3306,
         user: 'root',
         password: 'rootflzx3QC',
         db: 'deadline'
     );
     return await MySQLConnection.connect(settings);
  }
}
```

因为 conn 方法是类方法，使用时可以直接通过类 Db 调用。

17.3 编写服务端

43min

使用 shelf 包创建项目，并命名为 deadline_server。

17.3.1 实体类

43min

首先定义数据模型类 Plan，该类的属性与表 plan 的字段一一映射。示例代码如下：

```
//chapter17/deadline_server/bin/entity.dart
import 'package:json_string/json_string.dart';

class Plan with Jsonable{
  final int id;
  //计划的标题
  String title;
  //计划的备注信息
  String remarks;
  //计划完成时间
  DateTime plantime;
  //实际完成时间，默认
  DateTime endtime;
  //计划的标记主题
  String theme;
  //计划完成状态，默认未完成，0 表示未完成，1 表示完成
```

```
  int complete;
  Plan ({this. id, this. title, this. remarks, this. plantime, this. endtime, this. theme, this.
complete = 0});
  //将 Dart 对象转换为 JSON 数据
  @override
  Map < String, dynamic > toJson() {
    return {
      'id':'$id',
      'title':title,
      'remarks':remarks,
      'plantime':plantime,
      'endtime':endtime,
      'theme':theme,
      'complete':'$complete',
    };
  }
  //将 JSON 数据转换为 Dart 对象
  static Plan fromJson(Map < String,dynamic > json){
    return Plan(
        id:json['id'],
        title:json['title'],
        remarks:json['remarks'],
        plantime:DateTime.parse(json['plantime']).toUtc(),
        endtime:json['endtime'] == null ? null:DateTime.parse(json['endtime']).toUtc(),
        theme:json['theme'],
        complete:json['complete']
    );
  }
}
```

该类使用了 json_string 包提供的 Mixin 类 Jsonable，用于将 Dart 类转换为 Json 数据。提供的 fromJson 方法用于将 Json 数据转化为 Plan 对象的实例。

17.3.2 服务类

提供对计划进行增、删、改、查的服务类。示例代码如下：

```
//chapter17/deadline_server/bin/service.dart
import 'entity.dart';
import 'utils/dbconn.dart';

class PlanService {
  //创建新计划
  Future < int > create(Plan plan) async {
    var conn = await Db.conn();
```

```
    try {
      var result = await conn.query(
          'insert into plan(title,remarks,plantime,theme) '
          'values(?,?,?,?)',
          [plan.title, plan.remarks, plan.plantime, plan.theme]);
      //创建成功,返回插入 id
      if (result.affectedRows > 0) {
        return result.insertId;
      }
      //创建失败,返回 0
      return 0;
    } catch (e) {
      rethrow;
    } finally {
      await conn.close();
    }
  }

  //获取所有计划
  Future<List<Map>> getALL() async {
    var conn = await Db.conn();
    var list = <Map>[];
    try {
      var result = await conn.query('select * from plan');
      result.forEach((row) {
        //修正 MySQL1 包对于 UTC 时间的解析 Bug
        row.fields.update('plantime', (item) {
          if (item == null) return null;
          return DateTime.parse(item.toLocal().toIso8601String() + 'Z');
        });
        row.fields.update('endtime', (item) {
          if (item == null) return null;
          return DateTime.parse(item.toLocal().toIso8601String() + 'Z');
        });
        //将单行数据添加到 List 集合
        list.add(row.fields);
      });
      //返回所有数据
      return list;
    } catch (e) {
      rethrow;
    } finally {
      await conn.close();
    }
  }
```

```
//获取指定 id 的计划信息
Future<Map<String, dynamic>> getById(int id) async {
  var conn = await Db.conn();
  try {
    var result = await conn.query('select * from plan where id=?', [id]);
    //如果结果不为空,则返回单条数据
    if (result.isNotEmpty) return result.single.fields;
    //否则返回 null
    return null;
  } catch (e) {
    rethrow;
  } finally {
    await conn.close();
  }
}

//更新指定 id 的计划信息
Future<bool> update(Plan plan) async {
  var conn = await Db.conn();
  try {
    var result = await conn.query(
        'update plan set title=?,remarks=?,plantime=?,theme=? where id=?',
        [plan.title, plan.remarks, plan.plantime, plan.theme, plan.id]);
    //如果更新成功,则返回 true
    if (result.affectedRows > 0) return true;
    //如果更新失败,则返回 false
    return false;
  } catch (e) {
    rethrow;
  } finally {
    await conn.close();
  }
}

//更新指定 id 的计划信息的状态
Future<bool> updateStatus(int id, int complete) async {
  var conn = await Db.conn();
  try {
    var endtime;
    if (complete == 1) {
      //完成状态时确定实际完成时间
      endtime = DateTime.now().toUtc();
    }
    var result = await conn.query(
```

```
          'update plan set endtime = ?,complete = ? where id = ?',
          [endtime, complete, id]);
      //如果更新成功,则返回 true
      if (result.affectedRows > 0) return true;
      //如果更新失败,则返回 false
      return false;
    } catch (e) {
      rethrow;
    } finally {
      await conn.close();
    }
  }

  //更新指定 id 删除计划
  Future<bool> delete(int id) async {
    var conn = await Db.conn();
    try {
      var result = await conn.query(
          'delete from plan where id = ?',[id]);
      //如果删除成功,则返回 true
      if (result.affectedRows > 0) return true;
      //如果删除失败,则返回 false
      return false;
    } catch (e) {
      rethrow;
    } finally {
      await conn.close();
    }
  }

}
```

17.3.3 时间转换类

在实际项目中需要对日期类型的数据转码为字符串。代码如下：

```
//chapter17/deadline_server/bin/utils/to_encodable.dart
dynamic myEncode(dynamic item){
  if(item is DateTime){
    return item.toLocal().toIso8601String();
  }
  return item;
}
```

17.3.4　路由器

采用路由注解编写路由器，用于响应客户端请求。示例代码如下：

```
//chapter17/deadline_server/bin/routers.dart
import 'dart:convert';
import 'package:shelf/shelf.dart';
import 'package:shelf_router/shelf_router.dart';
import 'package:json_string/json_string.dart';
import 'service.dart';
import 'entity.dart';
import 'utils/to_encodable.dart';
part 'routers.g.dart';
//定义路由器
class Dl{
  //实例化服务 PlanService
  final PlanService _planService = PlanService();

  //GET 请求,获取所有计划
  @Route.get('/plans')
  Future<Response> _getAll(Request request) async{
    var dls = await _planService.getALL();
    return Response.ok(jsonEncode(dls,toEncodable: myEncode));
  }

  //GET 请求,获取指定 id 的计划信息
  @Route.get('/plan/<id>')
  Future<Response> _getById(Request request) async{
    var id = int.parse(params(request,'id'));
    var dl = await _planService.getById(id);
    return Response.ok(jsonEncode(dl,toEncodable: myEncode));
  }

  //POST 请求,创建新计划
  @Route.post('/plan')
  Future<Response> _create(Request request) async{
    var plan1,plan2;
    await utf8.decoder.bind(request.read()).join().then((content){
      //将 Json 内容解码为 Employee 实例
      plan1 = JsonString(content).decodeAsObject(Plan.fromJson);
    });
    //创建计划并缓存插入 id
    var id = await _planService.create(plan1);
    //如果创建成功则通过插入 id 获取创建的计划
    if(id != null && id != 0){
```

```
    plan2 = await _planService.getById(id);
  }
  return Response.ok(jsonEncode(plan2,toEncodable: myEncode));
}

//PUT 请求,更新指定 id 的计划信息
@Route.put('/plan/<id>')
Future<Response> _update(Request request) async{
  var id = int.parse(params(request,'id'));
  var plan1,plan2;
  await utf8.decoder.bind(request.read()).join().then((content){
    //将 Json 内容解码为 Employee 实例
    plan1 = JsonString(content).decodeAsObject(Plan.fromJson);
  });
  //更新计划并返回结果
  var result = await _planService.update(plan1);
  //如果更新成功则通过 id 获取更新后的计划
  if(id != null && result){
    plan2 = await _planService.getById(id);
  }
  return Response.ok(jsonEncode(plan2,toEncodable: myEncode));
}

//PUT 请求,更新指定 id 的计划的完成状态
@Route.put('/plan/<id>/<status>')
Future<Response> _updateStatus(Request request) async{
  //解析请求中的参数 id
  var id = int.parse(params(request,'id'));
  //解析请求中的参数 status
  var status = int.parse(params(request,'status'));
  //更新状态
  var result = await _planService.updateStatus(id, status);
  var plan;
  //如果更新成功则通过 id 获取更新后的计划
  if(result){
    plan = await _planService.getById(id);
  }
  return Response.ok(jsonEncode(plan,toEncodable: myEncode));
}

//DELETE 请求,更新指定 id 的计划信息
@Route.delete('/plan/<id>')
Future<Response> _delete(Request request) async{
  var id = int.parse(params(request,'id'));
  //Map 对象,用于储存响应信息
```

```
    Map res = Map();
    if(id != null){
      res.addAll({'id':id});
      //删除计划并返回结果
      var result = await _planService.delete(id);
      //储存删除结果
      res.addAll({'delete':result});
    }
    return Response.ok(jsonEncode(res));
  }

  //拦截所有指向未知路由的请求
  @Route.all('/<ignored|.*>')
  Future<Response> _notFound(Request request) async{
    return Response.notFound('页面未找到');
  }
  //返回本路由器的处理函数
  Handler get handler => _$DlRouter(this).handler;
}
```

17.3.5 跨域中间件

在实际项目中服务端通常需要响应本地之外的请求，因此需要为服务端提供响应跨域请求的能力。创建中间件，用于配置跨域请求。示例代码如下：

```
//chapter17/deadline_server/bin/utils/middle_cors.dart
import 'package:shelf/shelf.dart';

//创建中间件并提供corsHeaders配置
final cors = createCorsHeadersMiddleware(
    corsHeaders:{
      'Access-Control-Allow-Origin': '*',
      'Access-Control-Expose-Headers': 'Authorization, Content-Type',
      'Access-Control-Allow-Headers': 'Authorization, Origin, X-Requested-With, Content
-Type, Accept',
      'Access-Control-Allow-Methods': 'GET, POST, PUT, PATCH, DELETE'
    }
);

Middleware createCorsHeadersMiddleware({Map<String, String> corsHeaders}) {
  //未提供corsHeaders时，使用默认配置
  corsHeaders ??= {'Access-Control-Allow-Origin': '*'};

  //请求处理程序
```

```
  Response handleOptionsRequest(Request request) {
    if (request.method == 'OPTIONS') {
      return Response.ok(null, headers: corsHeaders);
    } else {
      return null;
    }
  }

  //响应处理程序,将 corsHeaders 的配置应用于响应
  Response addCorsHeaders(Response response) => response.change(headers: corsHeaders);

  //创建中间件并返回
    return  createMiddleware  ( requestHandler:  handleOptionsRequest,  responseHandler:
addCorsHeaders);
}
```

17.3.6 适配器

本项目采用 shelf 框架提供的适配器来启动服务。示例代码如下：

```
//chapter17/deadline_server/bin/server.dart
import 'package:shelf/shelf.dart';
import 'package:shelf_exception_handler/shelf_exception_handler.dart';
import 'package:shelf/shelf_io.dart' as io;
import 'utils/middle_cors.dart';
import 'routers.dart';

const _hostname = 'localhost';

void main() async {
  //返回路由器的处理程序
  var routerHandler = Dl().handler;
  //连接中间件和处理程序
  var handler = const Pipeline()
      .addMiddleware(exceptionHandler())
      .addMiddleware(logRequests())
      .addMiddleware(cors)
      .addHandler(routerHandler);
  //启动服务
  var server = await io.serve(handler, _hostname,1024);
  print('服务地址 http://${server.address.host}:${server.port}');
}
```

打开编辑器中的命令行工具，执行如下命令：

```
pub run build_runner build
```

命令执行成功后，路由生成器会生成一个新文件。生成的代码如下：

```
//chapter17/deadline_server/bin/routers.g.dart
//GENERATED CODE - DO NOT MODIFY BY HAND

part of 'routers.dart';

//*****************************************************************
//ShelfRouterGenerator
//*****************************************************************

Router _$DlRouter(Dl service) {
  final router = Router();
  router.add('GET', r'/plans', service._getAll);
  router.add('GET', r'/plan/<id>', service._getById);
  router.add('POST', r'/plan', service._create);
  router.add('PUT', r'/plan/<id>', service._update);
  router.add('PUT', r'/plan/<id>/<status>', service._updateStatus);
  router.add('DELETE', r'/plan/<id>', service._delete);
  router.all(r'/<ignored|.*>', service._notFound);
  return router;
}
```

然后运行 server.dart 文件即可启动服务。

60min

17.4　编写客户端

创建 Angular 应用程序，命名为 deadline_web。将服务端的 entity.dart 文件复制到项目的 lib/src/plan 目录下。

59min

17.4.1　管道

编写模板中会用到的管道类 TimeIntervalPipe 和 CompletePipe，前者用于计算时间间隔，后者根据完成状态筛选计划列表。代码如下：

60min

```
//chapter17/deadline_web/lib/src/plan/date_pipe.dart
import 'package:angular/angular.dart';
import 'entity.dart';
//时间间隔计算管道
@Pipe('timeInterval', pure: true)
class TimeIntervalPipe implements PipeTransform{
  String transform(DateTime now,DateTime plantime) {
```

```
    if (plantime == null) return null;
    var planMills = plantime.millisecondsSinceEpoch;
    var nowMills = now.millisecondsSinceEpoch;
    //计划期限内
    if(nowMills < planMills){
      var mills = planMills - nowMills;
      return '${_str(mills)}';
    }
    //超时
    if(nowMills > planMills){
      var mills = nowMills - planMills;
      return '${_str(mills)}';
    }
  }
  //计算、天、时、分、秒
  String _str(int mills){
    var str = '';
    //相距天数
    var days = mills~/(3600 * 24 * 1000);
    //小时
    var hours = (mills % (3600 * 24 * 1000))~/(3600 * 1000);
    //分钟
    var minutes = (mills % (3600 * 24 * 1000)) % (3600 * 1000)~/(60 * 1000);
    //秒
    var seconds = (mills % (3600 * 24 * 1000)) % (3600 * 1000) % (60 * 1000)~/1000;

    if(days > 0) str += '$days天';
    if(hours > 0) str += '$hours时';
    if(minutes > 0) str += '$minutes分';
    if(seconds >= 0) str += '$seconds秒';

    return str;
  }
}
//根据 complete 值过滤集合
@Pipe('complete', pure: true)
class CompletePipe implements PipeTransform{
  List<Plan> transform(List<Plan> value, int complete){
   return value.where((plan) => plan.complete == complete).toList();
  }
}
```

17.4.2 服务

编写向服务器发出请求的服务类，本项目采用 http 包提供的方法向服务器发出请求，并处理响应。代码如下：

```
//chapter17/deadline_web/lib/src/plan/plan_service.dart
import 'dart:async';
import 'dart:convert';
import 'package:angular/core.dart';
import 'package:http/http.dart' as http;
import 'package:json_string/json_string.dart';
import 'entity.dart';

//请求服务器地址和端口
const _URL = 'http://localhost:1024';
//定义服务
@Injectable()
class PlanService{

  //从服务器获取所有计划
  Future<List<Plan>> getAll() async{
    var plans;
    await http.get('$_URL/plans').then((response){
      if(response.statusCode == 200){
        //解析服务器响应信息并转换为 Plan 对象集合
        plans = JsonString(response.body).decodeAsObjectList(Plan.fromJson);
      }
    });
    return plans;
  }

  //根据指定 id 从服务器获取单个计划
  Future<Plan> getById(int id)async{
    var plan;
    await http.get('$_URL/plan/$id').then((response){
      if(response.statusCode == 200){
        //解析服务器响应信息并转换为 Plan 对象
        plan = JsonString(response.body).decodeAsObject(Plan.fromJson);
      }
    });
    return plan;
  }

  //提交创建请求
  Future<Plan> post(Plan pl)async{
    var plan;
    await http.post('$_URL/plan',body:jsonEncode(pl)).then((response){
      if(response.statusCode == 200){
        //解析服务器响应信息并转换为 Plan 对象
```

```
            plan = JsonString(response.body).decodeAsObject(Plan.fromJson);
        }
    });
    return plan;
}

//更新计划
Future<Plan> put(Plan pl)async{
    var plan;
    await http.put('$_URL/plan/${pl.id}',body: jsonEncode(pl)).then((response){
        if(response.statusCode == 200){
            plan = JsonString(response.body).decodeAsObject(Plan.fromJson);
        }
    });
    return plan;
}

//更新计划的完成状态
Future<Plan> putStatus(int id,int status)async{
    var plan;
    await http.put('$_URL/plan/$id/$status').then((response){
        if(response.statusCode == 200){
            plan = JsonString(response.body).decodeAsObject(Plan.fromJson);
        }
    });
    return plan;
}

//删除计划
Future<bool> delete(int id)async{
    var result = false;
    await http.delete('$_URL/plan/$id').then((response){
        if(response.statusCode == 200){
            //解析服务器响应信息并转换为 Map 集合
            var map = JsonString(response.body).decodedValueAsMap;
            if(map != null) {
                //获取删除结果
                result = map['delete'];
            }
        }
    });
    return result;
}
}
```

17.4.3 添加计划组件

创建组件 PlanAddComponent,用于添加计划。代码如下:

```
//chapter17/deadline_web/lib/src/plan/plan_add_component.dart
import 'dart:async';
import 'package:angular/angular.dart';
import 'package:angular_components/angular_components.dart';
import 'package:angular_components/utils/browser/window/module.dart';
import 'package:angular_router/angular_router.dart';
import 'package:angular_forms/angular_forms.dart';

import 'plan_service.dart';
import 'entity.dart';
import 'package:intl/intl.dart';
import '../route_paths.dart';

@Component(
  selector: 'plan-add',
  templateURL: 'plan_add_component.html',
  directives: [
    coreDirectives,
    formDirectives,
    NgFor,
    NgIf,
    materialInputDirectives,
    MaterialDateTimePickerComponent,
    MaterialMultilineInputComponent,
    MaterialYesNoButtonsComponent,
    MaterialSubmitCancelButtonsDirective
  ],
  providers: [windowBindings,datepickerBindings,ClassProvider(PlanService)],
)



class PlanAddComponent{
  final PlanService _planService;
  final Router _router;
  Plan plan_add = Plan();
  //设置用户可选择的最小时间点
  DateTime minDateTime = DateTime.now();

  //定义日期格式
  DateFormat dateFormat = DateFormat("yy年MM月dd日");
```

```
  DateFormat timeFormat = DateFormat("HH 时 mm 分");

  //控制 yes 按钮待定状态
  bool pending = false;

  //注入服务_planService 和路由 _router
  PlanAddComponent(this._planService,this._router){
    //初始化计划时间
    plan_add.plantime = DateTime.now().add(Duration(hours: 1));
  }

  //响应 yes 按钮的单击事件
  void add() async{
    pending = true;
    var dl = await _planService.post(plan_add);
    if(dl == null){
      getWindow().alert('添加失败!');
    }else{
      getWindow().alert('添加成功!');
    }
    pending = false;
  }

  //响应 no 按钮的单击事件,前往计划列表
  Future<NavigationResult> gotoList() =>
      _router.navigate(RoutePaths.plans.toURL());
}
```

添加计划组件的模板代码如下：

```
//chapter17/deadline_web/lib/src/plan/plan_add_component.html
<form style="width:360px;border:1px solid #eee;margin:0 auto;padding:30px;">
    <div>
        <material-input label="事件*"
                        floatingLabel
                        required
                        requiredErrorMsg="此输入框必填"
                        blurUpdate
                        [(ngModel)]="plan_add.title"></material-input>
    </div>

    <div>
        <material-input multiline
                   floatingLabel
                   blurUpdate
```

```
                    rows = "2"
                    maxRows = "4"
                    label = "备注信息"
                    [(ngModel)] = "plan_add.remarks"></material-input>
    </div>
    <div>
        <text>截止时间</text>
        <material-date-time-picker [(dateTime)] = "plan_add.plantime"
                                    [outputDateFormat] = "dateFormat"
                                    [outputTimeFormat] = "timeFormat"
                                    [minDateTime] = "minDateTime"
                                    required>
        </material-date-time-picker>
    </div>
    <div>
        <text>标注颜色</text>
        <input type = "color" [(ngModel)] = "plan_add.theme">
    </div>
<material-yes-no-buttons yesAutoFocus
                        submitCancel
                        raised
                        reverse
                        noText = "返回列表"
                        [pending] = "pending"
                        yesText = "添加"
                        (yes) = "add()"
                        (no) = "gotoList()"
                        [yesDisabled] = "plan_add.title == null || plan_add.title == ''">
</material-yes-no-buttons>
</form>
```

17.4.4　编辑计划组件

添加组件 PlanEditComponent，用于更新计划。代码如下：

```
//chapter17/deadline_web/lib/src/plan/plan_edit_component.dart
import 'package:angular/angular.dart';
import 'package:angular_components/angular_components.dart';
import 'package:angular_components/utils/browser/window/module.dart';
import 'package:angular_forms/angular_forms.dart';
import 'package:angular_router/angular_router.dart';
import 'package:intl/intl.dart';

import 'plan_service.dart';
```

```
import 'entity.dart';
import '../route_paths.dart';

@Component(
  selector: 'plan-edit',
  templateURL: 'plan_edit_component.html',
  directives: [
    coreDirectives,
    formDirectives,
    NgFor,
    NgIf,
    materialInputDirectives,
    MaterialDateTimePickerComponent,
    MaterialMultilineInputComponent,
    MaterialYesNoButtonsComponent,
    MaterialSubmitCancelButtonsDirective
  ],
  providers: [windowBindings,datepickerBindings,ClassProvider(PlanService)],
)
//实现 OnActivate 接口
class PlanEditComponent implements OnActivate{
  final PlanService _planService;
  final Router _router;
  Plan plan_edit = Plan();

  //设置用户可选择的最小时间点
  DateTime minDateTime = DateTime.now();

  //定义日期显示格式
  DateFormat dateFormat = DateFormat("yy年MM月dd日");
  DateFormat timeFormat = DateFormat("HH时mm分");

  //控制 yes 按钮待定状态
  bool pending = false;

  //注入服务_planService 和路由 _router
  PlanEditComponent(this._planService,this._router);

  //实现 onActivate 路由生命周期函数
  @override
  void onActivate(RouterState previous, RouterState current) async{
    //解析当前路由的参数 id
    var id = current.parameters['id'];
    //获取需要编辑的计划
    if(id != null){
```

```
        plan_edit = await _planService.getById(int.parse(id));
      }
    }

    //响应 yes 按钮的单击事件,更新计划
    void update() async{
      pending = true;
      var plan = await _planService.put(plan_edit);
      if(plan != null){
        plan_edit = plan;
        getWindow().alert('更新成功!');
      }else{
        getWindow().alert('更新异常!');
      }
      pending = false;
    }

    //响应 no 按钮的单击事件,前往计划列表
    Future<NavigationResult> gotoList() =>
        _router.navigate(RoutePaths.plans.toURL());
}
```

编辑计划的组件模板代码如下：

```
//chapter17/deadline_web/lib/src/plan/plan_edit_component.html
<form *ngIf="plan_edit != null" style="width:360px;border:1px solid #eee;margin:0 auto;
padding:30px;">
    <div>
        <material-input label="事件*"
                        floatingLabel
                        required
                        requiredErrorMsg="此输入框必填"
                        blurUpdate
                        [(ngModel)]="plan_edit.title"></material-input>
    </div>

    <div>
        <material-input multiline
                        floatingLabel
                        blurUpdate
                        rows="2"
                        maxRows="4"
                        label="备注信息"
                        [(ngModel)]="plan_edit.remarks"></material-input>
    </div>
```

```
    <div>
        <text>截止时间</text>
        <material-date-time-picker [(dateTime)] = "plan_edit.plantime"
                                    [outputDateFormat] = "dateFormat"
                                    [outputTimeFormat] = "timeFormat"
                                    [minDateTime] = "minDateTime"
                                    required>
        </material-date-time-picker>
    </div>
    <div>
        <text>标注颜色</text>
        <input type = "color" [(ngModel)] = "plan_edit.theme">
    </div>
    <material-yes-no-buttons yesAutoFocus
                             submitCancel
                             raised
                             reverse
                             noText = "返回列表"
                             yesText = "更新"
                             [pending] = "pending"
                             (yes) = "update()"
                             (no) = "gotoList()"
                             [yesDisabled] = "plan_edit.title == null || plan_edit.title =
= ''"></material-yes-no-buttons>
</form>
```

17.4.5 计划列表组件

添加组件 PlanListComponent，用于展示所有的计划。代码如下：

```
//chapter17/deadline_web/lib/src/plan/plan_list_component.dart
import 'dart:async';
import 'package:angular/angular.dart';
import 'package:angular_components/angular_components.dart';
import 'package:angular_components/utils/browser/window/module.dart';
import 'package:angular_router/angular_router.dart';

import 'plan_service.dart';
import 'entity.dart';
import 'date_pipe.dart';
import '../route_paths.dart';

@Component(
  selector: 'plan-list',
  styleURLs: ['plan_list_component.css'],
```

```
  templateURL: 'plan_list_component.html',
  directives: [
    MaterialListComponent,
    MaterialListItemComponent,
    MaterialCheckboxComponent,
    MaterialIconComponent,
    MaterialButtonComponent,
    MaterialTooltipDirective,
    coreDirectives,
    NgFor,
    NgIf,
  ],
  providers: [popupBindings, windowBindings,ClassProvider(PlanService)],
  pipes: [TimeIntervalPipe, CompletePipe, AsyncPipe, DatePipe],
)
class PlanListComponent implements OnInit {
  final PlanService _planService;
  //储存完成状态的计划
  List<Plan> plans_complete = <Plan>[];
  //储存未完成状态的计划
  List<Plan> plans_uncomplete = <Plan>[];
  DateTime time_now = DateTime.now();
  Timer _timer;
  final Router _router;

  PlanListComponent(this._planService, this._router) {
    //定时器,每隔 1s 更新一次 time_now 的值
    _timer = Timer.periodic(Duration(seconds: 1), (timer) {
      time_now = DateTime.now();
    });
  }

  @override
  Future<Null> ngOnInit() async {
    //获取所有计划
    var plans_get = await _planService.getAll();
    if (plans_get != null) {
      //根据计划完成状态分组
      plans_get.forEach((item) {
        if (item.complete == 1) {
          plans_complete.add(item);
        } else {
          plans_uncomplete.add(item);
        }
      });
```

```
      //根据计划事件排序
      plans_uncomplete.sort(sortByPlantime);
      plans_complete.sort(sortByPlantime);
    }
  }

  //更新指定 id 的计划
  void edd(int id) async {
    var index = plans_uncomplete.indexWhere((d) => d.id == id);
    if (index != null) {
      var dl2 = await _planService.put(plans_uncomplete[index]);
      plans_uncomplete[index] = dl2;
    }
  }

  //删除指定 id 的计划
  void delete(int id) async {
    var index = plans_uncomplete.indexWhere((d) => d.id == id);
    var result = await _planService.delete(id);
    if(result){
      plans_uncomplete.removeAt(index);
      getWindow().alert('删除成功!');
    }
  }

  //更新指定 id 计划的状态
  void eddStatus(int id, int status) async {
    var index = plans_uncomplete.indexWhere((d) => d.id == id);
    var dl = await _planService.putStatus(id, status);
    if (dl != null && dl.complete == 1) {
      plans_complete.add(dl);
      plans_uncomplete.removeAt(index);
      plans_complete.sort(sortByPlantime);
    }
  }

  //响应复选框 checkedChange 事件
  void onChecked(int id, bool isChecked) {
    eddStatus(id, isChecked ? 1 : 0);
  }

  //根据计划完成时间排序的函数
  int sortByPlantime(a, b) {
    return a.plantime.millisecondsSinceEpoch
        .compareTo(b.plantime.millisecondsSinceEpoch);
  }
```

```
  //跳转
  void onSelect(int id) {
    _gotoDetail(id);
  }

  //跳转到指定 id 计划详细页
  Future<NavigationResult> _gotoDetail(int id) =>
      _router.navigate(RoutePaths.edit.toURL(parameters: {'id': '$id'}));
}
```

模板代码如下：

```
//chapter17/deadline_web/lib/src/plan/plan_list_component.html
<div>待完成</div>
<material-list>
  <material-list-item *ngFor="let plan of plans_uncomplete; let i = index">
    <div class="item-dl" [style.border-left-color]="plan.theme">
      <div>
        <material-checkbox [disabled]="false"
                           [checked]="false"
                           (checkedChange)="onChecked(plan.id, $event)"
                           materialTooltip="标记为完成"></material-checkbox>
        <span class="action">
          <material-icon *ngIf="" icon="{{time_now.millisecondsSinceEpoch > plan.
plantime.millisecondsSinceEpoch ? 'more_time' : 'timelapse'}}"
                           materialTooltip="{{time_now.millisecondsSinceEpoch > plan.
plantime.millisecondsSinceEpoch ? '超时' : '倒计时'}}"></material-icon>
          <material-button icon (trigger)="delete(plan.id)" materialTooltip="删除">
            <material-icon icon="delete"></material-icon>
          </material-button>
          <material-button icon (trigger)="onSelect(plan.id)" materialTooltip="编辑">
            <material-icon icon="edit"></material-icon>
          </material-button>
        </span>
      </div>
      <div>
        <span class="title">{{plan.title}}</span>
        <span class="current">
          {{time_now | timeInterval:plan.plantime}}
        </span>
      </div>
      <p class="remarks">{{plan.remarks}}</p>
      <div class="time">
        <span class="plantime">{{plan.plantime | date:'yyyy.MM.dd HH:mm'}}</span>
```

```
        </div>
      </div>
    </material-list-item>
  </material-list>
  <div>已完成</div>
  <material-list>
    <material-list-item *ngFor="let plan of plans_complete">
      <div class="item-dl" [style.border-left-color]="plan.theme">
        <material-checkbox [disabled]="true"
                              [checked]="true"></material-checkbox>
        <div>
          <span class="title">{{plan.title}}</span>
          <span class="current">已完成</span>
        </div>
        <p class="remarks">{{plan.remarks}}</p>
        <div class="time">
          <span class="plantime">{{plan.plantime | date:'yyyy.MM.dd HH:mm'}}</span>
          <span class="endtime">{{plan.endtime | date:' - yyyy.MM.dd HH:mm'}}</span>
        </div>
      </div>
    </material-list-item>
  </material-list>
```

样式代码如下：

```
//chapter17/deadline_web/lib/src/plan/plan_list_component.css
material-list{
    width:720px;
}

material-list-item{
    padding:0;
}

.item-dl{
    width:720px;
    padding:6px 10px;
    border-left: 3px solid #5dbe8a;
    border-bottom: 1px solid #eee;
}
.title{
    font-weight: bold;
}
.remarks{
    margin: 1px 0;
```

```
    font - size:small;
    font - weight: 300;
}
.time{
    font - size:12px;
    line - height:14px;
    font - weight: 100;
    font - style: italic;
}

.current{
    float:right;
    align - items: center;
    display: flex;
}

.action{
    float:right;
    align - items: center;
    display: flex;
}
```

添加未知页响应组件。代码如下：

```
//chapter17/deadline_web/lib/src/not_found_component.dart
import 'package:angular/angular.dart';

@Component(
  selector: 'not - found',
  template: '< h2 >页面未找到</h2 >',
)
class NotFoundComponent {}
```

17.4.6 路由

添加路由路径定义文件。代码如下：

```
//chapter17/deadline_web/lib/src/route_paths.dart
import 'package:angular_router/angular_router.dart';

class RoutePaths{
  //添加计划的路由路径
  static final add = RoutePath(path: 'plan/add');
  //编辑计划的路由路径
  static final edit = RoutePath(path: 'plan/edit/:id');
```

```
  //计划列表的路由路径
  static final plans = RoutePath(path: 'plans');
}
```

添加路由定义文件。代码如下：

```
//chapter17/deadline_web/lib/src/routes.dart
import 'package:angular_router/angular_router.dart';

import 'plan/plan_list_component.template.dart' as plan_list_template;
import 'plan/plan_add_component.template.dart' as plan_add_template;
import 'plan/plan_edit_component.template.dart' as plan_edit_template;
import 'not_found_component.template.dart' as not_found_template;

import 'route_paths.dart';
export 'route_paths.dart';

class Routes{
  //计划列表的路由定义,连接添加计划到组件
  static final plans = RouteDefinition(
    routePath: RoutePaths.plans,
    component: plan_list_template.PlanListComponentNgFactory,
  );
  //添加计划的路由定义
  static final add = RouteDefinition(
    routePath: RoutePaths.add,
    component: plan_add_template.PlanAddComponentNgFactory,
  );
  //编辑计划的路由定义
  static final edit = RouteDefinition(
    routePath: RoutePaths.edit,
    component: plan_edit_template.PlanEditComponentNgFactory,
  );

  static final all = <RouteDefinition>[
    plans,
    add,
    edit,
    //拦截首页并跳转到路由计划列表
    RouteDefinition.redirect(
      path: '',
      redirectTo: RoutePaths.plans.toURL(),
    ),
    //拦截未知页
    RouteDefinition(
```

```
      path: '.*',
      component: not_found_template.NotFoundComponentNgFactory,
    ),
  ];
}
```

17.4.7 布局

项目通常会利用根组件进行布局,并且会将其作为路由组件。代码如下:

```
//chapter17/deadline_web/lib/app_component.dart
import 'package:angular/angular.dart';
import 'package:angular_router/angular_router.dart';
import 'package:angular_components/angular_components.dart';
import 'src/routes.dart';

@Component(
  selector: 'my-app',
  styleURLs:
  ['package:angular_components/app_layout/layout.scss.css','app_component.css'],
  templateURL: 'app_component.html',
  providers: [routerProviders],
  directives: [
    routerDirectives,
    MaterialPersistentDrawerDirective,
    MaterialToggleComponent,
    MaterialIconComponent,
    MaterialButtonComponent,
    MaterialListComponent,
    MaterialListItemComponent,
    ],
  exports: [RoutePaths,Routes],
)
class AppComponent{}
```

根组件模板代码如下:

```
//chapter17/deadline_web/lib/app_component.html
<header class="material-header shadow dense-header">
    <div class="material-header-row">
        <material-button icon
                        class="material-drawer-button" (trigger)="drawer.toggle()">
            <material-icon icon="menu"></material-icon>
        </material-button>
        <span class="material-header-title">应用栏</span>
```

```
    </div>
</header>
<material-drawer persistent #drawer="drawer">
    <!-- 抽屉内容 -->
    <material-list>
        <!-- 不带标签的组元素 -->
        <div group>
            <material-list-item>
                <a [routerLink]="RoutePaths.plans.toURL()"><material-icon icon="list">
</material-icon>返回所有计划</a>
            </material-list-item>
            <material-list-item>
                <a [routerLink]="RoutePaths.add.toURL()"><material-icon icon="add">
</material-icon>添加</a>
            </material-list-item>
        </div>
        <!-- 带标签的组元素 -->
        <div group>
            <div label>Tags</div>
            <material-list-item>
                <material-icon icon="star"></material-icon>Favorites
            </material-list-item>
        </div>
    </material-list>
</material-drawer>
<material-content style="width:720px;margin:0 auto;padding:30px;">
    <!-- 内容 -->
        <router-outlet [routes]="Routes.all"></router-outlet>
</material-content>
```

样式代码如下：

```
//chapter17/deadline_web/lib/app_component.css
material-list-item a{
    width: 100%;
    align-items: center;
    display: flex;
    text-decoration:none;
    color:rgba(0,0,0,0.87);
}
material-list-item a:hover{
    text-decoration: none;
}
```

运行项目，将会在浏览器中展示计划列表，可以通过导航前往添加和编辑计划的页面。

图书资源支持

感谢您一直以来对清华大学出版社图书的支持和爱护。为了配合本书的使用，本书提供配套的资源，有需求的读者请扫描下方的“书圈”微信公众号二维码，在图书专区下载，也可以拨打电话或发送电子邮件咨询。

如果您在使用本书的过程中遇到了什么问题，或者有相关图书出版计划，也请您发邮件告诉我们，以便我们更好地为您服务。

我们的联系方式：

地　　址：北京市海淀区双清路学研大厦 A 座 701

邮　　编：100084

电　　话：010-83470236　010-83470237

资源下载：http://www.tup.com.cn

客服邮箱：tupjsj@vip.163.com

QQ：2301891038（请写明您的单位和姓名）

教学资源・教学样书・新书信息

人工智能科学与技术
人工智能|电子通信|自动控制

资料下载・样书申请

书圈

用微信扫一扫右边的二维码，即可关注清华大学出版社公众号。